Contact with Alien Civilizations

Contact with Alien Civilizations

Our Hopes and Fears about Encountering Extraterrestrials

Michael A.G. Michaud

COPERNICUS BOOKS

An Imprint of Springer Science+Business Media

© 2007 Springer Science+Business Media, LLC

Published in the United States by Copernicus Books,
an imprint of Springer Science+Business Media.

Copernicus Books
Springer Science+Business Media
233 Spring Street
New York, NY 10013
www.springer.com

Library of Congress Control Number:
2006923105

Manufactured in the United States of America.
Printed on acid-free paper.

9 8 7 6 5 4 3 2 1

ISBN-10: 0-387-28598-9
ISBN-13: 978-0-387-28598-6

The considerations of the intelligent always include both benefit and harm.
—Sun Tzu, fifth century B.C.

Contents

Introduction

The proper study of mankind is not merely Man, but Intelligence.
—Arthur C. Clarke, 1951[1]

In the long-running television series "The X-Files," the original Deep Throat said to FBI Special Agent Mulder that "there are those like yourself who believe in the existence of extraterrestrial life." Ah, but that's not the question.

If extraterrestrial life exists, most of it may be in simpler forms comparable to the one-celled organisms of Earth biology. Finding such life would be fascinating for scientists, but may be of only passing interest to the general public. What intrigues the average citizen is the possibility of contact with extraterrestrial intelligence. We want to communicate with other sentient beings, learning what they know and telling them about ourselves. We want to find out how they are like us and how they are different. Microorganisms don't have a lot to say.

There is another implication of contact that underlies this book: Intelligent extraterrestrials might have an impact on our future. The information they send us—if any—might change our cultures. They could have conscious intentions toward us, and possibly the technologies to reach us directly. Their intentions may be benign—or not.

Our interest in alien minds is not new. The idea that intelligent beings exist beyond the Earth has been part of the Western intellectual tradition for more than 2000 years. Sometimes this belief was widespread; at other times, it was out of fashion.

Over the millennia, our thinking about extraterrestrials has evolved from metaphysical speculation to scientifically testable hypotheses. Yet we still have no proof.

At the philosophical and scientific levels, the most durable feature of the controversy has been the split between "Copernicans," who argue that Humankind does not occupy a special place in the universe, and "Aristotelians," who believe that we do. During the Enlightenment, philosophical speculations wrestled with another fundamental question. If there are intelligent beings beyond the Earth, some of them might be more intellectually advanced than we are. How, then, could Man be the measure of all things?

The common thread, the underlying philosophical question, is the importance of Humankind. What is the status of humans in the cosmos? Are we unique, a chosen species, or are we but one of many examples of intelligent life? "The answer calibrates our place in the universe," said astronomer Jill Tarter.[2]

The public is engaged. Opinion surveys since the 1960s have shown a steady rise in the percentage of North Americans who believe that extraterrestrial intelligence exists. A 2000 poll found that 82% of Americans thought there was intelligent life elsewhere in the universe.[3] Many people in other countries share these views.

The higher the education, income, and occupation level, the more likely a person is to believe in the existence of extraterrestrial intelligence. "What has to be explained is not that some people are interested in the subject," said Carl Sagan, "but that some people profess not to be interested in it."[4]

We are in the preconfirmation phase of this intellectual enterprise— the stage that allows the widest and most imaginative range of possibilities. Thinking about intelligent life elsewhere stretches our imaginations. We must conceive of gulfs of space and eons of time. We must visualize alien environments, in which evolution could take very different paths. We must imagine the histories of civilizations unlike our own, projecting their evolution far beyond our present level. Wide-ranging speculations are not only possible, they are required.

When we discuss extraterrestrial intelligence, we cannot limit ourselves to the narrow perspective of one science or academic discipline. Astronomy, biology, history, even climate studies contribute to the debate.

As philosopher David Lamb saw it, no other subject is as capable of bringing together the fragmented elements of contemporary science.[5] This applies not only to scientists but also to nonexpert citizens, for whom this topic may be the most intriguing entry portal into science's diverse attractions.

Speculations about contact often challenge conventional wisdom, from biology to religion. Imagining extraterrestrials, and what they might do, irritates those most sure of an established worldview. We must reconsider what we have taken for granted.

Our thinking about extraterrestrial intelligence also must contend with an innate human longing to feel important. To some people, that requires that we remain unique. The easiest solution available to them is to declare that intelligent aliens do not exist.

Hopes and Fears

Much of the debate about this question has been conducted at a high intellectual level, in the language of science and metaphysics. Scholars have given us fine histories of the idea, particularly Karl Guthke's elegant literary analysis in *The Last Frontier*, Michael Crowe's fascinatingly detailed study *The Extraterrestrial Life Debate, 1750–1900*, and Steven Dick's paradigm-setting book *The Biological Universe*.[6]

There is another level in this debate—the level of emotion and personal experience. Many people look forward to contact because they expect more than scientific and philosophical discussions. They hope to learn about other arts, other moralities, other ways to organize societies. They hope for guidance that will help us to solve our problems. They want to be reassured about the future of intelligent life.

There is a yearning that goes beyond a thirst for knowledge. We seek encounters with things strange and wonderful and larger than ourselves. We want a more appealing and responsive universe than the lifeless, mindless expanse that science has so far revealed to us. This intellectual venture connects us with deep emotions.

In the recent past, we have relied on science fiction to disrupt the mundane. Now we hope that our scientific explorations will achieve the same end. "Across the seas of space lie the new raw materials of the imagination," Arthur Clarke told us. "Strangeness, wonder, mystery, and magic—these things, which not long ago seemed lost forever, will soon return to the world."[7]

Many hope that contact will change us for the better; others worry about the demoralization of Humankind, even its destruction. The cosmos is a vast canvas on which we can paint our most imaginative and radical theories, our hopes, and our fears. "Space is a sea without end which washes on countless strange and exotic shores," wrote space visionary Dandridge Cole, "where the conceivable forms of the living and the dead are greatly outnumbered by the inconceivable; where the known is lost in the unknown; where new dangers hide in undiscovered shadows in unimagined forms."[8]

How realistic are these hopes? How realistic are these fears? We will consider the arguments about both.

Searching

After Galileo, humans could begin testing theories about extraterrestrial life and intelligence by observation. Astronomers first studied the Moon; when the near side appeared lifeless, pluralists populated the other side.[9] The search then expanded outward to the other planets of our solar system, particularly Mars. Although some observers thought they had detected the works or signals of alien minds, their claims were discredited. Finding no confirmed evidence of other intelligences nearby, we have extended our search to the stars.

The way we search has been limited to the technical means available at the time. In each phase, observers have been operating at the limits of those technologies; initial results, and extrapolations from them, often proved to be incorrect.

This is not unusual in astronomy. Edwin Hubble, speaking in the 1930s, described the uncertainty of exploring the universe with telescopes. Although we know our immediate neighborhood rather intimately, our knowledge fades rapidly with increasing distance. "Eventually, we reach the dim boundary—the utmost limits of our telescopes. There, we measure shadows, and we search among ghostly errors of measurement for landmarks that are scarcely more substantial."[10]

At different times, scientific findings and theories have supported or diminished the probability of extraterrestrial life and intelligence. Stars with planets were assumed to be common, then rare, and again common. The complexity and uncertainty of biological evolution was seen by some as making alien life and intelligence highly unlikely; others argued that the universality of physical processes, including evolution, made them probable.

These differences are reflected in a prevailing division between those who believe extraterrestrial intelligence exists and those who do not: believers and deniers. As we do not yet have sufficient evidence to prove either case, these views are opinions. Another body of opinion, the agnostics, may lie between, but they are much less likely to be heard.

Many advocates of the search are eager to find extraterrestrial intelligence, as if contact were sure to change our future in a positive way; one detects a hint of transcendental aspiration. The searchers persist, even in the knowledge that the first detection may not take place in their lifetimes.

Others do not want scientists to search for alien intelligence. Particularly striking is the strenuous resistance by some critics to government financing of any search programs, even though the amounts actually sought have been a tiny percentage of public funding for science. Reacting to the publicity surrounding the first modern search for alien radio signals, astronomer Otto Struve concluded that this effort had divided his colleagues into two camps: those who are all for it and those who regard it as the worst evil of our generation.[11]

Since the early years of the Space Age, searching for alien life has been politically accepted as legitimate science. Searching for alien intelligence remains far more controversial. This has not been just a matter of degree, in which a compromise could be reached. More is at stake than a zero-sum budget rivalry.

Scientists in many other fields frequently propose exotic explanations of physical phenomena. Although hypotheses such as string theory have not yet been proven, their authors are taken seriously enough to receive taxpayer-funded grants. The search for extraterrestrial intelligence is far more precarious. As astronomers Iosif Shklovskii and Carl Sagan pointed out, there are unconscious factors operating in the arguments of both the proponents and their opponents.[12]

Those who most doubt that we will find evidence of alien minds sometimes seem to be motivated by a healthy skepticism about unproven theories. At other times, they seem driven by philosophies, religious or ideological beliefs, or emotions. We could make similar comments about those who are most confident that the search will succeed.

Despite the advances that science has made, no one can invoke authority on these questions. The widespread belief that intelligent extraterrestrials exist still rests on logic and intuition, not observations and experiments. That does not mean that this belief is wrong—only that it is unproven.

Consequences

The cardinal question of the actual outcome of the encounter of mankind with extraterrestrial civilization—whether it will be beneficial or harmful—has not been answered unanimously.
—Soviet astronomer S.A. Kaplan, 1969[13]

Why do we search for others? It may be more than curiosity, or an extension of normal science. In modern times, the search has been driven increasingly by a desire to introduce new and hopefully positive factors into human affairs.

During most of its long history, this debate did not address the effects of encountering alien intelligence. Some authors used imaginary meetings with aliens on the Moon (and the Sun!) to satirize our own species and to suggest utopian futures, but they rarely foresaw significant consequences arising from contact. The threat posed by a multitude of inhabited worlds was to the presuppositions of dominant thought systems, not to human safety.[14]

The debate went through a major turning point in the 1890s. Improvements in astronomical technology and technique had sharpened our perceptions of other planets, particularly Mars. Percival Lowell's theories about an advanced civilization on the red planet stirred widespread interest.

They also stimulated fictional statements about direct contact with Martians who traveled through space to the Earth, using their greater technological powers to colonize our planet. Kurd Lasswitz, in his novel *On Two Planets*, envisioned an initial conflict followed by reconciliation and the beginnings of a tutelary relationship that would advance Humankind to a higher state. H.G. Wells, in his *The War of the Worlds*, foresaw disaster. This division of opinion about the outcome of contact has continued into our own time.

Many speculations about the consequences of contact have rested on assumptions about what extraterrestrials will be like and how they will behave. Those assumptions rest, in turn, on analogies with the nature and behavior of humans. As later sections of this book will show, those analogies have been questioned forcefully.

The most intensely debated issues about the consequences of contact are not scientific. They concern social, political, philosophical, and religious questions of some magnitude, ranging from intellectual disorientation to the extinction of the human species. Many speculations about the societal implications of contact are outside science; they tread on other sensitivities.

Attitudes toward the consequences of contact have reflected a wide range of cultural and personal views, with many arguments gathered around the poles of millenarian optimism and catastrophic pessimism. Some extrapolate from our history as it has been, including its horrors. Others foresee a future history as they would prefer it to be, imagining that other intelligent beings do not share our faults. Deniers evade the debate about consequences entirely by claiming that we will never come into contact with intelligent aliens.

Scientists who firmly attach themselves to what can be observed and confirmed when they are doing science in their own fields often abandon that hard connection with reality when they speculate about the behavior of extraterrestrials. Many commentators leave out the possibilities that do not support their opinions, defying or ignoring precedents in our history that do not fit within their preferred visions.

Science fiction author David Brin found that "few important subjects are so data-poor, so subject to unwarranted and biased extrapolations—and so caught up in mankind's ultimate destiny—as this one." Many people have an emotional investment in the outcome.[15]

Arguments about the consequences of contact continue to evolve with new scientific findings and new technological capabilities, and with changes in our cultures. Although specific visions and worries may change with the times, the underlying split between optimism and pessimism remains.

Discussions of consequences have been democratized. Participation in this debate has widened beyond the astronomical community to include biologists, anthropologists, historians, psychologists, journalists, international lawyers, legislators, and many others. None of the participants can claim authority, as no human has actually studied an extrasolar civilization. All of us are speculating.

All the theorizing and experimentation that scientists do, proposed neuroscientist Christof Koch, depends on particular metaphysical background assumptions about the world. Nowhere is that more true than in theories about extraterrestrial intelligence. "Where facts are few," declared psychologist Carl Jung, "speculations are most likely to represent individual psychologies."[16]

Before and After 1960

The year 1960 provides a useful dividing line between eras in the debate about consequences. Scientific and technological advances since then have implied that contact could have a real impact on our lives. Signals from a more advanced civilization might contain information that would change our cultures and our opinion of ourselves. Direct contact could be a more intimate and possibly devastating experience.

Turning points cluster around 1960. Giuseppe Cocconi and Philip Morrison's seminal article in 1959 called attention to our new capability to search for alien radio signals at interstellar distances. Frank Drake undertook the first radio telescope search in 1960. A year later, he proposed an equation that allowed us to estimate the probability of detecting technological civilizations.[17]

Meanwhile, our species acquired a new potential visibility because of the radio, television, and radar signals we were emitting. For the first time, a remote alien observer with only the equipment of the kind already operating on Earth could detect human technological activity.[18]

The Space Age began at about the same time. The first artificial satellite went into orbit in 1957; the first human orbited the Earth in 1961. This technology fed a different set of expectations, including visions of exploration and colonization beyond our solar system.

Our new capabilities supported different paradigms: a slow exchange of radio messages between ourselves and a distant civilization, or direct contact with extraterrestrials (or their machines) that come to our solar system in interstellar spacecraft. As we will see, those paradigms collided in the 1970s. By then, speculations about the positive consequences of contact had become sweepingly enthusiastic among those who expected a remote encounter by radio. Their optimism provoked a counterreaction, whose spokespersons sometimes took equally extreme positions.

Both sides may be too sure of their answers. As cultural historian Jacques Barzun once noted, history gives no comfort to the many able, dedicated minds that crave finality and certitude.[19]

The Book

This book looks at the long debate from both sides. Under probabilities, we look at both the positive and negative arguments for the existence of extraterrestrial intelligence. Under consequences, we look at both the optimistic and pessimistic predictions of what contact might bring. The dream of cosmic concord has been tempered; contact may not lead to a Woodstock of the skies. It may not imply Armageddon either.

We begin with a condensed history of speculations about contact with extraterrestrial intelligence up to 1959, with an emphasis on foreseen consequences. After brief descriptions of the scientific searches for extraterrestrial life and alien signals, we consider arguments about the probability of finding other technological societies. Most of the multiple factors that have been proposed implicitly limit the probability of contact, although to what degree is highly subjective.

We then take a close look at a second major model of contact—one that has very different implications from detecting a faint radio transmission. This paradigm could significantly increase the probability of encountering extraterrestrials. That leads us to review what some see as the paradoxical implications of results so far. A review of possible explanations suggests that there may be no paradox.

In the second half of the book, we review the main predictions that have been offered about the consequences of contact. We critically examine underlying assumptions and consider what has been missing from this debate.

The book ends with some conclusions about the search and its consequences. We consider paradigm shifts that appear to be under way and suggest some directions for Humankind's future role. An annex describes steps that have been taken or proposed to prepare for contact.

An Apology

The greatest variety of speculations about the consequences of contact is to be found in science fiction. Although scientists searching for signals from many light years away have not given much weight to this literature, it must be taken into account, particularly when we consider the possibility of a more direct encounter.

The imaginative output of science fiction writers on this subject deserves a book of its own. This one includes only samples of the most relevant ideas.

A Belief in Other Minds

One can hardly be a complete human being without at least occasionally calling to mind the community of rational beings, as yet unknown, to which we presumably belong.

—Stanislav Lem, 1976[1]

An Ancient Idea

For much of their time on Earth, most humans have believed that they share their world with non-human intelligences, particularly spirits that animate natural phenomena. Those spirits could do good, providing rain for crops and granting survival to children. They could do harm, bringing droughts, floods, or disease.

Such animators were implicitly intelligent. Earlier humans may have believed the world around them to be vibrant with unseen spirits motivated by thoughts and emotions mirroring their own, speculated astronomer Edward Harrison. He proposed that this "magic universe" arose hundreds of thousands of years ago, when our ancestors began to acquire advanced linguistic skills.[2]

These animistic beliefs often evolved into the polytheism seen in many earlier societies. Monotheism responded to the same need, providing a God who designed, created, and directed the world around us. Whether plural or singular, animators were intangible, beyond human sight or reach.

This belief in sapient but unseen others has recurred in one form or another throughout our history. Their imagined natures and locations have evolved with cultural change, intellectual advance, and new conceptions of the cosmos. They have remained invisible, detectable only by the effects of their actions.

Here we enter into our first speculation. Is this belief in unseen intelligences a characteristic of *Homo sapiens*, a peculiarly human attribute? Or is it shared by other sentient beings that may exist elsewhere in the universe? The answers to those questions have powerful implications for the probability of contact with extraterrestrials, and for its consequences.

The prehistoric universe was an intimate one; for the vast majority of human beings and for most of their history, there were no other worlds than Earth. Not until the ancient Greek rationalists do we have documented evidence of a belief in intelligent beings on other worlds. Some of the Pythagoreans of the sixth to fourth centuries B.C. speculated that the Moon and stars were peopled with humans or similar beings.[3]

The later Greek Atomists gave us the first arguments from probability. Democritus argued that there must be an infinite number of worlds because space and time are infinite. "Nowhere exists an obstacle to the infinite number of worlds," wrote Epicurus in his Letter to Herodotus. "We must believe that in all worlds there are living creatures and plants and other things we see in this world."[4]

The Roman poet Lucretius (Titus Lucretius Carus) spread the Atomists' themes in the first century B.C., expressing an argument that still underlies much of the thinking about the probability of extraterrestrial intelligence:

It is in the highest degree unlikely that this earth and sky is the only one to have been created . . . You are bound therefore to acknowledge that in other regions there are other earths and various tribes of men and breeds of beasts. . . . Nothing in the universe is the only one of its kind, unique and solitary in its birth and growth."[5]

Speculation was not limited to Greece and Rome. The Vedic worldview of eighth- to twelfth-century India envisioned a hierarchy of planets encompassing 400,000 humanlike races and 8,000,000 other life-forms. Muslim natural philosophers speculated about a hierarchy of intelligences supposed to animate the successive planetary spheres of the pre-Copernican universe.[6] None of these visions—Eastern or Western—were confirmed by observation.

The Wisdom of the Buddha

Consider Siddhartha Gautama's vision of the universe in the sixth century B.C. The Buddha's concept embraced billions of "minor world systems," each of which resembles our own planetary system. In the infinite space of the universe there exist billions of suns, billions of moons, billions of inhabited regions. "It is amply clear,"astronomers Fred Hoyle and Chandra Wickramasinghe told us, "that Buddha viewed life and consciousness (which he thought to be associated with all life) as cosmic phenomena, linked inextricably with the structure of the Universe as a whole."

The Buddhist treatise known as The Lotus of the True Law depicts Bodhisattvas spread in all directions in hundreds of thousands of worlds. This cosmic plenitude of Buddhas-to-be does not seem to play a significant role in the faith.[7] However, the idea that life and intelligence are natural expressions of the universe now is advocated by many scientists.

The plurality of inhabited worlds that some took for granted was a belief, not a proven fact. Others assumed that only the Earth is peopled with intelligent beings, establishing a division of opinion that has survived into our own time.

Plato, while admitting that the habitability of the planets was an open question, believed that the Earth was unique. Aristotle argued more strenuously against plurality. In his geocentric universe, the planets could not be inhabited because they were made of completely different substances than ours; the physical laws governing heaven and Earth were not the same.[8] This idea that our world and the surrounding cosmos were composed of different elements remained dominant in the Western world until the sixteenth century and was not disproved scientifically until the nineteenth century.

The Closed Universe

Ptolemy (Claudius Ptolemaeus) codified the Aristotelian conception of the Cosmos in his Almagest in the second century A.D. The Ptolemaic universe—a series of concentric spheres centered on the Earth—was closed, its outer boundary a fixed sphere of stars.

Although we now think of it as wrong, Ptolemy's universe was elaborately worked out and, by the standards of the time, internally consistent. Ptolemy and his colleagues may not have believed that their model was completely accurate; they may have seen it as a model compatible with observations and as one that provided a basis for reasonable predictions.[9]

The early Christian church endorsed this geocentric universe. The Aristotelian–Ptolemaic worldview became dogma in Europe for more than 1000 years, as did the presumed centrality and uniqueness of Humankind. Some argue that the Church saw this as only the most plausible of the existing descriptions of the universe, but that subtlety was lost on most people.

The church's domination of the European intellectual world may have delayed the development of Western science for centuries. Not only had rational thought been suppressed by the fifth century, Charles Freeman found, but there had been a substitution for it of mystery, magic, and authority that drew heavily on pagan elements. The mystical theologian

Pseudo-Dionysius declared in the early sixth century that "it is most fitting to the mysterious passages of scripture that the saved and hidden truth about the celestial intelligences be concealed through the inexpressible and the sacred and be inaccessible to the *hoi polloi*."[10]

Why did this geocentric conception survive so long? Even if it was not scientifically accurate, the Christian world view of the Middle Ages gave an integrated, rational account of all creation. The medieval universe might be grounded in mystery, but its articulation and structure were comprehensible and logical; Harrison described it as the most satisfying and self-sufficient world system ever devised by the human mind.[11] By comparison with some Asian conceptions, this Western universe seems cramped. However, it was comfortable.

A dilemma arose when the teachings of Aristotle were reintroduced into Europe in the thirteenth century: the dogma of our world's uniqueness threatened belief in the omnipotence of God. The Condemnation of 1277, while ruling that the plurality of worlds *as a reality* was a pagan error, forbade teaching that God could not have created a plurality of worlds. This was not done in defense of plurality, but in defense of God's power.[12]

Cardinal Nicolas of Cusa (Nicolas Cusanus) made a bold statement of plurality in 1440. According to Nicolas, the universe was infinitely large; there was no essential difference between terrestrial matter and celestial matter; each star was a sun like ours with its own complement of planets, which probably were inhabited. The existence of humankinds on other worlds, wrote the Cardinal, is an absolute certainty.[13]

Pluralism had spokesmen in other cultures as well, although they had no more proof than Westerners. During the Sung dynasty in thirteenth-century China, Teng Mu restated the argument from probability: It would be unreasonable to suppose that, besides the Earth and sky we can see, there are no other skies and no other earths.[14]

The Common Man

Intellectual histories of this debate give great weight to speculations among the literate elite. What was the illiterate, uneducated European public thinking about non human intelligences? Their world view included angels and demons who inhabited the heavens, took human form, and visited the Earth.

Space historians Roger Launius and Howard McCurdy observed how popular images of mysterious others continued to evolve. In the spiritually conscious nineteenth century, people turned their attention to ghosts and sought methods for communicating with the dead. As science replaced superstition, people embraced extraterrestrials, especially those with advanced technologies.[15]

Copernicanism

Nicolas Copernicus, however unintentionally, gave plurality new life in 1543. His treatise "On the Revolutions of the Heavenly Spheres" proposed a universe in which the Earth and the other known planets revolved around the Sun, sending our world into motion and displacing the Earth and its inhabitants from the center of the cosmos. Copernicus did not invent this heliocentric conception of the universe; he knew that Aristarchus of Samos had proposed it long before. He refined and extended the idea through logic, mathematical calculation, and intuition—and very few observations.[16]

Copernicus has became a symbol, the man who ended one age and ushered in another. To many, he is the figure who led the West out of medieval obscurantism, giving us a more realistic conception of the universe. In our eagerness to praise him, we should recall that he did not dismantle the entire Ptolemaic scheme. He held that the Sun was the center of the universe, not just of the solar system. The Copernican cosmos remained closed, bounded by a sphere of fixed stars.[17] Long-established conceptions of the universe that frame our sense of order are hard to kill.

Nonetheless, Copernican theory had powerful implications for the debate about the multiplicity of inhabited worlds. It suggested that the other planets of our solar system were worlds rather than points of light on celestial spheres. Copernicanism also implied that there is no qualitative difference between one part of the universe and another. This did not prove that other planets were inhabited, but it did suggest that they might be inhabitable.

Many people today think that Copernicanism demoted humans by removing them from their central pedestal. In fact, the Ptolemaic universe had placed the Earth at the lowest point in the cosmos, the least perfect level of existence; Arthur Lovejoy described it as "the dim and squalid cellar of the universe."[18] To remove humans from the center of things actually raised them from their low estate.

Religious authorities found heliocentrism less troubling than the plurality of inhabited worlds. It was not the position of our planet in space that gave the Earth its unique status and its unique share in the attention of Heaven, but the fact that it alone was supposed to have an indigenous population of rational beings. By implying that there might be many Humankinds, plurality subverted our uniqueness.[18]

Some sought to reconcile plurality with Christian doctrine by invoking what later became known as the Principle of Plenitude. The Creator's omnipotence and freedom of action meant that whatever God can create will be created, including all possible worlds. Multiple worlds with sentient life-forms could be seen as demonstrating the wisdom and benevolence of God.[19] This theory did not resolve the underlying dilemma for Christian theology; if there were intelligent beings on other worlds, we might not be God's unique and central concern.

The Passionate Monk

Forty years after Copernicus published his great work, the itinerant monk Giordano Bruno was forcefully advocating a decentralized, infinite, and infinitely populated universe. Believing that the supreme characteristic of the cosmos is its homogeneity, the intellectually passionate Bruno argued that the physical composition of all heavenly bodies is essentially the same. He declared that the fixed stars are suns and supposed, by analogy, that they are orbited by planets. For him, those planets were worlds, every one inhabited. He ridiculed the assumed centrality of Humankind by proposing that those who inhabit the moon "without doubt" believe themselves to be at the center of the universe.

These radical ideas may have awakened the Catholic church to Copernicanism's heretical implications. Bruno was put on extended trial by the Inquisition for a variety of offenses, including attacks on some basic tenets of the Christian religion such as the uniqueness of the Incarnation.[20] His grim fate—being burned at the stake in 1600—may have been inevitable even if he had never campaigned for a plurality of inhabited worlds.

Meanwhile, Copernican theory had stimulated new ways of looking at the heavens. Tycho Brahe, watching the skies without a telescope, spotted a comet traveling within the planetary system, puncturing the assumed etherial spheres of the Ptolemaic universe. He also observed a new star (nova) that remained visible for over a year in the supposedly immutable realm of the fixed stars. The stage was set for the greater revolutionary, Galileo.

The Triumph of Observation

Galileo initiated a new age of astronomy when he scanned the sky with his simple telescope, particularly during the winter of 1609–1610. He observed that our Moon had mountains and plains like the Earth and thought he had seen seas as well. He reported planets as spherical objects; he saw that Venus went through phases as it would if it orbited the Sun. He detected moons orbiting Jupiter, proving that not all bodies revolved around the Earth.

Galileo's discoveries greatly strengthened the credibility of Copernican theory. Although they did not prove definitively that the Copernican model was right, they made it possible for the interested public to believe that it was real. Speculations about inhabited worlds could move beyond imagination to become extrapolations from fact.

The Galilean telescope came along at the right moment in history, argued astronomical anthropologist Anthony Aveni. Innovative ideas, fully formed, were ripe for empirical testing. After Galileo, the sky was no longer a place filled with moving lights that affect our destinies and passions; it became a vast domain populated by worlds like our own, with surfaces, atmospheres, mountains, cores, and mantles—worlds that whirl around one another under a different rule of law, one of unseen powers that are part of a natural rather than a divine order.[21]

Galileo gave new weight to a distinction that has been central in the debate about extraterrestrial intelligence, between what we believe to be true and what we can actually detect. He pointed out that there were no sure observations on the question of inhabitants of other worlds; the astronomer cannot affirm that something exists merely because it is logically possible. If life existed on the Moon, he thought, it would be extremely diverse and far beyond our imaginings.[22]

Meanwhile, the European age of discovery and conquest had added other dimensions to the debate. Vast realms of our own planet, previously unknown to Europe, were found to be inhabited by strange peoples and alien societies that had developed independently of Eurasia. Saint Augustine had condemned the belief in the very existence of such "Antipodeans."

Confronted with reality, Europeans had to ask if these creatures were as human as they were, or of a separate, alien nature. In 1537, the Church decided that Christians could regard the beings they encountered in the Americas as human.[23]

Fictional Aliens

The revival of the plurality of worlds enabled authors to explore the imagined features of other planets and the nature of their human or quasi-human inhabitants. Many used fictional aliens as critics of human follies—or as models for rectifying them.

Some exploited the fictional cosmic voyage as a literary vehicle. Johannes Kepler's *Somnium* (Dream), written in 1609 but not published until 1634, described an imaginary journey to the Moon, which he imagined to be inhabited by intelligent analogs to humans. Kepler made the bold proposal that life may have originated in a spontaneous, mechanical fashion on Earth; by implication, it might arise elsewhere. Another Keplerian speculation was to resonate loudly in the twentieth century: Space travel would enable people from "our species of Man" to settle other worlds.[24]

Other authors imagined meetings between humans and their counterparts in space, although they did not foresee significant consequences from such contact. Bishop Francis Godwin's *The Man in the Moone* (1638) envisioned aliens as being similar to humans, but their world was like Paradise.[25] This conception of extraterrestrials being not only better than us but also happier reappeared in later speculations; imaginary societies gave us models of utopia to aspire to, or visions of abasement to be shunned.

Some pursued the idea that we could initiate direct contact with extraterrestrial beings. In his *Discovery of a World in the Moone*, first published in 1638, Bishop John Wilkins speculated that future generations might invent means for our better acquaintance with the inhabitants of other worlds; the problem was merely one of transport. Pierre Borel, a counselor to the French court, proposed in 1657 that humans could learn the truth about the plurality of worlds by aerial navigation, what we now would call space travel.[26]

The skeptics had their say. To novelist Charles Sorel, an encounter with aliens was the height of absurdity. He introduced the idea of an invasion of the Earth from outer space, but only as a ridiculous concept.[27]

The Importance of the Moon

The Moon, hung high above us, was the natural first target for our extraterrestrial aspirations. Brightly visible, obviously round, its face marked by intriguing patterns of light and dark, it was a little world. Except for the Sun, it was the only body in the solar system whose disk we could see with the naked eye; it suggested to us the shape of the planet on which we live. Although other heavenly bodies seemed to be pure light, unchanging, and eternal, the appearance of our natural satellite varied as it went through its monthly cycle.

With the invention of the telescope, we saw the Moon's craters, its seas of hardened lava, even the shadows of its peaks. Would we have imagined other worlds if there had been no Moon, or no planet close enough to look different from a star?

The Moon offered a tantalizing destination. Once we combined technology with intent, we broke the barrier of distance with startling rapidity. Only 8 years after the first human ventured into space, others set foot on our natural satellite. That achievement encouraged visions of human journeys to the planets and beyond. Would those visions be in our minds if the Earth had no Moon?[28]

This is not just idle speculation. It helps us to imagine what extraterrestrial civilizations might do—or not do.

An Impersonal Universe

Rene Descartes in 1644 proposed that the universe is not a void, but is filled with atoms whose vortices form stars and planets. He suggested that an infinite number of creatures far superior to us may exist elsewhere. Recognizing the implied demotion of Humankind, Descartes claimed that our merits are not diminished by the fact that intelligent beings on other heavenly bodies have similar ones.[29]

Cyrano de Bergerac drew radically anti anthropocentric and anti theological conclusions from Descartes' cosmology. In satiric novels published in 1657 and 1662, he dismissed the idea that the universe exists for our benefit. Portraying an imagined meeting between humans and their counterparts in space, he described the aliens as being superior in intelligence. He drew irony from this situation; the extraterrestrials had believed the Earth to be uninhabited.[30]

As the Enlightenment gained ground, the educated public increasingly accepted the belief that we are not alone. Two works had a particularly strong impact in Europe by making the idea of plurality accessible to non scholars. Bernard Le Bovier de Fontenelle's *Conversations on the Plurality of Worlds* (1686) expanded on Copernicus and the Cartesian universe, spreading the idea of a plurality of solar systems. The Frenchman argued that the Moon must be inhabited because it is like the Earth; the planets must be inhabited because they are like the Moon. If our Sun gives light to the planets, every star might do the same for its planets. De Fontenelle imagined that life elsewhere could be radically different from life on Earth; aliens might have new senses and other capabilities unknown to us.[31]

In his *Kosmotheros*, published in 1698, astronomer Christian Huygens also proposed a plurality of solar systems, primarily on the analogy that the stars were other suns. He reasoned that all the planets must have plants, animals, and intelligent life adapted to their environments. We can not feel threatened and downgraded by the higher degree of reason possessed by other planet dwellers, as enlightened Man is the highest conceivable form of life.[32]

Meanwhile, Newton had published major works on physical laws and the motions of the planets, showing how gravity proved the truth of Copernican theory. William Herschel's later observations of double stars confirmed that celestial objects light-years away behaved according to Newton's laws of gravity and motion.[33] Physical laws seemed to be the same everywhere in the known universe; we humans could understand the cosmos by means other than faith or revelation.

Many people found the Cartesian and Newtonian universes cold and impersonal. "I feel engulfed in the infinite immensity of spaces whereof I know nothing and which know nothing of me," said Blaise Pascal; "the eternal silences of those infinite spaces alarm me." De Fontenelle too felt lost in an immense universe: "Our world is terrifying in its insignificance."[34]

What could one do to give purpose and human relevance to such a fearful void? "Perhaps just because the silent emptiness of Newton's infinite universe threatened to engulf and utterly lose so petty a planet as Earth," theorized historian William McNeill, "the 17th and 18th centuries were particularly prolific of new religious movements and sects, all of them emphasizing a direct, emotionally charged experience of God."[35]

There also was an implied tension between feeling isolated in a vast universe and being insignificant and anonymous in a cosmos infinitely inhabited by other beings. "I was afraid of being overlooked amidst the immensity of nature," wrote Joseph Addison, "and lost among that infinite variety of creatures."[36]

There were hints of other consequences. The jubilant feeling of cosmic brotherhood, suggested Guthke, can easily give way to fear of intelligent extraterrestrial beings that are no longer subject to the wise surveillance of a Creator-God. What if such aliens were more powerful than we are? De Fontenelle had one of his characters say "Tis no great matter whether we go to them, or they come to us, we shall then be the Americans," that is, the weaker party.[37]

Intellectual dilemmas persisted. The Age of Reason sought to leave behind the traditional Christian concept of human nature by making Man the measure of all things. The possibility of alien intelligences challenged this belief; we could not be the measure if extraterrestrial beings were superior to us.[38]

By the end of the eighteenth century, the dominant cosmology assumed that physical and evolutionary laws were the same throughout the universe, that the building blocks of life were plentiful everywhere, and that a universe teeming with intelligent life was overwhelmingly likely. The concept of a solitary planet in an immense ocean of space, commented Thomas Paine, was giving way to the cheerful idea of a society of worlds.[39] This vision was revived in the modern debate.

Pluralism seemed to be in the ascendant. However, eighteenth-century pluralists left two problems for their nineteenth-century successors: reconciling pluralism with Christianity, and the lack of astronomical evidence.[40]

Whewell's Challenge

William Whewell, Master of Cambridge's Trinity College, published a major anti pluralist treatise in 1853. Attacking the use of analogy to draw scientific conclusions, he may have been the first to look critically at the empirical evidence for extraterrestrial intelligence, as opposed to the philosophical or theological arguments. He was the first to argue that life could develop only on planets within the boundaries of certain conditions, a concept that evolved into today's zone of habitability. Whewell also recognized that the existence of life on another world does not necessarily imply intelligent life. He pointed out that Humankind was very recent; for most of Earth's history, there were no intelligent beings on our planet.

As a practicing Christian, Whewell believed that, were worlds other than ours populated with intelligent beings, they would have some kind of relationship with God. That would dilute any special relationship we humans claim to have.[41]

Whewell was not the only skeptic. American astronomer Denison Olmsted argued that "The telescope . . . has added nothing to the amount of evidence in favor of the doctrine that the planets are inhabited. It has, in fact, greatly diminished that amount, since the points of dissimilarity to the Earth . . . have increased faster than the points of resemblance."[42] More than a century later, other skeptics revived the idea of our Earth being a rare phenomenon.

Evolution and Sharpened Questions

Before the nineteenth century, observed McNeill, men had seldom taken seriously the obvious proposition that all things in the universe, and the universe itself, have a history.[43] Arguments about extraterrestrial life had taken place in a static context; humans and other living things were always as they are, created at one moment.

After Darwin and Wallace, the idea that life on Earth is the result of a miracle or of a rare chance event was replaced by an evolutionary narrative, giving plurality a new boost. However, the concept of natural selection highlighted the probability that the evolution of life elsewhere in the Universe would be quite different than on Earth.[44] Natural selection also implied competition, inspiring a harsh social and political doctrine: The fittest will survive, and the less fit will die out.

Evolutionary theory had a broader impact on our view of the cosmos. Instead of being a static structure with fixed properties, the universe became an entity whose characteristics changed over time.

Nineteenth-century science gave us new tools. Armed with devices for spectrum analysis, astronomers were able to prove that celestial bodies were made up of the same chemical elements as our Earth, both within our solar system and beyond it. If life could evolve from those elements here, it might evolve on other planets as well. Science popularizer Camille

Flammarion, writing of countless human like beings in space, called on us to greet our "sister humanities."[45]

Despite such enthusiastic visions, a new climate of skeptical and rigorous inquiry began to discourage wild speculation about the existence of extraterrestrial beings. Science removed the supports from early casual assumptions about the existence of other minds, commented Isaac Asimov. The scientific view of the universe admitted only those phenomena that could be observed. As science found no evidence of other intelligences, Humankind might stand alone.[46]

Lowell, and Alien Invaders

Not by its body, but by its mind, would it be known. Across the gulf of space it could be recognized only by the imprint it had made on the face of Mars.

—Percival Lowell, 1908[47]

In the latter nineteenth century, the techniques of planetary observation still provided the best hope for detecting evidence of extraterrestrial life and intelligence. As improved telescopes showed the Moon to be a bleak, cratered world, the search moved outward to Mars—the only planet whose surface can be seen in any detail through telescopes on Earth.

In 1877, Italian astronomer Giovanni Schiaparelli described *canali* (channels) on the red planet. He speculated that these might be canals, the works of an alien civilization.[48]

American astronomer Percival Lowell, stimulated by Schiaparelli's observations and by Flammarion's 1892 book *Mars*, began conducting his own astronomical studies in 1894. Like Schiaparelli, Lowell saw the *canali* as straight lines that could not be of natural origin. Concluding that they were canals bringing water from the Martian polar caps to drier terrain, he developed a powerful theory: Mars was an aging, drying planet whose inhabitants had been forced to use massive engineering works to save their civilization.[49] The plight of the Martians was a preview of our own future.

As they had preceded us and were by implication an older and wiser species, Lowell expected the Martians to be more socially and politically advanced than we humans. Schiaparelli imagined collective socialism on Mars, a "paradise of societies." Wars would be unknown; society's efforts would be directed against the common enemy of "penurious nature."[50] The extraterrestrial utopia was back, but in quite a different form.

Lowell described his detection of an alien civilization as a second Copernican shock. The presence of Martians ousted us from any unique or self-centered position in the solar system. Humans were merely this

Earth's highest production to date; we are certain to be excelled by others.[51] Lowell recognized that encountering something unknown and superior is liable to induce fear and uncertainty. However, he argued, it is time for us to grow up. These ideas too were revived in the modern debate.

Lowell developed a concept that later gained credibility among scientists: planetology, the comparative study of worlds. He may have been the first to pursue an interdisciplinary approach to studying a purported alien civilization. Recognizing that this involved far more than astronomy, he invited a zoologist and a sociologist to participate in the debate.[52]

Stimulated by such theories, others imagined that more advanced aliens could directly affect our lives and even threaten our survival. In his 1897 novel *On Two Planets*, Kurd Lasswitz envisioned Martians colonizing the Earth. Like some Earthly imperialists of that time, they regard it as their duty to set up a protectorate over the indigenous intelligent species. After a human rebellion and a truce, the Martians launch a program to help our species advance morally. As Guthke saw it, Lasswitz expressed the hope of more rapid evolution through the educative intervention of extraterrestrial guardian angels.[53] This hope remains widespread in our own time.

H.G. Wells made grimmer assumptions in his *The War of the Worlds*, first published in magazine form in 1897. He foresaw that the life-forms most able to survive might not be the best from an ethical point of view; instead, they might be the most powerful and aggressive. His Martians are ruthless invaders intent on conquering the Earth. They treat humans like inferior animals, destroying any who get in their way. Humankind is saved only when the Earth's humblest organisms, the bacteria, infect and kill the Martians.[54] This division of opinion between seeing aliens either as enlightened guardians who reshape our future or as unfeeling technological monsters who threaten our existence persists to this day.

Historian of astronomy Agnes Clerke, responding to Lowell's theories, wrote in 1904 that evidence of extraterrestrial life was not at hand. She went beyond that objective statement of fact to argue that we could not search for such evidence.[55]

As an historian, Clerke should have known better. French philosopher August Comte had declared around 1835 that, although we may learn the forms, distances, sizes, and motions of stars, we can never know their chemical composition. Yet Fraunhofer already had discovered that spectrum analysis could be used to determine the chemistry of astronomical objects (although his work did not become widely known until years afterward). Bunsen and Kirchhoff developed a practical device for spectroscopy in the 1840s, bringing the chemistry of stars within our intellectual reach. What seems unknowable now may become knowable later.

Expansion and Skepticism

Twentieth-century astronomers, by revealing the unsuspected vastness of the universe, displaced us even farther from a central position. Harlow Shapley reported in 1918 that our solar system is at the periphery of the Milky Way galaxy. By 1924, Edwin Hubble had shown that there are many other galaxies, each containing multitudes of stars. Meanwhile, Einstein's theories denied the existence of any fixed point or any absolute space in the universe; the cosmos has no center.[56]

This enormous extension of scale powerfully strengthened the argument from probability. These advances in astronomy, declared science historian Steven Dick, destroyed the argument for an anthropocentric universe.[57]

Human life and history were dwarfed by the immensity of geological and biological time. As McNeill put it, astronomers coolly presumed the formation and snuffing out of innumerable stars, casually assumed the existence of other solar systems in all stages of formation and dissolution, asserted without qualm the indefinite reduplication of the galaxy, and speculated freely on superorderings of galaxies.[58]

Despite these findings, the plurality of worlds suffered a temporary setback when the nebular theory for the origin of our solar system was overtaken by another proposal. In the new model, the formation of planets depended on rare close encounters between two stars. British astronomer James Jeans, a leading advocate of this concept, argued that our solar system seemed to be very exceptional. Astronomy, he wrote in 1923, "begins to whisper that life must necessarily be somewhat rare" in the universe.[59] Although we humans no longer could think of ourselves as central, we still could think of ourselves as unique.

In the meantime, improving observations discredited the idea of an advanced civilization on Mars. Lowell's theory ultimately failed the Galilean test; the straight lines had been an illusion, a product of the human hunger for recognizable patterns. Other research showed that Mars was a frigid desert. If any life existed there, it might be limited to simple, hardy forms such as lichens. A few scientists claimed to have detected spectroscopic evidence of plant life, but most rejected or ignored those findings.

The search for extraterrestrials became so burdened with a "giggle factor" that many astronomers turned away from studies of Mars and the question of intelligent aliens. "The revulsion at this miscarriage of the scientific enterprise brought discredit not just to planetary science," wrote astronomer Frank Drake later, "but to the whole idea that there might be life elsewhere in space."[60]

Too many astronomers had tried to resolve a problem beyond the limits of the science of their time. This frustration still may be with us today as we search for signals from more distant civilizations.

Conditions on the other planets of our solar system proved to be even harsher than on Mars. The attractive vision of Venus as a younger, more tropical Earth faded away as that planet's blistering heat was revealed; by 1958, observers were reporting a surface temperature of 600 degrees.

Thwarted hopes for finding intelligence beyond Earth were kept alive in science fiction. The transformation of extraterrestrial life into one of the universal themes of literature, Dick observed, "is some measure of how deeply felt and firmly ingrained was the alien concept in the human mind." Guthke saw the encounter with the alien, as either an enemy or a guardian, as the philosophical and imaginative adventure of modern times.[61]

Skeptics had their say. "It is to be regretted," said astronomer James Keeler, "that the habitability of the planets, a subject of which astronomers profess to know little, has been a chosen theme for exploitation by the romancer. . . . The result of his ingenuity is that fact and fiction become inextricably intertwined in the mind of the layman."[62]

Lowellism

It tells us something that popular interest in life beyond the Earth remained high even after Lowell's ideas had been discredited scientifically. He had evoked not only an intellectual response, but an emotional one as well.

Lowellism inspired many of those who pioneered the Space Age, especially the modern search for extraterrestrial life and intelligence. Without Lowell and those he provoked, there might have been no *Viking* missions to Mars. As Edward Edelson put it, the Viking spacecraft went to Mars because of a human vision of what Mars is like.

This grand speculation set us up for a great disappointment when modern science found that Mars and the other worlds of our solar system were uninhabited. However, Lowellism also showed how big ideas can stimulate thinking for generations, even when the immediate example is proven wrong. They change people's conceptions of what may be possible.

Some who rejected Lowell's theories kept open the possibility of extra-terrestrial intelligence, but displaced it to interstellar distances. Astronomer W.W. Campbell spoke in 1920 of "other stellar systems . . . with degrees of intelligence and civilization from which we could learn much, and with which we could sympathize."[63]

Meanwhile, another profound change in our view of the universe was taking place. The ancient conception of the celestial sphere as a region of tranquility and harmony was giving way to a dynamic universe dominated by violent events.[64] The cosmos was not just a neutral background; it was potentially threatening.

Ubiquitous Life

Life but waits in the wings of existence for its cue, to enter the scene the moment the stage is set.

—Percival Lowell, 1908[65]

The long debate about inhabited worlds had taken another turn by the 1940s. Astronomers found that smaller, cooler stars were rotating more slowly than expected, suggesting that some of their angular momentum had been transferred to planets. By 1958, the idea of numerous planetary systems had regained scientific credibility.[66]

Meanwhile, early research into the origins of life on Earth had strengthened the idea that living things were a natural product of physical and chemical evolution. Russian biochemist Aleksandr Oparin contended in 1924 that there was no fundamental difference between a living organism and lifeless matter; the complex combinations and properties of life must have arisen in the process of physical evolution. British biochemist J.B.S. Haldane independently published similar ideas, describing how the action of ultraviolet light on the Earth's primitive atmosphere might have formed a "primordial soup."[67]

Other scientists began testing these concepts with experiments, finding that precursors to living matter are surprisingly easy to make. Stanley Miller proved in the 1950s that trace quantities of organic compounds that are life's building blocks could be formed through the action of electric discharges and ultraviolet light on the probable constituents of Earth's early atmosphere. If such an evolution happened on the Earth, it might occur elsewhere as well. Harvard biologist George Wald proclaimed that the Oparin–Haldane process would be an inevitable event on any planet similar to the Earth in size and temperature.[68]

Astronomer Otto Struve, who played a significant early role in supporting the scientific search for extraterrestrial intelligence, argued in 1955 that life is an intrinsic and inseparable property of certain aggregates of very complex organic molecules. Most sun like stars have planetary systems, he believed; the total number of planets with some form of life could be in the billions.[69]

Three years later, Shapley told us that "whenever the physics, chemistry and climates are right on a planet's surface, life will emerge, persist, and evolve." He reasoned that there could be millions of opportunities for life, including at least 100,000 life-bearing planets in our galaxy.[70]

Amid all of this optimism about life, there was a psychic cost as science increasingly excluded metaphysical considerations. In Shapley's words, Man was exposed as a recent and perhaps ephemeral manifestation in the unrolling of cosmic time.[71]

Under the scrutiny of science, our own species became peripheral. That provoked a strong reaction, one that was to become increasingly visible after 1970.

A New Era

Life is, after all, just a state of matter, albeit a weird one.
—Physicist Paul Davies, 2005[1]

Searching for Life

Before the Space Age, searching for life and intelligence on the other planets of our solar system always left room for doubt. We knew that our telescopes might not be able to detect evidence of living things or the artifacts of another civilization at interplanetary distances. If we had confined ourselves to peering through our atmosphere, we might have remained forever uncertain; debates could have continued without any prospect of final resolution.

The Spaceflight Revolution gave us new means of searching for life in our solar system. Our machines could observe planets and moons from close orbits, even land to conduct research on their surfaces. This boosted a new interdisciplinary science known as astrobiology or exobiology (astronomers often call it bioastronomy).

In 1962, the Space Science Board of the U.S. National Academy of Sciences set the search for extraterrestrial life as the prime goal of space biology, concluding that "it is not since Darwin—and before him Copernicus—that science has had the opportunity for so great an impact on man's understanding of man. The scientific question at stake in exobiology is in the opinions of many the most exciting, challenging, and profound issue, not only of this century but of the whole naturalistic movement that has characterized the history of Western thought for three hundred years." Harvard evolutionary biologist George Gaylord Simpson commented dryly that this was the first time a scientific discipline had been started before any evidence of its subject matter had been found.[2]

The raw materials for life seem to be abundant; the elements needed for biology exist throughout our galaxy. Discoveries since the 1950s have confirmed that space is rich in carbon-based compounds, including the basic

25

building blocks of life's chemistry; scientists proved in 1970 that a meteorite unambiguously contained extraterrestrial amino acids.[3] Researchers have demonstrated that the spontaneous generation of amino acids in interstellar space is possible.

Many scientists once assumed that not even the earliest steps in the chemistry of life could take place in space because ultraviolet radiation from stars would break up organic molecules. Now it appears that such chemistry can begin in the dust grains of interstellar clouds, at least those that are sheathed in ice. Higher carbon clusters, formed in the outflows of carbon stars, can survive passage through the interstellar medium. Crucial early processes appear to take place in space long before planet formation, making it possible for asteroids, comets, meteorites, and interplanetary dust to deposit complex organic material on early planet surfaces.[4]

Disappointment and Revival

Hopes for finding simple life on Mars still were high in 1965 when the *Mariner 4* spacecraft began sending back our first close-up images of the red planet. Our expectations suffered a blow when those pictures showed a cratered, Moon-like desert. We later learned that *Mariner 4* had imaged some of the bleakest parts of Mars; subsequent missions revealed more intriguing features. Nonetheless, the disappointment ran deep.

There remained the possibility that microorganisms might survive such harsh conditions. The *Viking* landers that reached the Martian surface in 1976 carried on-board laboratories to look for evidence of biological processes. Gilbert Levin, the designer of one of the experiments, believed that *Viking* had found evidence of life, but most scientists disagreed.

Some drew sweeping conclusions from our failure to find confirmed evidence of life on Mars. "It is now virtually certain that the earth is the only life-bearing planet in our region of the galaxy,"wrote biologist Norman Horowitz. "We have awakened from a dream. We are alone."[5]

Others have questioned such negative declarations. Since *Viking*, we have discovered that billions of years ago—when life on Earth was getting its start—enough water pooled on the surface of Mars to allow the possibility of living things. Many forms of Earth life could have survived and even thrived under those conditions. However, the Martian environment took a different path than Earth's, one too stressful for any known form of life to survive on the planet's surface. Mars life may have appeared early, only to be extinguished within the planet's first billion years.[6]

Might there still be life under Martian rocks and dust, drawing on subterranean water? Biologist Hubertus Strughold (sometimes called the father of space medicine) proposed 40 years ago that living things might survive in a subsurface "hydrosphere." Others too believe that organisms may still cling to life, hibernating during the cold spells and thawing out when climate conditions improve.[7]

Mind-Stretcher. Life may be a temporary phenomenon on many worlds, limited to a particular eon of a planet's history. Living things may arise from inanimate matter, survive for millions of years, and then die out.[8] We may find only their traces, frozen, scorched, or crushed by climatic and geological change.

The question of life on Mars—past or present—remains open. Three-quarters of the planetary scientists informally polled in 2005 believed that Mars once had conditions hospitable to life. One-quarter thought that it still does. Others argue that life might exist in a dark ocean under the icy surface of Jupiter's moon Europa.[9]

Finding an independent origin of biology on another world in our solar system would have momentous implications. As planetary scientist Christopher McKay put it, the search for life on Mars is a search for a second Genesis.[10] Discovering a second origin in our own small solar system would imply a fertile universe, one that might produce other minds.

Detecting Remote Life

How can we detect alien forms of life when they are very far away? By the effects they have on planetary atmospheres. "Except by a purposeful act of camouflage," wrote Gaia theorist James Lovelock, "any life system will reveal its presence through the chemical disequilibria caused by its contrivances." In the case of Earth, the simultaneous presence of molecular hydrogen, methane, and ammonia in significant quantities would be strong evidence for a dynamic system—life.

Would we recognize other signs? If we encounter alien forms of life, Simpson warned, we might not perceive them as living—or we might have to revise our conception of life. On Earth we make a distinction between living matter and nonliving matter. What of life as we do not know it? If we recognized it at all, we might have to place it in a third category.[11]

The boundary between life and nonlife has become blurred, even on Earth. In the case of terrestrial biology, viruses often are thought of as being in a gray area between living and nonliving. Scientists and medical experts have debated for years whether viruses can be called life, as they apparently can exist only as parasites on living cells.[12] The border may be even less clear when we encounter truly alien biology.

NASA has asked the National Academy of Sciences to look at possible alternative chemistries for life. Some researchers hope to create living things based on such different chemical processes.[13] There may be mechanisms for life that we have not yet imagined.

A Metadiscipline

Cosmology and biology are not separate disciplines, since life cannot be understood without tracing the origin and evolution of the universe; nor can the universe be comprehended without considering the life residing within it.

—Astronomer George Seielstad, 1989[14]

We have accumulated many supporting arguments for a biological universe—a concept that has been widely accepted despite the lack of evidence. What had been a borderland field is increasingly accepted as a science—a very interdisciplinary one, resting on many areas of research.[15]

Astronomer and science writer David Darling suggested that biological science may now be at a stage analogous to physics when Newton discovered the law of gravity, more than 200 years ago. We already can make out, amid the tumult of claims and counterclaims, the beginnings of a general theory of biology. In its attempt to understand how life springs from the evolution of the universe, astrobiology aims for a grand unification, bringing biology together with cosmology.[16]

Like other new sciences, this one remains open to innovative thinking. Nobel Prize winner Baruch Blumberg, the first head of NASA's Astrobiology Institute, thought that this new field would have the advantage of being "unencumbered by hypothesis." We should let our quest for life be guided by our theories, said McKay, but we should try not to be constrained by them. They may be wrong.[17]

We still do not know what is contingent and what is necessary in the evolution of life, what are the "universals," and what are the "parochials." Dick found that only two conclusions are warranted by the evidence at this stage. First, if life exists beyond the Earth, it will be astonishingly diverse. Second, the abundance of extraterrestrial life will be inversely proportional to its complexity; microbial life will be more abundant and intelligent life less so.[18]

So far, the search for extraterrestrial life has been frustrating. However, this is a long-term effort. Physicist Gerald Feinberg and chemist Robert Shapiro compared the search to the European exploration of the Western Hemisphere that began in the fifteenth century and took hundreds of years to complete. The human will to carry out projects over long periods of time cannot be dismissed lightly.[19]

Astronomy has demonstrated to the satisfaction of most that physical law is universal. Biological law also might be universal, suggested Dick, but that has not yet been proven. Even if the Earth is not physically central, the question of whether it is biologically central remains unanswered.

The widespread assumption that alien biology exists still is an expectation, not a proven fact. But what an expectation. Just one success in proving an independent origin of life, and the biological Copernicans win.

The interested public seems at ease with this, as the consequences are assumed to be largely scientific. Finding nonintelligent alien life would not resolve the question of greatest concern to nonscientists: whether or not the universe has produced other minds equal or superior to ours.

Panspermia

We must gradually accustom ourselves to think that living beings have endured for eternity, and thus have no origin in time, that they originate through germs which come from other celestial bodies, that they die out when conditions have become unfavorable, but that they then live on elsewhere in the universe.

—Svante Arrhenius, 1903[20]

Twenty-five hundred years ago, the Greek philosopher Anaxagoras hypothesized that all life came from the combination of tiny seeds pervading the cosmos, a "panspermia." Swedish chemist Svante Arrhenius revived this concept, visualizing that spores spreading through interstellar space could reach favorable planetary environments such as ours.

Seventy years later, biologist Francis Crick and chemist Leslie Orgel suggested that technological societies—which might have arisen elsewhere before the formation of the Earth—might deliberately spread life through the galaxy in a "directed panspermia." We ourselves might be descendants of life forms deliberately or accidentally transferred to Earth's surface.[21]

Researchers have shown that some amino acids in meteorites can withstand collisions with the Earth. "There's just no doubt that some of the amino acids survive the impacts," said biochemist David Deamer. A molecule, literally dropped from the sky, could have jump-started or accelerated a simple chemical reaction key to early life.[22]

Hoyle and Wickramasinghe argued in a series of books that life on Earth stemmed from a piecing together of prebiotic molecules from outer space. Instead of being the biological center of the Universe, our planet is just an assembly station; no great innovation in biology ever happened here. Even viruses and bacteria that cause disease may come from beyond the Earth. "Attacks of a viral disease represent the final stage in an attempted matching process," they claimed, "a process that, in the minority of cases where it succeeds, is responsible for directing the evolution of species. Diseases are foiled evolutionary leaps."

According to Hoyle and Wickramasinghe, evolution proceeds in bursts of activity stimulated by the arrival of new genes from the sky. They cited in particular a single event 570 million years ago, perhaps a collision with comet fragments or a molecular cloud, that may have deposited organisms on the Earth that provided most of the genetic information that now characterizes Earth life. They described mutations, gene doublings, and recombinations as no more than fine-tuning superposed on the much greater cosmic assembly process.

Mind-Stretcher. Instead of seeing life as a collection of isolated pockets, Hoyle saw it as a coherent whole developed out of a single aggregate of cosmic genes. If all life-forms in our galaxy share a common genetic heritage, that could make life everywhere vulnerable to the same biological threats to their health—including diseases from space.

Earth continues to receive ready-formed living structures such as bacteria and viruses, these scientists argued. In their conception, the primitive Earth may have been transformed into a habitable condition mainly due to contributions from comets, which created the primordial oceans and atmosphere. Wickramasinghe noted that 100 tons of cometary material reaches Earth every day.

Hoyle suggested that peptide chains might reproduce in interstellar space, possibly becoming clumps that consume others and expel waste. He proposed that microorganisms set the right physical conditions within clouds of interstellar gas so that suitable stars and planets form. Hoyle and Wickramasinghe even claimed that cosmic bacteria, or superstructures built from them, may be in control of our galaxy. The surest way for such bacteria to prosper would be by maintaining a firm grip on the interstellar magnetic field.[23]

Wickramasinghe still was advocating these theories in 2002, after Hoyle had passed on. He argued that we should extend the boundaries of Darwin's warm little pond to encompass the largest possible amount of carbonaceous material in the cosmos. "Life," he concluded, "would inevitably develop on every habitable planet, descended from the same all-pervasive genes."[24]

Most scientists remain skeptical of these theories. However, Deamer found space delivery of amino acids billions of years ago a plausible companion to organic synthesis in the atmosphere.[25]

The Mars Rock

The transfer of life to the Earth might be local. The famous physicist Lord Kelvin suggested in 1871 that an astronomical body might strike a planet with enough force to blast debris out of its gravitational field. As a result, "many great and small fragments carrying seed and living plants and animals would undoubtedly be scattered through space. . . . If at the present instant no life existed upon this earth, one such stone falling on it might. . . . lead to its becoming covered with vegetation." Acknowledging that this hypothesis may seem wild and visionary, Kelvin maintained that it was not unscientific.[26]

Kelvin's vision gained credibility during the 1980s, when scientists discovered that some meteorites found in Antarctica actually were pieces of Mars, blasted off the red planet by impacts. More than 5 billion Martian rocks capable of carrying living microbes have fallen to Earth in the past 4 billion years.[27] We have found so many of them that mail-order houses offer samples for sale.

A group of researchers led by geologist David McKay reported in 1996 that a chunk of Martian rock contained possible chemical signs of ancient life. Most provocatively, McKay's team spotted shapes that might be micro-fossils of tiny bacteria, resembling some found on our own planet.[28] By implication, Earth and Mars were not quarantined from each other; if life started either on Earth or on Mars, it could have spread to the other.

NASA officials saw the McKay group's findings as so important that they asked President Clinton to announce them to the public, a scene borrowed by the film "Contact." The McKay team's interpretations have been challenged by other researchers, leaving the issue unresolved.[29]

Recent research suggests that life may have started earlier on Mars, which cooled off more quickly and stabilized its surface earlier than our own planet. (Percival Lowell, who believed that evolution on Mars was further along than evolution on the Earth, would have enjoyed hearing that.) Some scientists believe that tiny fossils—mineralized communities of microorganisms—still might offer the best hope for finding evidence of past life on the red planet.[30]

Could the ancestors of life on Earth be Martian microorganisms, transported to our planet in chunks of rock? That might explain why life on Earth is separated into three distinct lineages, suggested Stanford's Norman Sleep; these might represent three distinct sowings of Martian seed.[31]

There is a downside to this theory. If Earth and Mars can contaminate each other, it may be very difficult to determine if there were two independent origins of life.

Was Arrhenius Right after All?

If hardy "nanobacteria" exist, relics of ancient life such as those found in the meteorite from Mars might be the descendants of interstellar colonists. A group of scientists who studied this question reported that no meteorite from a planet of another solar system has ever landed on Earth. However, other researchers have found interstellar dust grains in our solar system;

they can survive their journey through the interstellar medium. Interstellar particles have been detected in the Earth's atmosphere; some reach the surface without vaporizing.[32]

We may have borrowed material from our nearest stellar neighbor. If the Alpha Centauri system is surrounded by a cloud of icy planetoids as our own solar system is believed to be, the two clouds may overlap. Science fiction author Ben Bova speculated that some of the comets we see in our skies originated in the Alpha Centauri system.[33]

Physicist Freeman Dyson cited research showing that substantial numbers of objects from the star Beta Pictoris—ranging from dust grains to kilometer-sized bodies—pass through our solar system. A fraction of these objects will be captured into orbits around our Sun. These findings suggest that life adapted to vacuum has the potential to spread not only from world to world within our solar system, but far and wide through the galaxy.[34]

Searching for Intelligence

I have no doubt that there are many other inhabited worlds, and that on some of them beings exist who are immeasurably beyond our mental level. We would be rash to deny that they can use radiation so penetrating as to convey messages to the Earth. Probably such messages now come. When they are first made intelligible a new era in the history of humanity will begin.

—Bishop Barnes of Birmingham, 1931[1]

The means for contact with extraterrestrial intelligence are potentially within our hands.

—Carl Sagan, 1973[2]

What a noble endeavor SETI is! A bountiful cosmos, advanced technology, and dedicated researchers: everything is in place.

—Andrew J.H. Clark and David H. Clark, 1999[3]

Radio Days

Searching our solar system through optical telescopes failed to provide persuasive evidence of extraterrestrial intelligence. To pursue the question further, searchers broadened the quest with new technologies and extended it outward to the stars. This expanded enormously the volume to be searched.

What evidence can we look for at interstellar distances? Although intelligent life is likely to be much less common than nonintelligent life— perhaps orders of magnitude less common—it can be much easier to detect. We can observe the consequences of its actions, meaning its use of technologies.

Asimov proposed a definition of intelligence within the context of such a search: A species is intelligent if it can develop a complex technology. This definition makes it unnecessary to delve into psychology and philosophy. Instead of trying to divine the inner being and its inner thoughts, one merely looks at what is being accomplished. This definition also means that we don't have to bother with the intelligence of individuals or even minor groups; the unit of definition is the species.[4]

Jill Tarter refined the definition: The search for intelligent life is a search for a technology that is detectable by our technology. We are looking for a species' ability to technologically modify its local environment in ways that can be detected over interstellar distances.[5] This may be only a tiny sampling of the Galaxy's minds; more numerous examples may be beyond the reach of our detectors.

Since the discrediting of Martian canals, the primary medium for detecting evidence of alien technology has been radio—first at the interplanetary scale, then at the interstellar. Inventors Nikola Tesla and Guglielmo Marconi not only foresaw the use of radio technology for communication beyond Earth; they believed they had actually detected signals of intelligent origin. Tesla was the first to publish this claim, in 1901.

Marconi said that he often had received strong signals that seemed to come from some place outside the Earth and might have originated from the stars. He listened for signals from Mars during a transatlantic voyage on his yacht in 1922 and believed that he heard them.[6]

Astronomer David Todd, who had suggested in 1909 that Martians might communicate with Earth by radio waves, organized amateur radio operators to listen during Mars' close approach to the Earth in August 1924. Although the U.S. Army and Navy declined his request for periods of radio silence, they did instruct radio operators to monitor and report any unusual signals. None were heard.[7]

In 1932, Karl Jansky of Bell Telephone Laboratories detected strange radio static that he could not attribute to any known source. A year later, he announced his interpretation: The radio emissions were coming from beyond the solar system. Although Jansky's discovery made the front page of *The New York Times*, most professional astronomers ignored his pioneering work.[8]

Radio engineer Grote Reber, who built the first radio telescope dish in his back yard in the late 1930s, conducted the earliest systematic survey of cosmic radio waves. He reported in 1940 that virtually the entire Milky Way was a source of radio "noise." Again, there was no immediate impact; for nearly a decade, Reber was the world's only radio astronomer.[9]

Mind-Stretcher. What if Reber had detected a signal from extraterrestrial intelligence during that decade and sought to publish his findings in an appropriate journal? Given the lack of scientific interest in Jansky and Reber's other findings, his discovery might have been ignored—even if a journal had agreed to print it.

Wartime advances in radio and radar technologies made much more powerful radio telescopes possible, instruments capable of conducting extensive, detailed studies of radio signals from the sky. By the early 1950s, astronomers were actively observing such features as hydrogen clouds that outlined our galaxy's spiral arms.[10] Radio astronomy in the United States received public support with the founding of the National Radio Astronomy Observatory at Green Bank, West Virginia. Other nations built their own radio observatories, the most famous being the giant dish at Jodrell Bank in England.

Spotting the potential of these systems, physicists Philip Morrison and Giuseppe Cocconi began discussing the feasibility of interstellar communication. Cocconi calculated that the 250-foot radio telescope at Jodrell Bank could detect Earth-like radio signals from the nearest star. In a paper published in *Nature* in 1959, he and Morrison pointed out that such instruments made it feasible to communicate with other civilizations—and to search for their signals. They concluded their paper with a now classic statement: "The probability of success is difficult to estimate; but if we never search, the chance of success is zero."[11]

Astronomer Frank Drake realized independently that searching for interstellar radio signals might allow us to detect technologically advanced extraterrestrials. Drake launched the modern radio search with his Project Ozma in 1960, using a radio telescope at Green Bank. His brief observation of two stars detected a false positive coming from a secret military facility, but no evidence of extraterrestrial technology. Drake persisted in his quest, becoming a major figure in the scientific search.[12]

By 1962, scientific attitudes toward extraterrestrial intelligence were shifting from slightly amused neglect to more open-minded inquiry. As so often happens in science, observed astronomer Robert Rood and physicist James Trefil, the transition occurred because someone pointed out that technological developments had brought within reach a goal previously thought to be unattainable.[13]

A new, controversial scientific field was emerging—a search for electromagnetic evidence of alien *technologies*, rather than for living beings. For radio searches, that meant detecting beacons or broadcasts, intercepting beamed transmissions, or eavesdropping on local communications.

A Dissenting Voice

In a 1960 paper that did not attract much attention at the time, radio astronomer Ronald Bracewell presented a different model of interstellar contact. He proposed that, instead of searching for radio signals from many light-years away, more advanced civilizations would send out robotic probes to the most interesting stars. Those machines could report back to their launching civilizations and could use radio to contact any technological civilization they found.[14] Bracewell later made an important point about this model: We should focus on those technological civilizations that can reach us.

Most radio astronomers interested in interstellar communication either dismissed or ignored Bracewell's arguments. His direct contact scenario came back to haunt them 15 years later.

From Russia with Theories

In 1962, Soviet astronomer Iosif Shklovskii published a book in Russian called Universe, Life, Mind, popularizing the question and providing a broad context for the search. Another Soviet astronomer, Nikolai Kardashev, called the first USSR-wide meeting on this subject in 1964. Kardashev proposed three levels of alien technology reflected by the power of signals they could emit: Type I, a planetary technology comparable to that of Earth; Type II, a technology exploiting the energy of a star; Type III, disposing of energy comparable to that of an entire galaxy. That theory supported a Soviet strategy of looking for powerful signals from a few civilizations vastly more advanced than our own. This approach required far fewer transmitting societies than the American strategy, which assumed abundant civilizations with a modest radio transmission capability.[15]

There were false alarms. The Soviet news agency TASS announced in 1965 that Soviet astronomers had detected rhythmic fluctuations in a powerful radio source called CTA 102 that might be the beacon of a supercivilization. That source turned out to be a recently discovered phenomenon known as a quasar.[16]

A 1971 conference of American and Soviet scientists at the Byurakan Observatory in the USSR endorsed the idea of CETI (Communication with Extraterrestrial Intelligence), declaring that recent discoveries had transferred this subject from the realm of speculation to a new realm of experiment and observation. The Byurakan resolutions laid out proposed research directions and suggested the types of instruments that would be needed. Recognizing the interdisciplinary character of CETI, the conferees agreed that "a wide circle of specialists, from astrophysicists to historians, should participate in the planning of this research." Sagan commented at the time that the Byurakan conference made the subject of communication with extraterrestrial intelligence scientifically respectable.[17]

American Initiatives

Meanwhile, news about the search began to spread beyond the radio astronomy community. The New York Times science editor Walter Sullivan's 1964 book We Are Not Alone brought the search to the attention of a broader audience in the United States.

Sagan collaborated with Shklovskii on an expanded version of his book, published in English in 1966 as Intelligent Life in the Universe. That textbooklike work was a landmark in establishing the credibility of a scientific search; scientists could claim that their research was based on extrapolations from established fact and theories.[18]

Ambitious Automata

Another, less widely known book published in 1966 suggested a different model of sentience beyond the Earth. In their work *Intelligence in the Universe*, early electronic computing specialist Roger MacGowan and space expert Frederick Ordway proposed that the most advanced alien minds would be intelligent machines, which they called automata.[19] Although this theory has been revived by others, it has lacked the emotional resonance of encountering biological intelligences like ourselves.

One of the hotbeds of activity was the NASA Ames Research Center in California, which had been doing experiments on the origins of life in our solar system. John Billingham, a medical doctor who worked on life sciences research at Ames, persuaded center director Hans Mark to authorize a study of making the radio search an official project.

In 1971, Billingham and electrical engineer Bernard Oliver brought together a group of scientists and engineers to do a design study of a system for detecting radio signals from extraterrestrial intelligent life. The result was a report proposing Project Cyclops, an expandable array of radio telescopes that would grow in size as needed. (This report expanded on a concept proposed by Oliver in 1966.) At its maximum extent, Cyclops would aim 1000 great dishes at the heavens, a giant multifaceted eye.

The scale of the full Cyclops array would have been staggering, with projected expenditures comparable to those of the Apollo program. The report's authors even visualized a city called Cyclopolis, where the observatory's staff and their families would live.[20]

Mind-Stretcher. Cyclops might be seen as the scientific equivalent to the medieval cathedral, built incrementally by generations of believers as an act of faith. The study's authors explicitly admitted that the premises behind their conclusions were beliefs.

Although NASA never sought funding for Cyclops, the science and engineering concepts in the study proved to be a rallying point for people interested in the radio search. "Project Cyclops was really the start," Billingham said years later. "That was the thing that launched SETI (Search for Extraterrestrial Intelligence)."[21]

Momentum was building within the scientific community. The Astronomy Survey Committee of the National Academy of Sciences took a visionary stance in 1972, stating that "our civilization is within reach of one of the greatest steps in its evolution: knowledge of the existence, nature, and activities of independent civilizations in space." In a Saganesque phrase, the report suggested that "at this instant, through this very document, are perhaps passing radio waves bearing the conversations of distant creatures—conversations that we could record if we pointed a telescope in the right direction and tuned to the proper frequency."[22]

Sagan and others reached out to the general public through interviews, television appearances, and works like Sagan's 1973 book *The Cosmic Connection*.[23] The pro-search lobby mixed idealistic hopes for an epochal discovery with the salesmanship necessary to get support for an actual program. As their optimistic vision of contact spread through the interested public, expectations began to soar.

A series of workshops on interstellar communication chaired by Morrison reported in 1977 that, within the previous two or three decades, we had entered a new communicative epoch. A signal sent from an existing radio dish on Earth could be detected with ease across the galaxy by a similar dish—if it were pointed in the right direction at the right time and were tuned to the right frequency. The report concluded that the present "climate of belief" made it timely to search for extraterrestrials. A significant program with substantial potential secondary benefits could be undertaken with only modest resources; large systems of great capability could be built if needed. Such a search was intrinsically an international endeavor in which the United States could take a lead.

This report coined the term SETI—the Search for Extraterrestrial Intelligence—distinguishing that effort from communication with such intelligence (CETI). As we will see, that distinction remains an issue within today's debate. Jill Tarter and Stanford's Christopher Chyba commented later that, instead of calling this search SETI, it might be better to call it SET-T, a search for extraterrestrial technologies.[24]

NASA established a small SETI program office at Ames in 1976, headed by Billingham. The Jet Propulsion Laboratory in Pasadena formed its own office the next year. Jill Tarter, the first astronomer to devote her career to SETI, became an increasingly important player in the scientific preparations for a search program.

Senator Proxmire set back the project in 1978 by giving it his Golden Fleece Award (an example of the government wasting the taxpayer's money). Proxmire later retreated from active opposition, at least partly because of a conversation with Sagan. That skilled science popularizer included contact with alien civilizations in his highly successful 1980 television series and book Cosmos.[25]

Saganism

We can't avoid Carl Sagan when we look at the modern debate. He was the great popularizer of the search, reaching beyond the astronomical community to the interested public. He spread the word through his frequent publications, interviews, and television appearances, often drawing criticism from more conventional scientists.

Sagan was an optimist in two senses: about the probability of detecting another civilization and about the outcome of that encounter. When he addressed the consequences of contact, he drew on the most positive analogies from our history while often dismissing the most negative. This optimistic model has great power, despite the fact that it remains unproven.

Many of Sagan's specific predictions may be disproved (perhaps inevitably, as he spoke on both sides of some issues). Like Lowell, he will be seen as influential even when he got the details wrong.

After skeptics questioned the scientific validity of such a search, Sagan put together a pro-SETI petition in 1982, published in the letters column of the journal *Science*. The petition proposed that, instead of arguing about the issue, we should look: "We are unanimous in our conviction that the only significant test of the existence of extraterrestrial intelligence is an experimental one. No *a priori* arguments on this subject can be compelling or should be used as a substitute for an observational program." A U.S. National Academy of Sciences Committee on the future of astronomy published a report the same year recommending funding for SETI.[26]

After initial failures to get money from Congress, NASA obtained modest funding to develop instrumentation for a radio search. Much of the effort focused on data processing; the growing power of computers made it possible to survey much larger numbers of stars in an expanded range of wavelengths and to extract more useful data.[27]

SETI became an approved NASA project in 1990, with total funding estimated at 100 million dollars through 2001. This despite the opposition of Congressman Ronald Machtley, who stated that "we cannot spend money on curiosity today when we have a deficit."[28]

This government-sponsored effort to detect alien intelligence reached its high point in October 1992, when a two-pronged search effort got under way: an all-sky survey using the Deep Space Network and a targeted search using large radio telescopes to study about 1000 sunlike stars. This High Resolution Microwave Survey was a major advance on earlier searches, as it was optimized for the detection of technological signals.

A U.S. Senator intervened to cancel the NASA program at the end of its first year. The American Congress was enthusiastic about supporting the search for extraterrestrial life (SETL), but found the search for extraterrestrial intelligence (SETI) too speculative—or potentially unsettling.

The search did not die with the end of the official program. By then, it was bolstered by a small social movement inspired by genuine idealism—and by the euphoric hopes some of its leaders had raised in their efforts to gain support.

The Politics of SETI

NASA historian Stephen Garber, after studying the program's history, concluded that the major factors leading to its cancellation were fervor over cutting the federal budget deficit, lack of support from other scientists, a history of unfounded associations with nonscientific elements (presumably meaning UFOs), and bad timing. The small size of the program ($12.5 million for its last year) actually was a disadvantage, as the program had few contractors and was easy to attack.

Although SETI involved truly fundamental science, it did not fit neatly into any existing scientific discipline.[29] Yet some other interdisciplinary programs—notably environmental research—are funded much more generously than SETI ever was.

Privatizing SETI

The nonprofit SETI Institute, which had been founded in 1984, successfully sought private funding for a revived targeted search. Among the contributors were prominent entrepreneurs of the information revolution, including David Packard and William Hewlett of Hewlett-Packard, Paul Allen of Microsoft, Gordon Moore of Intel, and Mitchell Kapor of Lotus.[30]

Acquiring much of NASA's SETI equipment, the Institute launched Project Phoenix in 1995, focusing on sunlike stars within 200 light-years of our solar system. Phoenix depended on part-time access to radio telescopes used for more conventional astronomical research, including instruments at Arecibo and Green Bank, and in Australia. Funded at about $4 million a year, the program ended early in 2004 after examining 710 nearby stars; no credible evidence of extraterrestrial technology was found.[31]

Although Phoenix was the most sensitive SETI experiment ever, it could not pick up the kind of ordinary leakage that the Earth releases into space such as TV carrier waves, even at the distance of the closest star beyond our Sun. Military radars would have been detectable dozens of light-years away if they were aimed in the right direction at the right time.[32]

Argentinian astronomer Guillermo Lemarchand commented during Phoenix's run that we can detect beacons, but are less able to detect "long-distance calls." Shklovskii and Sagan acknowledged that the probability of accidentally picking up interstellar long-distance communication signals is very small. Astronomer Sebastian Von Hoerner calculated that intercepting long-distance calls would be possible only if each advanced civilization were to converse simultaneously with 1300 neighbors.[33]

The SETI Institute continues to develop new technologies and search strategies. In cooperation with the University of California at Berkeley's Radio Astronomy Laboratory, the Institute drew up plans for the first radio observatory dedicated from its beginning to the search for extraterrestrial intelligence. This group of 350 small interlinked dishes—a mini-Cyclops with more sophisticated electronics—is named the Allen Telescope Array after its primary funder, former Microsoft executive Paul Allen. (Individuals can sponsor the construction of a dish for $100,000.) To be completed by 2010, the ATA will be able to broaden stellar reconnaissance from 1000 stars to 100,000, to resolve details three times better than the Arecibo telescope (the largest in the world), to operate 1000 times faster, and to observe multiple spectral windows simultaneously. The array will be able to conduct microwave searches on a continuous basis, improving the odds for detection. A small portion of the ATA was in operation as of early 2006.[34]

A report on the future of SETI explicitly acknowledged its dependence on rich contributors, concluding that the magnitude of the search should be scaled so as to be commensurate with the philanthropic capabilities of the world's visionary individuals of great wealth.[35] This is consistent with the financing of American astronomy from the second half of the nineteenth century to the mid-twentieth century, when several major new instruments and observatories were funded by a few rich men.

That philanthropy sometimes produced spectacular results. The 100-inch telescope on Mount Wilson, financed by wealthy Los Angeles businessman John Hooker, enabled Hubble to confirm the nature of other galaxies and the expansion of the universe. The more recent Palomar 200-inch telescope and the Keck I and II telescopes atop Mauna Kea in Hawaii, funded by private foundations, have made important contributions to astronomy including the discovery of quasars.[36]

Other nongovernment search programs have been underway for years. The nonprofit Planetary Society, founded in 1980 by Carl Sagan, Jet Propulsion Laboratory Director Bruce Murray, and JPL engineer Louis Friedman, spread its financial support for SETI among several projects. The Society began sponsoring astronomer Paul Horowitz's wide-sky surveys in the 1980s through Project META, using a radio telescope in Massachusetts and later one in Argentina. A new effort called Project BETA, with more powerful data processing capabilities, was switched on in 1995.[37]

The Planetary Society and the SETI Institute support Project SERENDIP, an innovative low-cost approach invented by astronomers Stuart Bowyer and Jill Tarter in the late 1970s. SERENDIP piggybacks receivers on radio telescopes conducting more conventional astronomical research. Although SERENDIP can not choose its targets, it can listen 50–70% of the time. Originally a Northern Hemisphere search, this project initiated a Southern SERENDIP in Australia in 1998. The program also is supported by the Friends of SERENDIP, a fund-raising group headed by Arthur C. Clarke.[38]

The search involved the public in an unprecedented way with the SETI at Home project, in which private citizens make their computers available to process data acquired by radio telescopes. SETI at Home went public in 1999 and claimed 5 million participants as of 2004. The volunteers seemed to have three predominant motivations: wanting to use their computers and the Internet productively, wishing to participate in an intriguing and worthwhile scientific project, and desiring to be connected to something bigger than themselves.[39]

Astronomer John Kraus and electrical engineer Robert Dixon began using Ohio State University's "Big Ear" radio telescope to search for signals in 1973; this became the world's first telescope dedicated to SETI. By the time the Big Ear was shut down in 1998 to make room for a golf course, the Ohio State program had become the longest-running search. One of its goals was to develop the technology for a proposed phased array known as Project Argus that could image the entire sky at one time.[40]

The nonprofit SETI League, founded in 1994, organizes search efforts by amateurs using small dishes, hoping to network these capabilities into a global SETI system of 5000 observing stations. The League's long-term goal is the realization of Project Argus. SETI Institute physicist Kent Cullers, pointing out that there are some good frequencies left for prospecting, raised the possibility that an amateur astronomer might be the first to detect evidence of extraterrestrials.[41]

Transient Events

Some believe that the Wow! signal recorded by Ohio State University's Big Ear in 1977 met most of the criteria for a signal generated by extraterrestrial intelligence. This strong flash of energy was not noticed until an astronomer examined a printout later, and the signal was never reacquired.

Few scientists would consider a one-time detection to be proof of a broader phenomenon; verification requires that the signal be found again. The Planetary Society's META searches, for example, detected a small number of tantalizing signals that met all the criteria of an extraterrestrial intelligent origin, except for the essential element of repeatability.

SETI astronomers might see alien beamed communications—especially those not meant for us—as flashes of energy that dim within seconds. SETI has focused on continuous or repeated signals such as radio beacons because transient signals are more difficult to pick up, and much tougher to find a second time. "Our present systems are very ineffective in dealing with transient signals," said Drake.[42] New observing technologies will have enhanced abilities to record very brief signals.

In the meantime, the technology for radio searches continues to improve. The Arecibo telescope has been upgraded, raising its highest usable frequency while lowering system noise. Radio astronomers are planning more powerful instruments, although these are not dedicated to SETI. First would be an expansion of the Very Large Array of radio telescopes in New Mexico, to be completed by 2010. The Square Kilometer Array that may be built by 2015 could have 50 to 100 times the sensitivity of any existing array. Also on the agenda is a Low Frequency Array.[43]

Experts meeting with SETI Institute sponsorship recommended developing and building a One Hectare Radio Telescope, using a phased array of many small dishes for surveys in the galactic plane. This could be a prototype for the Square Kilometer Array. The group also recommended building and operating an Omnidirectional SETI System, an all-sky, all the time microwave search for transient beacons. In addition, these experts called for support of optical searches for rapid pulses and data mining of existing records for optical signals. The estimated annual cost of the recommended programs was 8 million dollars per year.[44]

Meanwhile, computer technology for data processing is improving rapidly. Given the continuing reduction in computation costs, said SETI Institute astronomer Seth Shostak, we can expect that the speed of the reconnaissance will double every 18 months.[45]

Radio astronomers face increasingly serious interference from satellite and terrestrial transmitters. The 1982 SETI petition pointed out that, because of the growing problem of radio-frequency interference, the search will become more difficult the longer we wait. If we do not protect certain wavelengths from interference, said Bernard Oliver, we may doom humanity to galactic isolation.

Advancing technology may have changed the game. Chyba declared in 1996 that searches had reached a level of technical maturity where all interference can be recognized and excluded.[46]

SETI researchers have studied space-based observing systems that could open up new opportunities at frequencies that cannot be observed from the Earth's surface. Unfortunately, the high cost of such observatories makes early implementation unlikely. Horowitz proposed an interim step that resembles SERENDIP: adding a SETI component to a planned NASA mission, the Terrestrial Planet Finder.[47]

Other Wavelengths, Other Technologies

If *there are beings with radio out there,* if *they are willing to transmit messages into the unresponsive void (instead of just listening as we are), and* if *we listen in the right direction at the right time and the right frequency with the right bandwidth and the right detection scheme, then the radio approach to SETI will make a significant contribution to our knowledge.*
—Physicist Robert L. Forward, 1994[48]

For more than 40 years, the attraction of searching in the radio sector of the electromagnetic spectrum has been compelling. As Drake saw it, radio has two enormous advantages for interstellar communication: The galaxy is transparent to radio waves, and radio does not need to be aimed accurately. Radio also is more economical; it is cheaper to send radio photons than visible-light photons.[49]

Early observing programs adapted radio astronomy equipment to SETI tasks. Shklovskii offered a pragmatic reason for choosing radio frequencies: No one could accuse searchers of wasting money, as the equipment could be used for conventional radio astronomy.[50]

Whatever its advantages may be, searching by radio is not a comprehensive approach; we may be limiting our attention to a very restricted subset of all technological civilizations. By 1979, even Drake was worrying that the choice of radio had been unduly influenced by our booming expertise in radio technology.[51]

As early as 1961, laser pioneers Charles Townes and Robert Schwartz proposed that we use masers—including lasers—for interstellar communications. It may have been an historical accident that lasers were not discovered before radio as a means of long-range communication.[52]

Optical beacons now are almost as easy to detect as radio beacons. We could see a powerful pulsed laser a few dozen light-years away; a tightly focused laser beam could greatly outshine a planet's host star at a particular wavelength.[53]

Some astronomers already have conducted limited searches for laser signals. Stuart Kingsley, who pioneered these efforts in 1990, has argued that only lasers have the ability to probe interstellar space free of the significant distortion that smears radio signals. However, there is a disadvantage. Although optical SETI does not face terrestrial interference, it does require that the extraterrestrials deliberately target us with their lasers.[54] The Planetary Society initiated a new search for laser signals in 2006, using a telescope in Massachusetts.

There have been more exotic proposals for optical detection. Morrison raised the possibility of interstellar communication by modulating the visible light of a star. Extraterrestrials might attract attention by placing large objects in star-hugging orbits, proposed Luc Arnold of France's Haute-Provence Observatory; from the perspective of other civilizations, those objects would transit the stars and produce light curves whose artificial nature would be recognizable. Astronomer Martin Harwit suggested that other civilizations might transmit information by twisting light rather than using other encoding methods.[55]

The radio compulsion is yielding to a more eclectic view as SETI becomes multispectral. The authors of the SETI 2020 report recommended broadening the frequency range of searches to include optical and infrared wavelengths, looking for both continuous wave and pulsed signals. Detecting optical pulses may be limited to a few thousand light-years, but infrared pulses may be detected all the way to the galactic center.

The "holy grail" of an all-sky, all-frequencies, all the time search can be discerned on the horizon of radio technology, according to this group of experts. Its optical counterpart cannot be far behind.[56]

Above the Clouds

Thanks to rocket technology that has enabled us to place robotic observatories in orbit, astronomers now can study much wider ranges of the electromagnetic spectrum. In addition to the U.S.–Europe Hubble Space Telescope, NASA and other space agencies have deployed major instruments observing in the infrared, X-ray, and gamma-ray regions of the spectrum. Although none are designed to search for extraterrestrial technology, their findings may have significant implications for SETI.

Experts have concluded that all wavebands—radio/microwave, infrared/optical, X-ray/gamma-ray—are worthy of consideration for SETI searches. As X-ray and gamma-ray observations must take place above the atmosphere, searches undertaken just for SETI will remain too costly in the near future.[57]

Bursts of Energy

Gamma-ray bursts (GRBs), which may be caused by the collapse of massive stars or the merger of neutron stars, are among the most powerful energy sources in the cosmos. The rays from the brightest events are so highly collimated that they can be seen across the observable universe.

Because bursters concentrate their energy in beams, we only see the one in every several hundred of them that is pointed in our direction. The true rate of GRBs is much higher than it appears.[58]

That also could be true of brief, targeted communications between advanced societies, or among the scattered worlds of a star-faring civilization. Only rarely would the Earth pass through a communications beam. If we detect one, there may be many more.

We now have the means to detect ultrahigh-energy neutrinos that were not even known to exist in 1990 (scientists had proposed using neutrinos for interstellar communication at least 11 years earlier). Astronomers have produced the first sky survey based on detecting positrons, the antimatter equivalent of electrons.[59]

As our potential means of detection expand, the universe becomes more transparent to us. Like spectroscopy in the nineteenth century, new methods may evolve unexpectedly from conjunctions of technological developments and scientific insights. Such new capabilities will be focused on conventional astronomy rather than on finding evidence of extraterrestrial technology. However, serendipity might apply.

A Warning. Windows opened can become windows closed. Orbiting telescopes have finite lifetimes, often only a few years. If they are not replaced, the portion of the electromagnetic spectrum that they study may once again become invisible to us.[60]

Looking for the Astroengineers

Body is the last thing we are likely to know of them. Of their mind as embodied in their works, we may learn much more.
                                                    —Percival Lowell, 1908[61]

Science fiction author Olaf Stapledon suggested in the 1930s that we might discover alien civilizations by searching for signs of astroengineering.[62] More advanced civilizations may be able to transform natural objects and natural energies to suit their purposes. Although these works would not be intended as means of communication, they might be detectable by their effects.

Searching for radio signals is a fine idea, Dyson observed, but it only works if you have some cooperation at the other end. He suggested that we look for passive signals from uncooperative targets—evidence of intelligent activity without anything in the nature of a message. As any high technology must radiate away waste heat, we could search in the infrared, which does not assume anything special about the nature of extraterrestrial intelligence except that it be technological and carrying out activities on a large scale. The largest-scale activities will be the ones most likely to be found.[63]

Mind–Stretcher. Computer scientist Marvin Minsky questioned the assumption that advanced technological civilizations would radiate more infrared emission. Because radiation at any temperature above the cosmic background level is wasteful, the higher the civilization, the lower the infrared emission.[64] This implies that more advanced technological civilizations may be undetectable.

Dyson proposed in 1960 that we search for emissions in the far infrared from artificial biospheres that advanced civilizations may have built around their parent stars. Searches performed before 2005 have found nothing like "Dyson spheres" out to 80 light-years. A 25-year search of the sky for astronomical objects that might be artificial in origin found a number of very peculiar objects, but none appeared likely to be the product of alien masterminds. Within about 10,000 to 20,000 light-years around the solar system, no highly advanced extraterrestrial civilizations *intend* to reveal themselves through such objects (emphasis added).[65]

The James Webb Space Telescope, which may be deployed in space some time after 2010, will be designed to work best in the infrared part of the spectrum. This huge instrument, originally planned with a segmented mirror 6.5 meters (about 20 feet) across, is designed for the study of very old and distant galaxies, not to search for astroengineering projects.[66] Again, serendipity might apply.

Farside

The far side of the Moon is a symbol of remoteness and inaccessibility. The *Apollo* astronauts who looked down on Farside as they orbited the Moon were cut off from Earth, in a cone of radio silence. They were, for those minutes, the loneliest men alive.

It is that very insulation from the Earth that makes Farside of unique value to us. There we could locate the most powerful tools of our astronomy, undisturbed by the Earth's atmosphere, hydrogen geocorona, light, or radio signals. In the Moon's weaker gravity, radio antennas could grow to giant size. Astronomers or their robotic assistants could work in a carefully protected stillness, facing outward into the celestial deep. Great radio ears might strain to hear the whispers of the stars, and perhaps the distinctive patterns of intelligent communication.

Farside Station could be Humankind's unblinking eye on the cosmos, our most advanced listening post for energy and intelligence. Building it would extend human civilization around the Moon, enclosing a new world in our grasp.

French astronomer Jean Heidmann led an effort within the International Academy of Astronautics and the International Astronomical Union to reserve Farside for astronomy, including SETI. After Heidmann's passing, Italian physicist Claudio Maccone picked up the torch; the Academy's study of the concept might provide a basis for formal international action to protect Farside from lunar satellite or lunar base communications.

Farside's use would not be limited to radio astronomy. One scientist proposed setting up a battery of large infrared telescopes there to look for evidence of oxygen in the atmospheres of planets orbiting nearby stars. Others argue that technological advances make deployment in space a more attractive option for such instruments, although perhaps not for very large radio telescopes.[67]

Exotic Means

Dyson proposed in 1963 that we search for gravitational radiation from what he called "gravitational machines" that might be used as interstellar propulsion systems. Two decades later, astronomer John Kraus suggested that gravity waves could be used as a means of communication.[68] The first large gravity wave observatories are now in operation.

Lemarchand listed what he called "extravagant methods" for finding evidence of extraterrestrial technological activity. In addition to detecting infrared radiation from a Dyson Sphere, we might discover optical radiation from rare isotopes in stellar spectra or observe X-ray pulses from nuclear explosions or from material dropped on to neutron stars. We might look for anomalous gamma-rays generated by matter–antimatter propulsion systems or artificial neutrino beams. We might discover a coded message capable of replicating matter in suitable environments. Or we might find space probes or other artificial objects in our own solar system. Lemarchand described other methods—wormholes, faster-than-light particles, or the use of physical laws we have not yet discovered—as "exotica" from science fiction.[69]

Mind-Stretcher. "Star Trek"'s transporter may not be totally fanciful. William Reupke of the Computer Sciences Corporation thought it might be possible to transmit enough information over interstellar distances to recreate physical objects—including human beings. Matter transmitters deployed elsewhere in the Galaxy could provide new targets in the search for extraterrestrial intelligence.[70]

We do not know which of these techniques, if any, will reveal the presence of alien minds. What we are looking for, Shostak reminded us, is an uncertain manifestation of a hypothetical presence.[71]

SETI researchers continue to push back the horizon of possibility, with admirable dedication. Their enthusiasm is contagious, although their underlying beliefs are not yet proven.

Sending Our Own Signals

We ourselves are now capable of producing at will various phenomena. . . .
which could be observed from distant planets. . . . We must, therefore,
revise our thinking and incorporate in our theories possible effects of the
free will of other living beings.

—Otto Struve, 1962[1]

Active SETI

Some of those interested in searching for extraterrestrials have proposed
that we initiate contact by sending our own signals, in the hope that another
civilization will detect them and respond. Calling attention to ourselves by
such signaling has been described as "active SETI," in contrast to passive
listening.

The proposed means of sending signals have evolved with technological
advance and with the remoteness of the target. Early ideas focused on
creating geometric patterns on the Earth's surface that would be visible to
astronomers on other bodies in our solar system.

Mathematician Karl Gauss allegedly proposed in 1826 that we signal
intelligent beings on the Moon by clearing vast lanes of forest in Siberia to
show the Pythagorean theorem by means of areas surrounding a right tri-
angle. Austrian astronomer Joseph Von Littrow reportedly suggested in
1840 that we signal Mars by digging long canals in geometric shapes in the
Sahara desert, filling them with water and kerosene, and setting them on
fire at night.

Other concepts employed the transmission of light. Gauss recommended
that we signal the Moon with an array of 100 mirrors; Charles Cros proposed that huge mirrors be used to focus light on Mars or Venus, sending
messages by periodic flashes. American astronomer Henry Pickering,
who suggested signaling Mars with mirrors during the opposition of 1909,
thought that a mirror one-half square mile in area would be dazzlingly
conspicuous to Martian observers. As recently as 1941, James Jeans
proposed that we flash prime numbers toward Mars.[2]

The radio spectrum has gradually emerged as the preferred channel. Tesla, who built a radio transmitter in Colorado in 1899, believed that his signals had reached Mars. He predicted that interplanetary communication would become the dominating idea of the century that had just begun.[3]

Radar systems developed during World War II gave us a new means for transmitting powerful signals over interplanetary distances. The US Army Signal Corps "bounced" radar signals off the Moon in 1946. Twelve years later, scientists at the Massachusetts Institute of Technology sent a signal to Venus and received the echo. The California Institute of Technology achieved similar results with signals sent to Mars in 1963.[4] Military and astronomical radars remain the Earth's most powerful emitters of electromagnetic signals.

Plaques and Records

Rocket pioneer Robert Goddard suggested another way of sending messages in 1920: Interplanetary spacecraft might bear metal plates inscribed with geometrical shapes and astronomical objects. His proposal became reality five decades later with the launch of *Pioneers 10* and *11*, the first human-made machines destined to leave our solar system. Those robotic craft carried plaques designed to tell aliens who might find them about the nature of our species and our location in the galaxy. The later *Voyagers 1* and *2* carried "records" containing greetings from Humankind.

These plaques are destined to be the longest-lived works of our species, Sagan and Drake declared. They will survive virtually unchanged for hundreds of millions, perhaps billions, of years. They are like messages in bottles, cast into the sea of space in the hope that some beachcomber of the stars will find them—and note our existence.[5]

Such small machines are unlikely to be found in the vastness between suns. Sending these messages was more symbolic than practical; we are likely to be the only recipients.

The Arecibo Blast

The most famous of our deliberate interstellar transmissions was a radio signal sent in 1974 from the largest radio telescope in the world, at Arecibo in Puerto Rico. This was the strongest man-made signal ever transmitted; on its wavelength, the signal would have made our Sun appear to be by far the brightest star of the Milky Way.[6]

The Arecibo transmission was a shout into the cosmic dark, demonstrating that humans want to be noticed—or at least to leave some evidence of their existence. As astronomer Donald Goldsmith put it, we seem to enjoy announcing our existence to the universe. "If we did not," he argued, "surely someone by now would have pointed out the dangers of continually emitting large amounts of electromagnetic radiation."[7]

In fact, several people have pointed out potential dangers from transmitting powerful signals. British Astronomer Royal Sir Martin Ryle, who believed that calling a more advanced civilization's attention to our existence could be dangerous, reacted to the Arecibo signal by asking the International Astronomical Union to ban any further transmissions to the stars unless they were internationally agreed upon.[8]

Sagan responded defensively to Ryle's critique. The Arecibo message was clearly not intended as a serious attempt at interstellar communication, he claimed, but rather as an indication of the remarkable advances in terrestrial technology. The staff of the Arecibo Observatory saw the signal as a demonstration that radio telescopes were entirely adequate for interstellar communication over immense distances. Nonetheless, such demonstrations might prove to be the most likely cause of our being detected—if another civilization is listening at radio wavelengths. Sagan later seemed to have reconsidered his views, favoring listening instead of sending.[9]

The NASA workshop report of 1977 established a conventional wisdom. There is an immediate payoff if we receive a signal; transmitting requires that we wait out the round-trip light time before we can hope for any results. Transmission should be considered only in response to a received signal or after a prolonged listening program has failed to detect any signals. Twenty-five years later, a group of SETI experts reached the same conclusion.[10]

Deliberate transmission does not yet make sense; the Earth already is quite bright with radio leakage. Transmitting enough to improve this significantly is expensive, and a transmitting strategy cannot pay back for many years.

While our own technology is still leaking distinctive signals, passive listening remains the most cost-effective strategy for discovering extraterrestrials. "Sending on our own remains a delayed option," Morrison wrote in the prologue to the SETI 2020 report, "perhaps to be considered at the close of a century of search."[11]

Bernard Oliver, noting that there are people who are fearful about announcing our presence, said "You have to convince these people or enough of the populace that it is a good idea before you go ahead and do it in a democratic society. On the other hand, listening has no such hazards in anybody's mind." If we are worried about calling attention to ourselves, others argue, we can choose not to respond to a signal from extraterrestrials. There is no way for the transmitting civilization to determine if its message was received and understood on Earth.[12]

Despite his central role in the Arecibo blast, Drake stated that "we will not send signals until we have received them, because we do not know in which direction to send. It would be far too expensive to construct enough radio telescopes to bathe the sky in detectable signals."[13]

The future may be different. As our technology improves, the balance tilts in favor of transmitting for two reasons: The Earth will produce less leakage, and transmitting becomes easier and less expensive as time goes on. However, any active transmission strategy would have to be long-lived.[14]

Within 100 years, MacGowan and Ordway predicted, humans may be willing to undertake a more or less permanent transmitting program to selected stars. Until then, we must be content with planning on serious, extensive, and selective listening.[15]

Is This Research?

Transmission is a diplomatic act, an activity that should be undertaken on behalf of all humans.

—The SETI 2020 Report, 2002

The SETI Institute's Douglas Vakoch and others have argued that "active SETI" is an alternative search strategy that may increase our prospects for success in finding other civilizations. Vakoch added that it would provide more opportunity for funding by appealing to a new group of potential supporters and by emphasizing that at least one civilization is transmitting messages.[16]

Even if an extrasolar society does receive our messages, there is no assurance that it would be willing to acknowledge them. We can only hope, observed MacGowan and Ordway, that our signals will inspire worlds more advanced than ours to reveal themselves to us in a manner within our powers of comprehension.[17]

Goldsmith raised the possibility that we may have to compete for the privilege of getting detailed messages from extraterrestrials. Our most competitive strategy may be actively transmitting messages that are sufficiently intriguing to capture the interest of others.[18]

Private organizations and commercial enterprises already have sent their own signals. Canadian astronomer Yvan Dutil and physicist Stephane Dumas created a pictographic message that was transmitted in 1999 to four nearby Sun-like stars, with the support of a U.S.-based organization called Project Encounter. That group's Cosmic Call 2003 sent signals from a 230-foot Ukrainian radio telescope to five selected stars. According to the Project Encounter organization, such a call is important because its members "believe that we must do everything we can to try to make First Contact" and because humanity stands to gain much beneficial knowledge. The Command Center for this Cosmic Call was in the UFO Museum and Research Center at Roswell, New Mexico.[19]

Despite its name, Active SETI is not scientific research. It is a deliberate attempt to provoke a reaction from another technological civilization whose capabilities and intentions are not known to us. It is a cultural and political act whose consequences are not predictable.

Several people, including your present author, have argued that sending a deliberate, unusually powerful message is a decision that belongs properly with all Humankind. The SETI 2020 experts were divided on the issue; although most believed that transmitting now would be merely harmless and wasteful, a few felt that transmissions should not be carried out without international consultation and approval. Nonetheless, predicted Brin, we can expect more unilateral spasms as radio equipment becomes less expensive and more available to those who lack the patience or courtesy to respect the wishes of others.[20]

Probabilities

The ancient covenant is in pieces: man at last knows that he is alone in the unfeeling immensity of the universe, out of which he has emerged only by chance.

—Biologist Jacques Monod, 1971[1]

An intrinsically improbable single event may become highly probable if the number of events is very great.

—Astronomer Otto Struve, 1960[2]

Probability and Analogy

Is contact likely enough to hope for, or worry about? That depends on how probable we think other technological civilizations may be—and on what we think they might do.

Since their beginnings, arguments supporting the existence of extraterrestrial intelligence have rested on two kinds of logic: probability and analogy. Probability has been employed to determine how frequent and widespread extraterrestrial intelligence may be and, by derivation, how likely contact may be. Many believe that the argument from probability is sufficient to justify exploration. However, we cannot *prove* the existence of extraterrestrials this way.[3]

The case for alien intelligence also uses the analogy of our own presence; if we exist, intelligent beings may appear elsewhere as well. Yet we cannot prove the existence of alien minds by analogy any more than we can by probability. While our own existence shows that the development of life and intelligence is *possible*, Davies explained, we cannot use that fact to argue that the formation of intelligent life is *probable* (emphasis added).[4]

No law, theory, or worldview can be proven from a sample of one. Yet, although it may be risky to generalize from a single example, our analogies may not always be wrong.

The Drake Equation

At an informal meeting of scientists in November 1961, Frank Drake proposed his now famous equation as a way to estimate the number of communicating civilizations in our galaxy. Drake, who was trying to quantify SETI, thought that the equation could be used to justify and optimize searches. His formula also gave the subject a credibility that allowed other people to enter the field.[5]

Astronomy writer Govert Schilling later commented that the Drake equation, by breaking down a great unknown into a series of smaller, more addressable questions, made the search for alien civilizations more realistic and promising. The formula focused our attention on the really important issues.[6] Others put it more modestly, describing the equation as a way of organizing our ignorance.

An Evolving Equation

The original Drake equation read as follows, with N representing the number of communicating civilizations now in existence:

$$N = R\, f(p)\, n(e)\, f(l)\, f(i)\, f(c)\, L$$

The first factors were physical. R is the rate of star formation in our galaxy; $f(p)$ is the fraction of stars that have planets; $n(e)$ is the number of planets per star suitable for life. Later factors had more to do with biological evolution: $f(l)$ is the fraction of those planets on which life develops; $f(i)$ is the fraction of life-bearing planets with intelligent life. Then came the cultural factors. $f(c)$ is the fraction of intelligent cultures that develop radio communication we can detect; L is the average time spent by civilizations in a communicative state.

The Drake equation has been modified since then. Shostak, describing the formula as it existed in 1998, stated that $f(c)$ is the fraction of civilizations that have the technology *and the incentive* to communicate over interstellar distances.[7]

A version available from the SETI Institute in 1999 showed N as the number of *communicative* civilizations in the Galaxy *whose radio emissions are detectable*. R here means the rate of formation of *suitable* stars; $f(p)$ means the fraction of those stars with planets; $n(e)$ now means the number of "Earths" per planetary system. $f(l)$ is the fraction of those planets where life develops; $f(i)$ now means the fraction of *life sites* where intelligence develops; $f(c)$ now means the fraction of planets *where technology develops*.

A further revision was presented in the SETI 2020 report, published in 2002. Here, $n(e)$ is the number of planets per planetary system *with an environment suitable for life*; $f(c)$ is the fraction of civilizations that *develop a technology that releases detectable signs of their existence into space*; L is the average length of time such civilizations *release detectable signals into space.*[8]

Even as revised, the equation still focuses on civilizations that send out electromagnetic signals that we can detect. There could be many others that have not reached this stage or that left it behind long ago.

Although a calculation of probabilities sounds scientific, the way that we have quantified the factors in the Drake equation has been heavily influenced by opinions and beliefs. As Goldsmith observed, everyone who deals with the probability of extraterrestrial life has some bias for or against finding another civilization.[9] Different scientists, inserting different numbers for each factor in the equation, have come up with wildly varying conclusions.

Those most optimistic about finding extraterrestrials once dominated discussions of this question. Drake, Sagan, and their colleagues proposed in 1961 that there are somewhere between 1000 and 100 million advanced civilizations in our galaxy. Nearly 30 years later, Sagan still estimated about 1 million civilizations more advanced than our own.[10]

Asimov went farther, suggesting that technological civilizations may have developed on as many as 390 million planets in our galaxy and that nearly all of them are more technologically advanced than we are. MacGowan and Ordway estimated that 3 billion stars in the Milky Way have evolved intelligent communicating societies, although we need to know how many still exist.[11]

At the other extreme, deniers derive numbers as low as one communicating civilization—our own. They back up their pessimism about SETI by inserting lower estimates of probability for one or more factors. When isolationists refuse to support SETI because they think we are alone, their hypothesis becomes a self-fulfilling prophecy.[12]

One can visualize the evolution of communicating civilizations as passing through a series of bottlenecks, suggested Rood and Trefil. All that is necessary to get a pessimistic result is to make one of them very narrow. For the optimistic result to hold, all of the bottlenecks must be wide.[13]

Individual preferences determine the width of these bottlenecks. Botanist William Burger, inserting his own arbitrary numbers for a long list of factors, produced an estimate of 3 to 30 Earth-like planets with technological civilizations in our galaxy.[14] The derived conclusion is not absence, but scarcity.

As Brin saw it, everyone involved in this debate has a favorite factor that they love to suppress. Uniqueness partisans squelch the number of stable stars, or decent planets, or the likelihood of intelligence. SETI enthusiasts squelch interstellar travel, species life span, and contact cross section. They are all pessimists at some level.[15]

New knowledge, claimed Rood and Trefil, has greatly reduced our freedom to choose any numbers we like for terms in the Drake equation. Yet they admitted that there are too many parameters and ways of treating physical phenomena to place complete faith in one calculation.[16]

Given our lack of data for some factors, we should be cautious about jumping to conclusions. Mathematical models may be suggestive, even indicative. Without confirmation in the real world, they are not proof.

Being Trendy

Probability and analogy are not the only tools employed in this debate. Some disputants have relied on straight-line projections of current trends in our own civilization, particularly the future courses of scientific and technological advance. Yet, past projections of such developments in human societies have been woefully inadequate, often failing to foresee major changes.

There is no reason to think that current humans have overcome that limitation. All visions of more advanced civilizations that rely on the extension of uninterrupted trends must be regarded with suspicion. As we will see later, this could have intriguing implications for the probability of contact and for its possible consequences.

Probabilities:
The Astronomical Factors

Guthke may have been too dismissive when he wrote that the Drake equation is made up entirely of unknowns.[1] Since that equation was written, there have been significant scientific findings affecting some of its factors.

We now have circumstantial evidence lending credibility to a plurality of inhabited worlds: Extrasolar planets exist in large numbers, strengthening the arguments of believers. Research showing that life on Earth is tougher and more adaptable that we had believed also may help their case.

Deniers have presented more detailed arguments casting doubt on the probability of Earth-like environments elsewhere. Some research results imply that astrophysical disasters may be more frequent than we once thought, shortening the lifetimes of intelligent species and the civilizations they may create.

Each factor deserves a book of its own. Here we can only touch on indicative samples from the current literature, with a score card as of early 2006.

Stars

Fermi's classical question "Where is everybody?" may be answered with "where the late F stars and the early G stars are."
—Space visionary Krafft Ehricke, 1975[2]

Astronomers have developed fairly reliable estimates of the rate of star formation in our galaxy, although the rate for the universe as a whole may have been decreasing for the past 5 to 8 billion years.[3] The question now is how many of these stars could sustain conditions suitable for the evolution of life on their planets.

Conventional wisdom long assumed that only relatively recent third-generation stars like our Sun could have planets where life, intelligence, and technology might evolve, because only those planets would have the necessary heavier elements (known to astronomers as "metals") in sufficient abundance. Until recently, searches had focused on stars similar to the Sun (spectral types F, G, and K), about 10% of the stars in our galaxy.

Recent observations have challenged this view. The onset of star formation and related element production seems to have been very rapid after the Big Bang. Heavy elements, including those necessary for life as we know it, appeared much earlier in universal history than astronomers once thought. One analysis suggested that 30% of the stars harboring life in our galaxy are, on the average, 1 billion years older than our Sun.[4]

Mind-Stretcher. The peaking of the universe's stellar birth rate about 5 billion years ago roughly coincided with the birth of our own solar system. If the evolution of life and intelligence in our case is typical, this may imply a flowering of intelligence in our own era. Astronomer Martin Rees even suggested that the unfolding of intelligence is near its cosmic beginnings.[5]

We now have a broader view of the types of stars that might have planets with life. SETI researchers have extended the range to include the smaller and dimmer M-class stars that may make up 85–90% the stellar inventory—perhaps 300 billion of them in our galaxy.[6]

A recent census of stars within 10 parsecs (32.6 light-years) of our Sun raised the count to 341. These include 4 white A stars, 6 yellow F stars, 21 G stars like our Sun, 45 orange K dwarfs, 20 white dwarfs, 9 brown dwarfs, and a whopping 236 cool, orange-red M dwarfs like Proxima Centauri, our closest neighbor.[7]

At a minimum, this widens the scope for simple life. Drake saw even greater promise: The types of stars that might harbor civilizations are much more extensive than we used to think.[8]

Astronomers already have found a large dust disk around a nearby M-class star. As such circumstellar disks are signposts for extrasolar planetary systems, this discovery provides clues as to how the majority of planetary systems might evolve. "The habitable zone around a red dwarf is much thinner than around a sun-like star," observed astronomer Todd Henry, "but since their number is so much larger, I believe that the first exo-Earth will be found around a nearby red dwarf."[9]

In place of the traditional sharp dividing line between stars and planets, we now have a continuum, with the apparent gaps filling in. Astronomers have discovered hundreds of dim stars known as brown dwarfs. These bodies, which occupy the mass range between 10 and 75 Jupiters, share some characteristics with stars and others with planets. Astronomers have detected bodies orbiting some brown dwarf stars, suggesting that they too might have planetary companions.[10]

As far back as 1979, astronomer Virginia Trimble concluded that stars with both ages and heavy element abundances comparable with those of the solar system are quite common in our Galaxy, particularly in its inner regions. The fact that we have so far failed to find extraterrestrials cannot be explained in terms of stellar and galactic evolution.[11]

Bottom Line, Stars: Advantage, believers. Suitable stars are not the bottleneck.

Planets

There are certain determined definite centers, namely the suns, fiery bodies around which revolve all planets, earths, and waters, even as we see the seven wandering planets take their course around our sun.
— Giordano Bruno, 1584[12]

Throughout this long debate, the plurality of intelligent life has been implicitly connected with the frequency of planets. Until recently, the existence of such worlds beyond our solar system rested on an argument from probability, without observational proof.

We have had a breakthrough. As of mid-2006, almost 200 extrasolar planets had been discovered. Although our statistics still are limited, it appears that vast numbers of planets accompany other stars.

Habitable Planets for Humans

More than 40 years ago, Rand Corporation analyst Stephen Dole extrapolated from existing knowledge of our own solar system to calculate that our Galaxy might host as many as 640 million Earth-like planets with life. Dole postulated some basic limits for planets that would be habitable by an expanding human species: mass greater than 0.4 Earth, but less than 2.35 Earth; period of rotation less than 96 hours; age more than 3 billion years; illumination at low equatorial inclination between 0.65 and 1.35 Earth normal; orbital eccentricity less than 0.2; the mass of the star less than 1.43 Sun.

Dole collaborated with Isaac Asimov on a popularized version of his study called *Planets For Man*. Inserting numbers into a Drake-like formula, including different estimates for different classes of stars, Dole and Asimov concluded that the average distance between a habitable planet and its closest neighbor in our region of the Galaxy is about 24 light-years. That distance may be less in more densely packed regions of the Milky Way. If these estimates are roughly accurate, there may be 600 million planets in our Galaxy that would be inhabitable by human beings.

Dole and Asimov noted that the development of modified species of humans will inevitably broaden the concept of a habitable planet. "The

Galaxy might well . . . be inhabited by varieties of men who are not only of separate species but whose criteria of habitability in planets may not be the same." They added that interstellar migration may become a new form of evolutionary pressure—both with respect to the new environments to which Man will be exposed and to the new requirements made of his mind and character.[13]

These observations would apply to nonhuman intelligences as well. Extraterrestrials might be able to evolve and thrive under a wider range of conditions than those required by humans. If they expand beyond their biospheres of origin, they too might diversify, broadening their potential range.

The new era of planetary discovery began in 1983 when astronomers detected a cloud surrounding the bright star Vega, possible evidence of dust and larger objects in orbit. A year later, astronomers obtained the first image of a circumstellar disk, around the star Beta Pictoris.

Since 1990, observers have found that many young stars have protoplanetary disks—potential solar systems in the making. The Hubble and Spitzer Space Telescopes have spotted such disks around mature, Sun-like stars known to have planets. They also have been detected around smaller, cooler M-class stars, which are much more numerous. At least 15% of nearby stars are surrounded by dusty disks.[14]

Dust disks around many stars are replenished from repeated collisions of large rocky objects, the way our solar system's inner rocky planets were formed. Many disks have inner holes that may have been swept out by accreting planets. This may mean that Earth-like planets are fairly common around other stars.[15]

Researchers also have found that planets may form around young brown dwarfs. The dense, cool atmospheres of these stars are an ideal environment for producing molecules, including water and carbon dioxide. These cool dwarfs may be as numerous as all the other stars in our region of the Galaxy.[16]

The first confirmed discovery of an extrasolar planet was made in 1991, when astronomers indirectly detected bodies orbiting a rapidly spinning neutron star known as a pulsar. Tiny variations in the regular pattern of signals from the pulsar indicated three planets, two of them weighing about the same as the Earth.[17] Most experts regard the environment around a pulsar as extremely inhospitable to life.

Four years later, astronomers drawing on computer analyses of small irregularities in the Sun-like star 51 Pegasi's motion reported evidence of a planet. Observers also have detected dips in the luminosity of stars, which imply that a large planet has crossed their disks as seen from the Earth.[18]

A Mirror Image. Astronomer John MacVey suggested that alien scientists studying our Sun might regard it as planetless because of the lack of an appreciable wobble as it journeys through space.[19]

As the search developed, astronomers found planets around some 10% of Sun-like stars; the real percentage may be much higher. Planetary scientist Stuart Ross Taylor proposed that we may eventually find that planets forming from disks rotating around young stars will occupy all available niches within the limits imposed by the cosmochemical abundances of the elements and the laws of physics and chemistry.[20]

In 2001, astronomers discovered the first planetary system resembling our own, with a Jupiter-like body in a circular orbit at the right distance. Two years later, searchers found a star–planet system even more like ours. Known as 18 Scorpii, this star has a mass, diameter, rotation, sunspot cycle, surface temperature, and age very similar to those of our own Sun. The system also has a Jupiter-like body in a nearly circular orbit far enough from its star to allow smaller, rockier planets to survive closer in. 18 Scorpii is only 46 light-years away; planets found in that system may be prime targets for instruments that search for signs of life.[21]

Jupiter-like objects have ambiguous implications. Our own Jupiter may have played an important role in our evolution by diverting comets away from the inner solar system, where their impacts might have reduced the chances for life evolving on inner planets. According to one estimate, without Jupiter, the current impact rate of comets would be a thousand times higher, with catastrophic collisions occurring every 100,000 years. On the other hand, a system with only Jupiter-type planets might be unlikely to evolve intelligent life.[22]

Astronomers have obtained the first images of planets orbiting other stars, including one about twice as massive as Jupiter but 20 times farther away from its sun.[23] The fact that we can see it at the astonishing distance of 450 light-years tells us how quickly our search for planets has progressed.

In 2005, planet-hunters Geoffrey Marcy and Paul Butler detected the first rocky planet orbiting another star, a red dwarf 15 light-years away. This stony world, seven times as massive as the Earth, may be so close to its sun that life could not develop or survive on its surface. Astronomers who found another planet of comparable mass concluded that worlds smaller than Neptune may be common in the inner parts of many solar systems.[24]

The prevailing conventional wisdom among astronomers long assumed that planetary systems were unlikely to form around binary or multiple stars, an estimated 60% or more of stars in our Galaxy. Now we know that they can. "The neglected majority of double stars," wrote two German astronomers, "could fill the galaxy with planets."[25]

Parallel Programs, Different Fates

In 1988—the year that NASA formally endorsed the SETI program—the agency's Solar System Exploration Division established a working group on strategies for finding other planetary systems. Like the SETI pioneers, scientists interested in detecting extrasolar planets held workshops to identify priorities and to build consensus.[26] Although both programs were modest in scale, official support of the search for alien planets has survived, but official support of searching for alien technologies has not.

At this writing, Earth-sized planets are still beyond our ability to detect directly; once again, we are operating at the limits of our instruments. Some models of solar system evolution suggest that an Earth-like planet is likely to form near our world's position; if correct, such results imply that Earth-like planets may be a common feature of other systems.[27]

Some scientists challenge the assumption that our Sun and its family are typical. Having many worlds and moons in near circular orbits may be more the exception than the rule. Earth-like planets may not form in many systems; if they do, their orbits may be perturbed by Jupiter-sized bodies, ending up far out of the system's original plane.[28]

On the other hand, we may be misled by selection effects. The more massive a planet is and the more closely it orbits around its star, the bigger the wobble, making hot Jupiters the easiest planets to find.[29] There may be many smaller planets whose effects are beyond our present ability to detect.

If there are Earth-like planets, some may be very old. One study showed that such planets began forming in the Milky Way about 9.3 billion years ago and that their average age may be about 6.5 billion years—2 billion years older than the Earth. Charles Lineweaver of the Australian National University calculated that three-quarters of all "Earths" are older than ours.[30] Some may have had plenty of time to evolve life, and possibly intelligence.

What does Earth-like mean? Artists who visualized the surfaces of other planets in our own solar system before the age of planetary exploration drew on analogies with Earth to produce recognizable, attractive environments. The realities that our spacecraft discovered were very different from what we had imagined—and hoped for. We underestimated how alien other planets would be, even around our own star.[31]

We also may be underestimating the variety of planetary environments that can support life, particularly by assuming that they must have liquid water on their surfaces. Our definition of "Earth-like" may need to be a broad one.

A Mirror Image. P.E. Cleator, one of the founders of the British Inter-planetary Society, pointed out in 1934 that, from an extraterrestrial point of view, conditions on the Earth may be deemed unsuitable for life. Arthur Clarke later developed this idea in his "Report on Planet Three," a fictional document prepared by Martian astronomers. They concluded that the prospects of life on Earth appeared to be extremely poor because of Earth's poisonous atmosphere, strong gravity, and other factors. Although some form of vegetation might be possible on the third planet, the Martians resigned themselves to the idea that they were the only rational beings in the solar system.[32]

Scientists have suggested that some moons of giant planets might be suitable for life, expanding the range of potential life into systems with no Earth-like planets. Shapley broadened the search space for discovery, speculating that there may be many planet-sized objects that are not gravitationally bound to stars. Some could generate enough internal heat to support biological evolution; perhaps thousands of these self-heating planets would be suitable for life. "Many billions of dark Earths could be roaming the Galaxy," David Darling wrote 40 years later, "cut adrift from the star systems in which they were made."[33]

The discovery of young, isolated planetary mass objects in star-forming regions suggests that a majority of large planets, ranging from 3 times to 80 times Jupiter's mass, may not be associated with suns. Such free-floating objects may exceed, by orders of magnitude, the total number of stars; given their numbers, they could constitute the largest environment for life in the galaxy. One or more interstellar planets could be closer to our solar system than the nearest stars.[34]

Astronomy continues to reveal significant bodies that had been hidden from us, challenging our assumptions about our galactic environment. We have learned that our own solar system extends much farther into interstellar space than we had thought. Instead of ending at Pluto, the Sun's family is now known to reach out at least three times that far.

In 2003, astronomers discovered a planetoid called Sedna that ranges as much as 84 billion miles away from our star. Sedna's elliptical orbit may carry it into the inner section of the long hypothesized but previously undetected Oort Cloud, an array of icy objects.

There could be more than a million bodies larger than 100 kilometers (62.5 miles) in diameter in the Sun's extended scattered disk. Observers already have found more than a thousand trans-Neptunian objects, including one bigger than Pluto. Astronomer Alan Stern speculated that some would be as large as the Earth.[35]

Other solar systems also may extend far beyond the orbits of the planets we can detect. There is much more stuff out there than Newton would have predicted.

Mind-Stretcher. There could be alien planets in our own solar system. Billions of years ago, our Sun and another star may have swapped their most distant companions during an encounter that disrupted orbits in the outer parts of both systems. Each might have captured thousands of Sedna-like bodies from the other. One theory suggests that Sedna is a true alien planet, tugged away from the outer disk of a low-mass star that passed relatively close to our Sun.

Theorists predicted that objects whose orbits are highly inclined to the plane of our own solar system (the ecliptic) would have come from outside. In 2005, astronomers found a minor planet larger than Pluto, but twice as far from the Sun—presently 9 billion miles out. Tentatively named Xena, this object is in an orbit inclined 44 degrees to the ecliptic. The new planetoid could have been discovered much sooner if anyone had looked that far away from the narrow band in which the classic planets revolve.[36]

Observatories planned for the near future will improve our ability to detect bodies orbiting other stars, possibly enabling us to identify Earth-sized planets. The Large Binocular Telescope Interferometer, designed to image exoplanets and to study circumstellar disks, is scheduled to go into operation soon on Arizona's Mount Graham.[37] The European orbiting observatory Corot, to be launched this year, and NASA's Kepler satellite, scheduled for launch in 2008, will monitor thousands of stars for slight dips in brightness caused by planets crossing their faces. The first space-based observatories able to detect Earth-like bodies orbiting other stars, these missions may find thousands of planets.

NASA's Space Interferometry Mission, to be launched some years later, will survey 2000 nearby stars for indirect signatures of medium and large planets. For 200 closer stars, this mission should reveal hints of Earth-sized bodies. The proposed Terrestrial Planet Finder could have a good chance of finding terrestrial-type planets within 50 light-years of the Sun if a good fraction of stars in our neighborhood have such companions (planners assume that 10% do).[38]

The most important discovery that astrobiology can realistically make in the next couple of decades, suggested Darling, is to find an out-of-equilibrium atmosphere on an extrasolar planet.[39] We could use spectroscopic analysis of the atmospheric composition of planets orbiting other stars to establish the existence of even nontechnological forms of life. Oxygen at any appreciable abundance would almost certainly indicate biology comparable to that of modern Earth; methane also would suggest some form of life.

The first steps have been taken. Less than a decade after finding the first planet beyond our solar system, observers were able to study the atmosphere of a "hot Jupiter" that passed in front of its sun.[40]

Scientists have tested our ability to detect biosignatures by aiming the detectors of Mars missions at the Earth. They were able to identify several constituents of our atmosphere, although our ability to do this at interstellar distances will depend on the resolving power of our instruments.

This type of research might even reveal the presence of a space-faring civilization. Finding two objects in an alien solar system that show signs of life in their atmospheres would suggest that one of them has been "terra-formed."[41] That, in turn, would tell us that an intelligent species had expanded beyond its home planet.

The Specialness of Earth

Those who are optimistic about finding extraterrestrial life and intelligence tend to see our Earth as typical of a class of planets that may exist in many systems. Others question that assumption.

For the first time in 500 years, Rood and Trefil proposed in 1981, we are coming to see the Earth as something special. It is in the system of a single G-class star; its orbit is in that zone where water will neither boil nor freeze for billions of years; it has a large moon which produces tides on Earth that favor tidal pools where life might have evolved; the tilt of the Earth's axis is just enough to cause periodic changes in the climate.

Geologist Peter Ward and astronomer Donald Brownlee revived this theme in their book *Rare Earth*, claiming that the conditions needed for the evolution and survival of higher life are so complex and precarious that they are unlikely to arise in many other places. The key factors include the probability of planetary systems of the right kind, the persistence of oceans and moderate temperatures, and a large moon. If the Rare Earth Hypothesis is correct, they declared, then SETI clearly is a futile effort. "There probably are other civilizations in the galaxy that have radio telescopes," they acknowledged, "but the vast numbers of stars and the vast distances involved are barriers that may always keep SETI more an experiment than a large-scale scientific endeavor."

Planetary scientist David Grinspoon challenged the Rare Earth thesis, particularly its failure to fully recognize the role of life in creating the Earth's unusual character. However, he agreed that our Moon has influenced Earth in numerous ways, slowing its rotation, raising tides in its oceans, steadying its spin axis and climate. Without such a moon, our planet might have undergone radical changes in the tilt of its axis, provoking extreme excursions of climate (there is some evidence that Mars has gone through such traumas).

A few claim that life might not exist at all if the Earth had no moon, particularly because of its effects on tides. On the other hand, some computer simulations suggest that small planets with big moons are likely to be quite common. As many as one in three young Earth-like planets may be struck hard enough to make big moons.[42]

Habitable Zones

Strughold proposed in 1956 what he called an "ecosphere of the Sun," made up of a biotemperature belt, a liquid water belt, and an oxygen belt. These belts all lie at about the same range from the Sun, from the orbit of Venus to beyond Mars. This concept, he suggested, could be applied to other stars.[43]

Scientists further developed the concept of a "zone of habitability" in our solar system, often defining it as a band of orbits within which liquid water could exist on planetary surfaces. Researchers studying the unique role of water in the chemistry of life have concluded that no simple molecule can mimic all of water's biological functions; water is an anomalous liquid with characteristics that seem critical to its biological role. Yet, science editor Philip Ball warned against the idea that, because life on Earth requires water, life anywhere else must have the same requirement. Life, which is adaptive, may simply have found ways to exploit what water has to offer.[44]

Experts disagree about the breadth of habitable zones around other suns. Astronomer Michael Hart argued that only G stars would have continuously habitable zones; other scientists have extended the range to other classes of stars.[45]

Ward, Brownlee, and astronomer Guillermo Gonzalez proposed a galactic habitable zone—the most hospitable areas of the Milky Way, lying between its dangerous center and its heavy-element-poor outer reaches. In a statement with a certain anthropocentric flavor, they wrote that the galactic habitable zone appears to be an annulus in the disk at roughly the Sun's distance from the Milky Way's center. Galactic travelers, if there are any, would tend to roam around the annulus.[46]

Other researchers have defined the galactic zone of habitability as a ring of stars that emerged about 8 billion years ago at 25,000 light-years from the core—slightly closer than the Sun's distance today. This zone has slowly spread toward and away from the galactic center and now embraces about ten percent of the stars born in the Milky Way. Roughly three-quarters of the stars in the zone are older than the Sun—typically 1 billion years older.[47]

A suitable star may not be the only environmental factor. Astrophysicist Priscilla Frisch commented that it does not make sense to look for habitable planets unless you look at the way their stars interact with local environments. A star passing in and out of dense interstellar cloud fragments would not have a stable environment, nor would its planets.[48]

Bottom Line, planets. A big win for the believers. It could be even bigger if astronomers detect Earth-like planets. The long-assumed plurality of worlds has been proven, although not the plurality of *inhabited* worlds.

Probabilities: Life

Miracle, Rare Accident, or Probable Event?

All the analogies of nature lead us to believe that, whatever the process which led to life upon this earth—whether a special act of creative power or a gradual course of development—through that same process does life begin in every part of the universe fitted to sustain it.
—Astronomer Simon Newcomb, 1905[1]

Believers argue that life is likely to emerge wherever physical and chemical conditions allow, making it a widespread phenomenon in the universe. Deniers claim that the origin of life is exceedingly rare, perhaps unique to the Earth. How can we address this complicated issue?

Let's begin with philosophical viewpoints about the origin of life, as described by Paul Davies. First, it was a miracle; this traditional view would be challenged by the discovery of an independent origin of life, or by the creation of life in a laboratory. Second, it was an accident, a highly improbable event; finding an independent origin of life would challenge this view as well. Third, it was a natural process of high probability.

Davies laid out principles underlying the case for extraterrestrial life. First is the Principle of Uniformity of nature: As the laws of nature are the same throughout the universe, the physical processes that produced life on Earth can produce life elsewhere. Second is the Principle of Plenitude: that which is possible in nature tends to become realized; if there is no impediment to the formation of life, life will form. Third is the Copernican Principle: The Earth does not occupy a special position in the universe, but apparently is a typical planet orbiting around a typical star in a typical galaxy.[2]

Each of those principles has been challenged by skeptics. Many biologists believe that the origin of life was highly contingent, some putting the factor $f(l)$ close to zero. They argue that the evolution of life as we know it is the result of such an extremely unlikely combination of circumstances that Earth may be its only home.

Others observe that life on Earth did not wait, but took advantage of pretty much the first opportunity offered. To astronomer David Koerner and neuroscientist Simon LeVay, this suggested that there is nothing improbable about the first spark that ignites life from nonlife.

Evolutionist Stephen Jay Gould believed that life on the Earth evolved quickly and is as old as it could be. This is quite different from the long, slow, drawn-out scenario so deeply ingrained in the origin of life community since Darwin.[3]

The oldest rocks known on Earth show that single-celled organisms probably have existed since our planet's surface cooled enough for water to remain liquid; the oldest signs of terrestrial life yet found date back 3.75 billion years. Scientists have found evidence that photosynthetic organisms had evolved and were living in an ocean more than 3.4 billion years ago.[4]

Microorganisms first appeared on our planet in nightmarish circumstances; comets and rocky bodies were pounding the Earth. Living things may have arisen repeatedly only to be wiped out every time—except the last. If life did emerge independently several times, we would not be aware of most of these events; all life today would have descended from just one of them. If any of those earlier beginnings had survived, Grinspoon pointed out, the evolution of life on Earth would have produced very different results.[5]

Rapid emergence may not always mean rapid development of more complex forms; long delays occurred in the evolution of living things from one stage to another. It took over 2 billion years for life on Earth to evolve the capacity for complex multicellular development. Oxygen levels in the Earth's atmosphere may have remained very low for more than a billion years, postponing the emergence of animal life.[6]

On the other hand, there can be rapid change. Although long periods of time were needed for the early evolution of microorganisms, much less was needed for the diversification of later multicellular life. When animals did emerge on the land, they seem to have adapted much more quickly than evolutionists once thought.[7]

Many researchers now believe that life may arise whenever a suitable energy source, a concentrated supply of organic material, and water occur together. Shapley had foreseen this in 1958, when he wrote that life will emerge and evolve wherever the chemistry, geology, and climatology are right.[8]

Mind-Stretcher. Scientists are planning to create living organisms from raw materials, challenging two older schools of thought about the origin of life: that it was created by a superior being or that it evolved blindly from chance encounters of simpler substances.[9] We may find that life can be created by intelligent creatures like ourselves, not just by nature or by God.

Chancists Versus Convergionists

Demonstrating the complexity of a process is different from demonstrating that the end result is rare.

—Physicist Lawrence Krauss, 2000[10]

All physical processes are a combination of chance and necessity, Nobel Prize winner Jacques Monod wrote in 1971. If chance was the dominant factor, the probability of a known organism forming from random molecular shuffling is absurdly small. Others have challenged this view, arguing that Monod had elevated chance to the level of a metaphysical principle; any scientific explanation that uses the hypothesis of singular chance comes into conflict with scientific standards.[11]

Nonetheless, evolution could have gone in different directions at many stages, producing very different results. The evolutionary process is not directional, orthodox Darwinists declare, and does not produce predictable outcomes.

Biologist Leonard Ornstein spoke for many when he argued that Darwinian selection is unparalleled by other physical processes and is much less likely to repeat itself. Evolution on Earth can easily have generated many "inventions," perhaps including intelligence, which are unique in the universe. He then went beyond science to declare that no significant social expenditure on SETI is warranted.[12]

Nobel Prize winning biologist Francois Jacob proposed that the appearance of life on our planet was not the necessary consequence of the presence of certain molecular structures in prebiotic times. "In fact," he wrote, "there is absolutely no way of estimating what was the probability for life appearing on Earth." The living world as we know it is just one among many possibilities; its actual structure results from the history of our planet. The interplay of local opportunities—physical, ecological, and constitutional—produces a net historical opportunity that determines how genetic opportunities will be exploited. It is this net opportunity that controls the direction and pace of adaptive evolution. As for extraterrestrial life, the sequence of historical opportunities there could not be the same as here.[13]

Astronomer George Seielstad looked at this question in a different way. Could life have been a certainty sometime, simply because its ingredients were periodically reshuffled until appropriate conditions fell into place?[14]

Scientists have found that the course of evolution is characterized by a trend to greater flexibility in the execution of the genetic program. With each innovation improving the transmission of genetic information, argued zoologist Mark Ridley, the complexity ceiling rose. Genomes may have potential capabilities that are hidden in normal times; organisms may be able to draw on preexisting genetic variation in response to environmental stress. Given long time periods, suggested anthropologist Richard Lee, evolution is a means for generating highly improbable results.[15]

Mind-Stretcher. This debate has implicitly assumed that the basic principles of evolution we know—genetic variation and natural selection—will operate on other worlds. Genes and chromosomes may not be the only way that biological information can be organized.

Evolutionist Simon Conway Morris, a frequent rival of Gould, saw evolution as seeded with probabilities, if not inevitabilities. "Convergence is ubiquitous from molecules to social systems," he declared. "In fact, the study of convergence reveals a deep structure to life. This strongly suggests that what is true on Earth is true anywhere."

Morris challenged the belief that the immense number of possibilities confers an inherent unpredictability on evolution. The main principle of evolution, beyond selection and adaptation, is the drawing of new plans but relying on the tried and trusted building blocks of organic architecture. Life is full of inherencies.[16]

As Morris saw it, convergent evolution, broadly speaking, produces predictable evolutionary outcomes. Multicellularity has evolved repeatedly, warm-bloodedness several times. Mechanisms used by diverse organisms to see, smell, hear, echolocate, sense electrical fields, and maintain balance are often convergent. For all of life's plenitude there is a strong stamp of limitation, imparting not only a predictability to what we see on Earth, but by implication elsewhere. The evolutionary routes are many, concluded Morris, but the destinations are limited.[17]

Gould recognized that "architectural constraints" limit adaptive scope and channel evolutionary patterns. While emphasizing that evolution is contingent, Gould also stressed the importance of emergence—the idea that novelties are present in complex phenomena that can neither be found in, nor explained by, simpler processes.[18]

Molecular biologist Sean Carroll found a long history of support for the general notion of overall evolutionary trends toward increases in size, complexity, and diversity. Scientists have proposed two fundamentally distinct mechanisms to explain these trends. One is a random, passive tendency to evolve through an overall increase in variance. The other is a nonrandom, active or driven process that biases evolution toward increased size and complexity.

Multicellularity evolved independently many times, Carroll emphasized. Once it did, macroscopic forms arose with new body plans or physiologies and with greater degrees of morphological complexity. The emergence of new forms was often followed by periods of rapid diversification. Although global trends may be passive, there may be active, directional trends nested within the overall arc of evolutionary history. Assuming a cellular basis of life elsewhere, the passive trends toward increases in organismal size, complexity, and diversity are certain to prevail.[19]

In the case of Earth life, what we regard as complex is usually inherent in simpler systems. Although this process takes time, what was impossible billions of years ago may become increasingly likely.

Take multicellularity. Darling saw this as a convergent property that bestows such overwhelming advantages that we will see it implemented routinely wherever living things emerge. The same is true of mobility.[20]

Evolutionist Ernst Mayr, although skeptical about extraterrestrial life and intelligence, acknowledged convergent evolution on Earth. Eyes evolved independently at least 40 times in different groups of animals.[21]

Recent research has offered additional support for convergence. "A lot of studies are finding quite a lot of surprising replicability of evolutionary outcomes," said evolutionary biologist Richard Lenski. "There's more to repeatability than we had suspected a decade ago," concluded Loren Rieseberg, another evolutionary biologist. He and his colleagues found evidence that evolution can repeatedly produce the same species.[22]

The evolutionary route that led to life seems to have taken the way with the fewest obstacles and to have chosen the most abundant construction materials available, argued planetary scientist Armand Delsemme. The genetic code appears to be well adapted to maximize the probability that mutation will lead to an improvement in the resulting protein.[23]

On the spectrum between total chance and complete determinism, concluded Davies, only an exceedingly tiny window would correspond to sparse life. If the truth lies to the deterministic side of that window, life will be common; if it lies much to the chance side, it will have happened only once in the observable universe. The discovery of one life-form with a different physical or chemical basis would settle the issue.[24]

Catastrophic Punctuation

Some scientists see a powerful connection between asteroid and comet impacts and the evolution of life on Earth. Citing the theory of "punctuated equilibrium" developed by evolutionists Stephen Jay Gould and Niles Eldredge, they argue that massive extraterrestrial impactors may have been the cause of this punctuation, provoking vulcanism, tectonic movements, and mountain building as well as atmospheric change.

The great extinction of 250 million years ago, which may have been caused by an impactor, by massive vulcanism, or both, was catastrophic; some researchers believe that life was nearly exterminated. Yet, life not only survived, it flourished.[25]

Teleology and Self-Organization

Chance does not exclude inevitability.
—Biologist Christian de Duve, 2005[26]

Evolutionary biology, as an historical science, is particularly plagued by teleological explanations, observed Mayr.[27] Does the direction of evolution have a teleological purpose—an innate aim to create higher life forms?

The idea that life is the manifestation of a universal organizing principle is not new; it can be traced back to Aristotle. Darwin himself suggested that "the principle of life will hereafter be shown to be a part, or consequence, of some general law."[28] Yet, the dominant scientific paradigm has long been against teleology and any hint of progress toward a goal.

More recently, we have seen growing interest in emergent properties. Noting that self-organization abounds in physics and chemistry, Davies argued that it would be astonishing if self-organization did not occur in biology too. Yet, any suggestion that biological order might arise spontaneously is considered a dangerous heresy.

The biosphere as a whole is undeniably more complex today than it was 3 billion years ago. The most complex organisms today clearly have a much greater complexity than the most complex organisms in the remote past. "Life, when allowed to flourish, rides an escalator of growth," argued Davies, "filling out every available niche, exploring new and better possibilities, developing ever more elaborate forms."

If we detect the presence of an alien intelligence, it would suggest that there is a progressive evolutionary trend outside the mechanism of natural selection. This assumption, made by most SETI scientists, strikes at the very heart of neo-Darwinism.

Davies concluded that biological evolution is just one more example of a lawlike progressive trend that pervades the cosmos. This does not mean that Darwinism is wrong, only that it is incomplete.[29]

Biological determinists like Nobel Prize winner de Duve believe that life will inevitably emerge, given enough time and favorable conditions; they see life as a preordained consequence of the laws of nature. However, de Duve cautioned that a deterministic view of the origin of life does not necessarily imply that life is widespread in the universe. It only means that life is as frequent or as rare as the physical and chemical conditions under which it must arise.[30]

Reacting to attempts to create life in the laboratory, Wald argued that experimenters are not creating life; they are simply trying to establish the conditions in which life can emerge. Life is part of the order of nature. It has a high place in that order; it probably represents the most complex state of organization that matter has achieved in our universe.[31]

Other scientists also are starting to see life as a manifestation of a universal tendency toward self-organization. When complexity theory reveals that many forms of Earth life are governed by deep geometrical rules, observed Grinspoon, it suggests a universal geometry of life.[32]

Life as We Don't Know It

It is understandable that humans—or the inhabitants of other isolated biospheres—may conclude that their own environments are uniquely suited for life. The evolution of life on Earth has exploited the idiosyncracies of carbon and water, making us carbon and water chauvinists.

These materials seem special to us, Feinberg and Shapiro argued, after 4 billion years of adapting to and building on their peculiarities. However, living things on other planets may exploit other chemical flows. As Clifford Pickover of IBM's Watson Research Center put it, life will evolve into whatever embodiment best suits its purposes.

Sagan warned years ago that we tend to overlook the possibility that other conceivable laws of nature might also be consistent with life.[33] Earth may be a very special place—for living things of the kinds we know on Earth.

Gaia and Her Sisters

Every planet changes in the course of its history, Ernst Mayr argued, and the sequence of changes has to be just right.[34] That may not be the whole story.

The Gaia theory originated by chemist and biophysicist James Lovelock and further developed by biologist Lynn Margulis suggests that living things can drive changes in the biosphere that benefit the long-term evolution of life. Early photosynthetic organisms arose in an ecosystem that was fundamentally different from the one we know today. They enriched our planet's atmosphere with oxygen, enabling very different evolutions on land. Life has "Terraformed" the Earth for billions of years.[35] While individual species go extinct, life survives and evolves.

Long-term habitability may depend more on the establishment of a robust and resourceful global living organism than on having a lucky planet and a lucky star, speculated Grinspoon. Once life starts, it is only a matter of time before it develops, in concert with its evolving planet, into a global system of interacting and self-regulating cycles. Life is what a planet becomes.[36]

The question may not be the probability of life, according to biologist Norman Pace, but, rather, the probability that life, having arisen, survives and comes to dominate a planet. Yet we must be cautious about assuming that Gaia—a description of what happened on Earth—will predict events on other worlds.[37] The exact model we see might be peculiar to our planet.

Living things on Earth, including humans, alter their environments and thereby influence their own evolution in a process known as "niche construction." Organisms on other worlds also might modify their environments and the selection forces they experience.[38]

Extremophiles and Dual Biospheres

We encountered our first exotic biosphere in 1977, when researchers found strange forms of life clustered around vents on one of Earth's mid-ocean ridges. There, beneath thousands of feet of water, far out of the reach of sunlight, was an entire ecosystem living off heat and chemicals from the Earth's interior.

Some researchers speculate that we might find similar forms of life in the global ocean believed to underlie the icy surface of Jupiter's moon Europa. Extrasolar planets may evolve unexpected biospheres that are even more exotic.

Life can exist in a much wider range of conditions than we once believed. Research has shown that terrestrial microorganisms can survive in boiling water or freezing cold, even in the interstices of rocks far below the Earth's surface. Certain types of heat-loving bacteria (hyperthermophiles) are able to replicate at temperatures above that of boiling water. Cold-loving (psycrophilic) organisms are plentiful at temperatures below freezing in Antarctica. Others function in highly acid or base solutions (acidophiles and alkaliphiles), in intensely salty brines (halophiles), under crushing high pressures (piezophiles or barophiles), or in extremely dry conditions (xerophiles).

Living things have evolved to deal with both metabolic and external sources of DNA-damaging agents by developing elegant mechanisms that repair the damage. Some bacterial types thrive in the interiors of working nuclear reactors.[39]

Our definition of the Earth's biosphere has greatly expanded. Simpler organisms (prokaryotes) have been found 3 kilometers below the surface, with some evidence suggesting that this can be extended to at least 4 kilometers (2.5 miles). Researchers have discovered prokaryotic populations as deep as 800 meters below the sea floor; they occur in sediments throughout the world ocean and may constitute as much as one-third of Earth's total living biomass.[40]

Astronomer Thomas Gold proposed that the "deep, hot biosphere" could be more extensive than the surface biosphere we know best. The bulk of terrestrial life may dwell in the rocks, down to a depth of 10 kilometers or more.

We may have two separate biospheres on Earth. Planetary scientist Von Eshleman suggested that near-surface and subterranean life-forms are essentially independent; either could exist without the other. Life might have started independently at two different levels. Davies speculated that we may find a second origin of life on the Earth: pockets of microbes deep in the crust that survived the early asteroidal bombardment 3.9 billion years ago.[41]

Mind-Stretcher. Calling a phenomenon extreme depends on its context. What would be extreme in an alien environment? If life had originated and evolved under conditions very different from those on Earth, organisms that we call extremophiles might be in the mainstream of living things. As Shostak suggested, they might be the most frequent form of life in the universe. What is exotic and rare on our planet could be common elsewhere.

The assumption that biology required light as well as liquid water limited the prospects for finding life beyond Earth, observed Davies. Now that we know that life flourishes under Earth's surface, the possibilities for life elsewhere look more promising. Subsurface life may be widespread among planetary bodies, even among those with totally inhospitable surfaces.[42] The ability of living things to survive in extreme conditions broadens the habitable zone in our solar system.

Over interstellar distances, surface life may be the only kind that we can detect, because of its effect on planetary atmospheres. If we do find evidence of surface life, that may imply that many other forms of life also exist, but out of reach for our instruments.[43]

Scientists now talk about life under extreme conditions, tightly linked evolution of living organisms and the planet on which they arise, and catastrophism. All of those ideas had seemed marginal or even heretical 25 years ago.[44] We had underestimated the resiliency and opportunism of living things—the insistence of life.

Bottom Line, Life. A modest gain for believers. Though we still lack confirmed evidence of extraterrestrial biology, the variety of environments capable of supporting life appears to be much greater than we thought in the past; the constraints on what living things need to survive have loosened.

Probabilities: Intelligence

The immediate pressure of necessity has brightened their intellects, enlarged their powers, and hardened their hearts.
—H.G. Wells, describing Martians, 1897[1]

Inevitable, or a Fluke?

Intelligence may be the most controversial factor in the Drake equation. Schilling found that the question of whether intelligence is inevitable is currently what most polarizes the discussion about SETI; there seems to be no middle ground. Some scientists see it as the crucial bottleneck in the Drake equation.[2]

Those who think that intelligence is likely to emerge on other planets emphasize the early appearance of life on Earth, which suggests that life on other worlds may have a long time to evolve intelligent beings. Our planet produced an advanced technological civilization "only" 4 billion years after it formed; the Earth may have a billion good years ahead before life is destroyed by an expanding Sun.

Sagan and Drake were more upbeat, claiming that intelligence and technology have developed on the Earth about halfway through the stable period in the life of our star. By contrast, physicist Brandon Carter argued that the time that most home stars remain stable would not be enough to allow the evolution of intelligent beings. Earth, he declared, is not typical, but an exception.[3]

Those optimistic about the emergence of intelligence cite generalized processes such as parallel or convergent evolution, noting that some Earthly species other than *Homo sapiens* show signs of intelligence. The many biological parallels to those features that define the emergence of humans, Morris argued, suggest that something similar will emerge elsewhere. Some features are so adaptive that they are essentially inevitable—like the ability to see, and the intelligence and self-awareness that characterize us. Given such rampant convergence on Earth, the emergence of human-level intelligence is a near inevitability.

Engineer Brian McConnell, reviving the principle of plenitude, took this argument to an extreme. If we can't prove that it is impossible for intelligent life to evolve elsewhere, then we should assume that it has.[4]

Others strenuously disagree. Many experts in evolutionary biology contend that the evolution of intelligence elsewhere is extremely improbable, implying that the Earth may be unique in harboring sentient life.

There is no central line leading steadily, in a goal-directed way, from a protozoan to man, insisted Simpson. The existence of *Homo sapiens* depended on a very precise sequence of causative events through some 2 billion years or more; selection has taken place through long chains of nonrepetitive circumstances. It is extremely unlikely that anything enough like us for real communication of thought exists anywhere in our accessible universe.[5]

The trouble with this reasoning is that it neglects alternate routes that converge to similar results. It is not important that the particular sequence of events leading to intelligent life on Earth be repeated elsewhere, according to the authors of the Cyclops report, but only that some sequence occur that leads to a similar end.[6] By implication, intelligent life on Earth could have taken a very different form.

Gould believed that critics of SETI had conflated two different issues. One treats the detailed repeatability of any particular evolutionary sequence; all evolutionists have vociferously denied this specific claim. As for the general question of whether attributes that we could identify as intelligence might arise, no strong opinion can be entertained. SETI needs only the general argument to be worth doing.[7]

Later, Gould seemed to harden his position. Humans, he argued, arose as a fortuitous and contingent outcome of thousands of linked events, any one of which could have occurred differently and sent history on an alternative pathway that would not have led to consciousness. The vast majority of replays would never produce a creature with self-consciousness; the chance that this alternative will contain anything remotely like a human being must be effectively nil.[8]

Arthur Clarke put this in a more positive way. There still might be intelligence on this planet if the terrestrial experiment started all over again, but it wouldn't look like us. In the dance of the DNA spirals, the same partners would never meet again.[9]

If consciousness is just an insignificant accident, an incidental outcome of random mutational processes, searches for extraterrestrial intelligence are unlikely to succeed. However, posited Davies, if we accept that mind is an emergent phenomenon requiring a certain critical level of complexity, we can imagine that level of complexity being achieved, given long enough and given the inherent self-organizing tendencies that we find in matter and energy.

Davies concluded that consciousness, far from being a trivial accident, is a fundamental feature of the universe, a natural product of the outworking of the laws of nature. Others caution that by making intelligence a natural and inevitable feature of life in the universe, the principle of mediocrity raises the quality most prized by humans to a norm.[10]

Social Animals

Is intelligence an evolutionary advantage? Many see it as an adaptation for more complex behavior, adaptive only in an organism that has behavior patterns involving many alternative choices.[11]

Is complex social life a prerequisite for the evolution of intelligence? The selective pressures of complex societies may have favored cognitive skills that were evolutionary precursors to some components of human sentience. However, social structure is not necessarily associated with intelligence; although ants and bees live in highly structured societies, they are not very bright.[12]

Some argue that human intelligence reduced to its essentials is synonymous with improved communication—the transmission of more complex information from one individual to another. However, the capacity for language may not have emerged because of some selective advantage; several scientists have suggested that it may have descended from ancestral systems evolved for other purposes.[13]

Sagan versus Mayr

The debate about the probability of extraterrestrial intelligence reached a publicly visible high point in the mid-1990s when Ernst Mayr criticized the optimism of Sagan and others about finding alien civilizations. Mayr, who thought adaptations toward greater intelligence highly improbable, argued that "only one of the approximately 50 billion species that have lived on Earth was able to generate civilizations. Among these approximately 20 civilizations, only one developed electronic technology." If intelligence has such high survival value, he asked, why don't we see it in more species?

Mayr's attack provoked quick responses. Drake argued that the evolution of so many species in the terrestrial biota demonstrates its flexibility and its ability to exploit any characteristic, such as intelligence. Among many civilizations, one will be the first, and temporarily the only one, to develop electronic technology. Those who contemplate this issue should never underestimate the opportunistic nature of biological systems or the scale of cosmic time. NASA SETI program manager John Rummel declared that it is better to perform experiments than to be walled off from the real world by the opinions of experts.

Sagan responded to Mayr by pointing out that as microorganisms are our ancestors, they did evolve "smartness." Extrapolation from our example would suggest that there are enormous numbers of Earth-like planets stocked with many species, and that at least one of those species would develop high intelligence and technology in much less than the lifetime of its star. The selection pressure for intelligence may be lower on some worlds, but it may be higher on others.

As this debate continued, Mayr conceded that life elsewhere in the universe is probable. He had acknowledged 10 years earlier that the probability of the repeated origin of macromolecular systems with an ability for information storage and replication can no longer be doubted. Yet, he still challenged the likelihood of intelligence and "electronic civilizations." Although "physicalist" thinking may be appropriate for physical phenomena, he found it quite inappropriate for evolutionary events or social processes such as the origin of civilized societies.[14]

How Brains Evolve, or Don't

The one characteristic which has always improved with time is intelligence.

—Frank Drake, 1974[15]

Skeptics have been particularly dubious about other evolutions to brains like ours, emphasizing that it was by no means a sure thing on Earth. They cite the specific sequence of events that produced human intelligence, arguing that this sequence is highly unlikely to be repeated anywhere else.

Some find that encephalization (the ratio of brain to body size) has increased at a remarkably uniform rate for at least half a billion years. These studies, wrote Drake, "seem to shout loudly that we can expect to find intelligence wherever initial circumstances and time have been sufficient."[16]

Others argue that there have been considerable variations in the rate of encephalization among primates. Brain size in some hominid species remained static for a million years, anthropologist Richard Klein reported, as did their cultures. He saw human evolution as three or four sudden and profound events spaced between lengthy stretches of time when little happened. According to one theory, hominid brains were allowed to expand by a genetic change 2.4 million years ago that reduced the powerful jaw muscles they had shared with their primate relatives; this removed a constraint on encephalization.[17]

Klein suggested that the chance mutation of a gene 50,000 years ago may have led to the evolution of the fully modern human brain in Africa; the spread of modern humans throughout the world was tied to that "dawn of culture." Although the human form remained remarkably stable after that time, behavioral change accelerated dramatically. Humans were transformed from a relatively rare and insignificant mammal to a geologic force. The explosion of technology in the past 10,000 years shows that cultural factors can unleash limitless accomplishments, proposed anthropologist Carel van Schaik, all with Stone Age brains.[18]

Hominid brains did enjoy an extraordinary increase in size from about 600,000 years in the past to about 30,000 years ago. Although some scientists argue that larger brains are a highly unusual anatomical complex—and a reproductive liability—they have been maintained for hundreds of thousands of years.[19]

Researchers have confirmed the common assumption that a larger brain makes it easier to adapt to the challenges of a new environment. That larger brain may be more responsive to problems—and opportunities. S. Ohno, a Japanese biologist, suggested that our cave-dwelling ancestors were provided with intellectual potential that was in great excess of what was needed to cope with the environment of the time.[20]

These findings do not prove that the evolution of brains like ours was inevitable. Evolutionary psychologists argue that the human mind is a collection of special-purpose circuits produced by natural selection to solve the problems of survival and reproduction. The evolution of cognition is neither the result of an evolutionary trend nor an event of even the lowest calculable probability, declared Owen Lovejoy, but the consequence of selection for unrelated characters such as locomotion and diet.[21]

Nobel Prize winner William Calvin saw the evolution of intelligence as being an unforeseen consequence of neural machinery that has been selected for some other reason. Our intelligence arose primarily through the refinement of some brain specialization, such as that for language. The transition to a cognitively fluid mind was neither inevitable nor preplanned, archaeologist Steven Mithen found; evolution simply capitalized on a window of opportunity that it had blindly created by producing a mind with multiple specialized intelligences.[22]

The Debate Goes on

De Duve supported the believers when he wrote that increasingly complex polyneural networks appear to be strongly favored by the fact that a more effective brain is advantageous under any circumstance. Others challenge this view; as many species on Earth have survived longer than humans, intelligence is not a necessary adaptation. Intelligence must first arise by chance and in degrees, commented science educator John Mauldin, then have its value tested in the environment. High intelligence is not automatic.[23]

Calvin argued that drastic climate changes in the past had profound effects on human evolution, including the development of the brain. SETI Institute Principal Investigator Emma Bakes drew optimistic implications from such events. A lack of environmental and thermodynamic stability may actively spark the formation of life and trigger the accelerated formation of intelligent and robust life-forms that no longer depend on their environment, but shape it to their advantage. The more unstable the environment is, the greater the driving force toward diversity and high adaptive intelligence.[24]

Advanced life and civilization are here courtesy of disaster, devastation, and worldwide freeze-ups, commented Darling. Somewhere between the extremes must lie just the right level of environmental stress to push life forward at its maximum possible rate.[25] This suggests that the types of brain that evolve elsewhere may be very specific to the histories of their environments.

There may be a more ominous message: Growth in intelligence may have been driven by competition with other species. Shostak and science educator Alex Barnett pointed out that Old World monkeys have much larger brains relative to their body size than monkeys in South America because the African simians were challenged more by predators than their cousins in the New World.[26] Intelligence doesn't just happen through inevitable evolution; it is a response to pressure, and opportunity.

Our Intelligent Companions

In the past, *Homo sapiens* shared the Earth with other humanoid species. This claim once rested on the simultaneous presence of Neanderthals and modern humans until about 30,000 years ago. Now we know of others; an older branch of the hominid family known as *Homo erectus* still existed in Indonesia as recently as 27,000 years ago. The present phenomenon of a solitary human species on Earth may be more the exception than the rule.[27]

Most startling were the dwarfed hominids whose remains were found on the Indonesian island of Flores. These miniature people shrank because of environmental circumstances until their average height was only a little over 1 meter. Although their brains were chimpanzee-sized, they walked erect and continued to employ the simple technologies of tool-making used by their larger ancestors. The little people remained in cultural and technological stasis for thousands of years, until they were wiped out by unknown causes.[28]

These hobbit-sized creatures existed for millennia alongside *Homo sapiens*, possibly surviving until only 12,000 years ago. We have no evidence for the nature of their interaction with our ancestors; it may have involved competition or predation, or no direct contact at all.

This discovery has intriguing implications. We tend to assume continuing progress in our own evolution, including the growth of our intellect. Now it appears that the expansion of the brain can be reversed under environmental pressures. The fact that these mini-humans made the same stone tools as their larger ancestors and may have been able to perform advanced cognitive tasks raises questions about the relationship between brain size and intelligence.

Human brains (and bodies) have shrunk about 10% on average during the past 50,000 years. The reasons are unknown, but might be connected with our abandoning the mentally demanding foraging ways of life and settling down to be farmers.[29]

A Close Call

The survival of a particular intelligent species can be chancy. Some anthropologists, geneticists, and population biologists have concluded that humans were squeezed down to a small population (possibly as few as 10 thousand breeding men and women) some time during the past 400,000 years. "Our ancestors survived an episode where they were as endangered as pigmy chimpanzees or mountain gorillas are today," said anthropologist Henry Harpending. That genetic bottleneck might have had a different outcome.

About 73,000 years ago, a huge volcano on the island of Sumatra blew its top in a massive eruption. Scientists estimate that this volcano, now known as Toba, ejected 670 cubic miles of material into the atmosphere, more than 500 times the amount produced by Mount Pinatubo in the Philippines in 1991. Sulfuric acid particles scattered and absorbed sunlight, cooling the planet's surface and reducing photosynthesis. Fine ash rained from the sky, penetrating animal lungs and fatally immobilizing birds. That eruption may have imposed a 6 year "volcanic winter" and a 1000-year ice age.

Studies show a bottleneck of genetic diversity among humans at roughly the same time as Toba's eruption.[30] *Homo sapiens* was then a young species, sharing the Earth with Neanderthals and other hominids. The survival of this upstart may not have been a sure thing.

Dolphins

It is as if once the early human beings learned language they killed off all members of the next lower level, at least those that were on land.

—Neurophysiologist John C. Lilly, 1961[31]

We humans still share our planet with at least one other form of life widely regarded as intelligent: certain species of dolphins. To some researchers, dolphin intelligence and communicativeness strongly suggest that these traits are not limited to humans. If intelligence has evolved more than once on our planet, the value of f(i) in the Drake equation may be higher than a unique evolution to intelligence would imply.

Medical researcher John Lilly argued more than 40 years ago that the existence of large-brained dolphins showed that more than one intelligent species had evolved on Earth. "Within the next decade or two the human species will establish communication with another species," he predicted, "nonhuman, alien, possibly extraterrestrial, more probably marine." He hoped that his book would spark public and private interest in time for us to make some preparation before we encounter such beings.[32]

Genetically, we are no closer to dolphins than we are to large cats or rodents; dolphins emerged from a much older evolutionary branch. For millions of years, dolphins and their relatives far exceeded the intelligence of all other animals. The sudden enormous increase in their brain size was not seen again until humans began to emerge.[33]

Mind-Stretcher. Physicist and science fiction author Gregory Benford hypothesized that an alien visiting Earth before humans emerged might have found dolphins the obvious evolutionary path for high intelligence. Who would have taken the time to scour African forests for elusive tribes of tool-using primates?[34]

Behavioral biologist Lori Marino thought it significant that dolphin brains, which metabolically are ruinously expensive, have been maintained for as long as 20 million years, far longer than in hominids. Dolphins and primates are not closely related, yet they both have large brains and similar behavior capacities. This supports the idea that convergent evolution can lead to intelligence.[35]

The convergent evolution of cognition was not built on convergent evolution of brains. Complex cognitive abilities have evolved in distantly related species with vastly different brain structures.[36]

Dolphin researcher Diana Reiss, finding evidence that there might be similar strategies for processing, storing, and using information in widely divergent life-forms, argued that intelligence cannot be conceived only in human terms. Scientists must find a delicate balance between being anthropomorphic (assigning human traits to other animals) and anthropocentric (assuming that we are unique in our abilities and that only our kind of intelligence is "real" intelligence).[37]

Some researchers believe that dolphins display such traits as advanced social behavior, an extended childhood, self-awareness, and language. Others disagree, arguing that we may be seeing what we want to see. Sea mammal researcher Ian Boyd warned of a danger that biologists who become closely and emotionally associated with their study animals begin to imbue them with a sense of self. Lilly reportedly cautioned the Green Bank conferees that we may have been duped by the mimicry we taught dolphins.[38]

Dolphin sentience casts doubt on the widespread assumption that tool-manipulating appendages are vital to the development of superior intelligence. However, large-brained cetaceans seem to have reached an evolutionary impasse. Because dolphins are highly adapted to the sea, they are unlikely to develop advanced technology.[39]

Jill Tarter extended this idea to SETI. "If the universe is teeming with beautiful worlds of water, populated by the super intelligent analogs of terrestrial dolphins and whales, lacking manipulative organs and any technology, then no passive search strategy will detect them directly. Their presence will only be deduced, if at all, by inference from an observed non-equilibrium chemistry in the biospheres of those planets."[40]

A Mirror Image. Geophysicist Norman Sleep suggested that intelligent deep-sea organisms might come up with many reasons why intelligence could not arise on land.[41]

Evolutionary biologist Bruce Fleury argued that dolphins and whales can serve as useful analogs for our hoped-for communication with an extraterrestrial intelligence. They live in a medium with physical properties different from our own. They navigate, communicate, and may even think in terms of a different sensory modality.[42]

Communications between humans and dolphins may involve a transformation between their auditory world and our visual world. Dolphins might have an active intellectual life, but one based on different processes than our own; their thinking may be opaque to us.

Despite decades of research, we still are unable to communicate in a substantive way with dolphins, an evolutionary product of our own biosphere. "If we fail here," predicted zoologist N.J. Berrill, "we will fail everywhere else as well."

Lilly was more optimistic, foreseeing that the space sciences would benefit from our having established contact here on Earth with alien creatures that had an evolutionary development separate from ours. At most, though, we will have graduated from the kindergarten of interspecies communication.[43]

Brother Apes

The continuity of the human mind with the animal mind is the most important question in human evolution.

—William L. Abler, Field Museum, 2005[44]

Our closest genetic relatives are other primates, particularly chimpanzees. We differ from chimps by 1.2% in single nucleotide changes; duplications and rearrangements of larger DNA segments add another 2.7%. The implications of those statistics lie in the mind of the beholder. Some scientists emphasize that chimpanzees share nearly all of our DNA; others see the genetic differences as significant.[45]

Researchers have found that nonhuman primates show signs of emergent intelligence, suggesting that there are many different levels of intelligence rather than a rigid division between intelligent and nonintelligent. Chimpanzees may be in a transitional stage. They can simulate or image actions, suggesting a preexisting capacity for later development that appears only in humans. They can represent what they perceive; humans can represent what they imagine.[46]

We judge chimpanzee intelligence on our terms, particularly through their ability to understand our languages. After years of effort, researchers have been able to teach chimps only very simple linguistic concepts. Even when taught analogies, they cannot use them to form words.

Psychology professor David Premack noted the contrast between human intelligence, which is flexible, and animal intelligence, which is specialized. Individual species such as bees and beavers are well adapted to certain specialized behaviors, but they are imprisoned by their adaptations. Nonetheless, the potential for intelligence may exist; some monkeys appear to have precursor areas of the brain that process species-specific vocalizations that may have evolved into language areas in human precursor species.[47]

Some believe that the dividing line between humans and other primates has grown increasingly blurry; in his book *Significant Others*, anthropologist Craig Stanford wrote about what he called "the ape–human continuum." Yet, most humans do not consider chimps to be intelligent; some even eat them. Researcher Jane Goodall warned in 2004 that this practice, which she described as a crime bordering on cannibalism, could drive chimpanzees to extinction within 10 to 15 years.[48]

A Mirror Image. The threshold between intelligence and nonintelligence may lie in the mind of the beholder. What if that beholder is an extraterrestrial whose intelligence far surpasses our own? Superior aliens might regard us as precursors of advanced species, as we look upon higher primates other than ourselves.

Our efforts to converse with dolphins and chimpanzees are tests of our ability to communicate across species lines. So far, we have shown only a modest talent for doing this.

Future Minds

Nothing in natural selection safeguards human beings from vanishing in the future or evolving into creatures of less intelligence.
—Edward Harrison, 1989[49]

We are not the ultimate achievement of evolution, observed de Duve, only a transient stage. It would be surprising if, in the future development of life on Earth, vertical evolution toward greater complexity did not continue to take place, perhaps leading to more intelligent beings.

Such beings could arise by further extension of the human branch of life, but they do not have to; there is plenty of time for a humanlike adventure to start all over again from another branch and perhaps go further. "Waiting in the wings of the theater of consciousness," Morris proposed, "are other minds stirring, poised on the threshold of articulation."[50]

The complete disappearance of humans from the Earth would permit the eventual evolution of some other species having a human level of intelligence, although this might take tens or hundreds of millions of years. Brin suggested that the interval between the Cretaceous catastrophe and the present is a reasonable estimate for the time it takes to develop a civilization from ancestors as modest as the first mammals: 65 million years.[51]

A dinosaur about the height and weight of a human once may have been well on its way to developing intelligence. Known to science as *saurornithoides*, these creatures stood on their hind legs and had "hands" with opposable thumbs, permitting a precision grip. Their brain mass was a little smaller than that of a human infant.[52] Like most other dinosaurs, *saurornithoides* vanished in the extinction event now believed to have been caused by an asteroid striking the Earth.

Dale Russell of the Canadian National Museum posed a hypothetical question: If the dinosaurs had survived, what would they look like now? He produced a scale model of what a dinosaur called *Stenonychosaurus* might have become: humanoid, with an upright posture, a large head, and intelligent eyes.

If we reran history without the great extinction of 65 million years ago, Earth's masters would not be human. However, high intelligence still might exist in what Darling called "this slightly disturbing form."[53]

Humanizing Monkey Brains

Future humans might keep themselves in front by self-initiated improvement. Our growing command of the genome could lead to the enhancement of our mental capabilities.

In the meantime, some researchers are modifying monkey brains by adding human neural stem cells. The resultant brains may have some human characteristics.

An ethics committee of scientists, recognizing that a nonhuman primate research subject might attain such human mental capacities as language and rationality, claimed that such changes are a potential benefit to the animal. To the extent that the primate acquires those capacities, that creature must be held in correspondingly high moral standing.

These scientists dealt with the ethical question by maximizing the interspecies distance, proposing that grafting into the brains of our most distant monkey relations is less likely to raise concerns about significant cognitive effects. The committee dismissed ethical objections based on crossing species boundaries, arguing that fixed boundaries are not well supported by scientific findings.[54]

Intelligence, it seems, is very much a matter of definition. There is no sharp dividing line between intelligent and nonintelligent life. The position of a particular species may be determined not by any particular quality or number, but by its locations on several different continua of abilities. Intelligent extraterrestrials might have different combinations of those abilities—and perhaps others that are unknown to us.

If intelligence has emerged in other biospheres, it may be quite unlike ours. Environmental pressures, the structures of nervous systems, the inheritance of instincts, and cultural influences will cause perceptions, associations, and memories to vary, leading to different ideas about self, species, and the surrounding world. Doris and David Jonas showed how different sensory equipment would shape not only perception, but also brains, motivations, cultures, and social structure.[55]

Bottom Line, Intelligence: Unresolved. This situation could change overnight. Just one confirmed detection of an alien technology, and the sentience Copernicans win.

Probabilities: Civilization, Technology, and Science

Civilization

Although the term *civilization* is used freely in the debate about the probability of detecting extraterrestrial intelligence, there is no agreed definition of what it means in this context. Our opinions about being civilized are highly subjective. Yet, we think that we will know the signs of alien civilizations when we see them.

In a sense, we have ducked the issue by defining a civilization in terms of technologies that we can detect. That definition may exclude most civilizations elsewhere in our galaxy.

Historically our term *civilization* came from ancient Rome, where to be a citizen meant to belong to the *civitas*, the Roman civil community—a very specific context.[1] In everyday usage, we think of civilization as a developed or advanced state of human society, or a particular stage or particular type of it. We often apply measures of societal complexity, economic, scientific, and technological development, religious or cultural distinctiveness, and standards of behavior.

Many prehistoric societies did not become what are generally termed civilizations. However, civilizations did arise independently in many different parts of the world—Mesopotamia, the Indian subcontinent, China, Egypt, Mesoamerica, Peru. Some early civilizations developed in isolation, unaware of others.

This was a very late event in human history. Yet, there clearly was a selective advantage for civilizations once they had arisen, because they continued to expand and to replace other ways of life—a process still going on.[2]

Mind-Stretcher. Are civilizations as we know them the highest form of social evolution? Or does some greater form lie beyond?

We have become accustomed to assuming that certain steps are required for the evolution of civilizations. In our case, the first step may have been language, which provided the novel inheritance system that allowed cumulative cultural and technological evolution.[3] Simple technologies were followed by agriculture, urbanization, and occupational specialization. Our recorded history has seen the development of industry, science, more complex technologies, greater command of energy, and the emergence of larger political units.

Would all these steps be required for intelligent beings living in very different environments? Would they occur in the same order? We don't know; once again, we are extrapolating from our own example. We probably cannot imagine all the possible routes to civilized societies.

Some argue that convergences are a fact in the emergence of cultures and civilizations as well as in biology. Toynbee, among other historians, saw recurrent patterns in the rise and fall of many human civilizations.[4]

Others warn us against assuming that there are universally valid "laws of history." Those who proclaim that ancient cultures must have developed in a particular way, in obedience to iron laws of human development, often insist that their own country should pursue an untrammeled course free of the dictates of overweening neighbors.[5]

Anthropologist Ben Finney hoped that learning about extraterrestrial societies might someday enable a "science of civilizations." In the meantime, we should be cautious about extending our concepts of how sociological and economic development, science and technology, or the evolution of language, art, and religion evolved on Earth to other civilizations; they may follow different evolutionary courses.[6]

Again, we suffer from the fact that we are extrapolating from a single example. Our thinking about alien civilizations is constrained by what happened in our own history, or by currently fashionable interpretations of those events.

Past examples of civilizations were confined to specific areas of the Earth's surface and may not be good analogs to the global civilization we know today. It is important to distinguish between the rise and fall of particular civilizations and the fate of civilized humans as a whole. Civilizations may come and go, noted Robert Wright, but civilization flourishes, growing in scope and complexity.[7]

"Progress," as that concept developed in the Western historical tradition, may not be inevitable. Civilizations can remain at certain levels of development or in given niches for millennia, if they are not disturbed by outside forces.[8]

Civilizations have histories, with many branching points; they are the products of particular, contingent events. Intelligent, civilized beings can make choices that alter their courses. Extraterrestrial civilizations may be as young as ours, or far older; simpler than ours, or far more complex; more technological, or less.

We are very far from knowing general laws governing the development of civilizations—the fundamental problem of what Soviet scientists called "exosociology." We must assign any particular model of an advanced society a very low probability of being a typical end result.[9]

Historians have presented us with many speculative alternative histories of human affairs, in which events went a different way.[10] Those speculations stimulate our thinking, but they are not predictions on which we can rely.

Cultures

Civilization is not synonymous with culture, another vague word used in many different ways. Prehistoric, precivilization societies all had cultures.

If we define culture as socially transmitted behavioral patterns, we must recognize that it has not been limited to humans. Researchers have found that chimpanzees can learn tool use by observation; that use varies among geographically separated groups of chimps, implying different cultures. Other investigators have discovered similar variations among geographically separated orangutans.

Great ape cultures may have existed for 14 million years or more, predating the arrival of hominids.[11] Although long established, those cultures lack the human ability to shape the material world on a large scale.

Technology and Science

The adaptive value of intelligence and manipulative ability is so great—at least until technical civilizations are developed—that if it is genetically feasible, natural selection seems likely to bring it forth.
                                        —Iosif Shklovskii and Carl Sagan, 1966[12]

The signs of alien minds that we are best able to detect are not those of civilization, but of only one of its possible attributes: technology. How likely are we to find it?

Many of those who support the search for extraterrestrial societies have assumed that the development of technology by intelligent life-forms is nearly inevitable. When beings having sufficiently high intelligence evolve, argued MacGowan and Ordway, they will sooner or later develop a technological understanding. Others have seen a reproductive advantage in the ability to store knowledge and build technology.[13]

A NASA workshop report made a forceful declaration:

Almost certainly once a species with the requisite intelligence, manipulative ability, and complex social organization has evolved, technological civilization will develop.... To go from a stone age culture to our present level of technological development required no biological evolution. All that was needed was the development of ideas, and their testing by trial and error.[14]

At its simplest, anthropologist Bernard Campbell reminded us, technology is older than reason. Some researchers have suggested that the ability to make and use simple stone tools is a primitive behavioral capacity that may have been "discovered" many times and utilized by more than one type of hominid. "If we hadn't walked out of Africa (with tools)," Morris proposed, "then probably sooner rather than later, our analogues would have strolled out of South America holding tools."[15]

Others argue that technological development is not inevitable; the motive must be present as well as the potential. One driving force for technological progress may be competition. Much human technology is motivated by the desire for efficient weapons.[16]

Although some see technology as a negative force that damages our environment, others argue that it is precisely because humanity has learned to control some aspects of nature that civilization has advanced. Iconoclastic historian Felipe Fernandez-Armesto proposed that a society is civilized in direct proportion to its distance from the unmodified natural environment, by its taming of climate, geography, and ecology.[17] The most civilized do what is condemned by some environmentalists: They reshape the world around them to suit their purposes.

Intelligence—or knowledge—may not necessarily lead to technological societies. Human civilizations with highly developed forms of sociopolitical organization did not all have highly developed technology.

In his book *Guns, Germs, and Steel*, physiologist Jared Diamond argued that human societies achieved different levels of technology not because of differences in culture or ability, but because of different sets of geographical advantages and disadvantages.[18] Intelligent beings that evolved in alien environments also would have different combinations. In less advantaged environments, technological development might be slower than on Earth; in more advantaged environments, it might be faster.

The rapid technological development that we have been experiencing is a very recent phenomenon. It may be driven by complex combinations of factors that might not be duplicated elsewhere. We cannot assume that the human case is typical, that our pattern of development will prevail on other worlds; there is nothing universal or necessary about its history.[19]

We must draw a distinction between technology and science; one does not automatically imply the other. Science is not a convergent phenomenon in our history, several scholars have argued, but a cultural development unique to post-Renaissance Europe, only recently adopted by most remaining cultures—and even then rather reluctantly. Technology can develop independently of science.[20]

Our models of alien technological civilizations are almost certainly too narrow, argued biologist Jack Cohen and mathematician Ian Stewart; they can only come from predictions of our own future technology. Our imaginations cannot conceive of anything truly alien.[21]

Is Science a Universal?

Many projections of how extraterrestrial civilizations will develop have been written by scientists extending trends far into the future. Yet, our own history casts doubt on straight-line projections.

A growing scientific enterprise and continuous technological development may not be inevitable. Even when civilizations possess the technical skill, social or psychological forces may deter them from advancing further, though that advance may seem obvious to outsiders and successors.[22]

Human interest in science may be rare, conceivably unique. Even if the concept of scientific inquiry exists in some extraterrestrial societies, its implementation may be limited or episodic.

Science gives us a powerful lever on fate. That lever that may not be equally available to all civilizations.

Until we have confirmable evidence, attaching a number to the fraction of intelligent cultures that use technologies we can detect—or that does our kind of science—rests on shaky analogies. Intelligence—even intelligence with many forms of technology—could be abundant in the universe without our being able to find it.

Bottom Line, Technological Civilizations: Unresolved. This too could change overnight.

Probabilities: Longevity

The search for extraterrestrial intelligence is principally a test to determine L.

                        —David Schwartzman and Lee J. Rickard, 1988[1]

How Long Do Technological Civilizations Live?

Our ability to answer this question is extremely limited. Soviet astronomer V.A. Ambartsumian summarized the problem in 1971: We do not have even a single example of the lifetime of a technical civilization.[2] As we know of only one technological species—our own—we are in effect trying to predict our own future.

Just as there is as yet no predictive theory of biology, so there is no predictive theory of history.[3] We do not know how long our own civilization will survive. We have only opinions, making L the most politically loaded factor in the Drake equation.

In the classic SETI paradigm, only a large value of L leads to significant numbers of civilizations coexisting at a given time. As Sagan saw it, L is strongly biased toward the small fraction of technological civilizations that achieve very long lifetimes (implying that most do not).

That does not necessarily mean that they would be sending signals; the lifetime of transmitting projects may be far shorter than the lifetime of a technological society.[4] In the original Drake equation, L meant the average lifetime of *communicating* civilizations, which might be a small subset of the total number of technological societies.

The debate about the longevity of civilizations often blurs what we mean by their deaths. Many implicitly assume that extraterrestrial civilizations often end with the extinction of the intelligent beings who belong to them—a drastic outcome. Spoiling a civilization is one thing, and perhaps not too arduous, commented Shapley; complete annihilation is quite another, and vastly more difficult.[5]

Others see the end as the collapse of a civilization, although they describe that differently. Diamond, for one, defined collapse as a drastic decrease in human population size over a considerable area for an extended time.[6] That is quite different from the annihilation of a sentient species. Other definitions refer to cultural or political collapse, a loss of morale, or absorption by other civilizations.

The Shadow of the Bomb

Our speculations about the longevity of civilizations—ours and others—have been influenced more by the circumstances of our time than by a long view of human history. The catalog of potential civilization-ending disasters is tied to our present or near future.

Since our transition point of 1960, many commentators have adopted a Standard Pessimist Model, resting on a belief that the problems of our time are uniquely threatening. More advanced technological civilizations, it is assumed, must have passed through the same crisis.

From the 1950s through the early 1990s, the threat of nuclear war weighed heavily on public opinion in many countries, provoking deep pessimism about our near-term future. This strongly influenced those who speculated about extraterrestrial societies. Several argued that interstellar communication and transportation arise simultaneously with the means of extinction; the universe might be littered with the remnants of civilizations that failed to resolve that dilemma. For this rationale to work, commented Brin, there has to be an easily triggered mechanism for destroying civilizations, such as a nuclear winter.[7]

Sagan claimed that if we find no other civilizations, the most likely explanation is that they destroy themselves before they are advanced enough to establish a high-powered radio-transmitting service. Shklovskii turned away from optimism about contact because he felt that nuclear war was inescapable. As Drake pointed out, this was a political calculation, not a scientific one.[8]

Sagan and astrophysicist William Newman claimed that weapons of mass destruction force on every emerging society a behavioral discontinuity. If they were not "aggressive," they probably would not have developed such weapons; if they do not quickly learn how to control that aggression, they rapidly self-destruct.

Those who are as destructive as we are never make it to interstellar expansion, argued Von Hoerner. "They destroy each other before they can go anywhere. We have to solve the question of peace first." Those who are aggressive would have killed each other or "blown up their planets;" those who survive would be more "gentle, peaceful, and reasonable."[9]

The commentators most dubious about long civilizational lifetimes often have been preoccupied with "stability." Oliver suggested that a major benefit of contact would be learning about advanced social behavior and political forms that are stable and encourage longevity; we might learn how they survived World War III or prevented it from happening. Bracewell thought it a matter of chance whether we would succeed in stabilizing the political situation as it existed in 1963.[10]

Stability also has been seen as a requirement for entry into a galactic society. "The achievement of great stability and serenity would seem a prerequisite for a society willing to make the expensive and enormously prolonged effort required to contact another world," asserted Walter Sullivan. Bracewell proposed that the achieving of stability for a long-enough period was a prerequisite to formation of, and membership in, a galactic chain of communities.[11]

For some, the absence of evidence of other technological societies drives a conclusion that our own civilization may be doomed. Von Hoerner thought that scientific and technological civilizations do not endure for long; if they did, we would have found them. Jill Tarter, endorsing the view that there can be many civilizations only if they are relatively stable and long-lived, concluded that we have no other indication that there is any possibility that we can somehow get through this stage of potential for nuclear destruction.[12]

Others reject the pessimistic scenario. Charles Seeger, the first astronomer to work full time on SETI, said that he did not share the "paranoid fear of the imminent end of our species."[13]

Perceptions of risk change over time. Sullivan commented in the 1994 revision of his book *We Are Not Alone* that the prospects for survival of technological civilizations seemed better than they had 30 years before.[14]

Other Means of Self-destruction

Creatures with a higher cognitive intelligence, like shooting stars that suddenly flash across the black vault of night, come into being from time to time, then quickly fade away.

—William Burger, 2003[15]

Worries about population growth, resource exhaustion, and environmental degradation have driven some pessimists to conclude that technological societies are inevitably short-lived even if they do not exterminate themselves with nuclear weapons. Burger, speculating that other technological civilizations would experience uncontrolled growth and foul their environments, concluded that "the present drama unfolding on planet Earth makes it seem highly likely that energy-guzzling technological societies have only a short life span."[16] Here again, many have blurred the distinction between a setback for a civilization and its destruction.

Some of the most extreme predictions by cultural pessimists in the 1960s and 1970s already have been disproved. The authors of *The Limits of Growth*, published in 1972, claimed that we would run out of oil, gold, zinc, and mercury in 20 years. Conservation biologist Paul Ehrlich wrote in *The Population Bomb* that "the battle to feed humanity is over; in the course of the 1970s, the world will experience starvation of tragic proportions, and hundreds of millions of people will starve to death.[17] Although many suffered from malnutrition, the number dying of starvation was orders of magnitude smaller than Ehrlich predicted.

Apocalyptic Punishment

Modern predictions of doom bear some resemblance to the traditional Christian apocalypse, in which the world as we know it is destroyed. Some see that destruction as just punishment for corrupt civilizations, with only the righteous being saved. As we will see later, this theme has appeared among the predicted consequences of contact with a more powerful civilization.

The failure of the doomsters' predictions showed that cultural pessimists had underestimated the human ability to deal with problems by adapting and inventing. Human behavior does change in response to pressures, including population growth and damage to the environment.

The failures of these predictions also illustrate the danger of assuming continued exponential growth in an index. This, warned journalist Gregg Easterbrook, is the Fallacy of Uninterrupted Trends.[18]

It is understandable—even admirable—that concerned people would put forward doomsday scenarios to deter behavior that could lead to a disaster. However, to make our self-destruction an assumption is something else again. A well-intended political purpose does not make predictions of short civilizational lifetimes a fact that we can plug into the Drake equation.

We may not be looking far enough into our future; there could be later thresholds as well. If those thresholds applied to alien societies, they might add up to a winnowing process that would reduce the number of civilizations that we can detect.[19]

Threatening Science

What science can do, it will do, some time, somewhere, whatever obstacles may be put in its way.

—Christian de Duve, 2002[20]

In his book *Our Final Hour*, astronomer Martin Rees issued a sharp warning about the dangers that science and technology might pose to our future. Technology has so highly leveraged the power of the individual or the small group that one person's act of irrationality, or even one person's error, could do us all in. Our timescale has contracted, claimed Rees; the odds are no better than fifty-fifty that our present civilization on Earth will survive to 2100. Our destiny depends on the choices we make during this century.[21]

Consider biological experimentation. Scientists are actively attempting to create life. The resulting artificial cells might be quite different from any existing or extinct form. The "top-down" approach aims to create such cells by simplifying and genetically reprogramming existing cells with simple genomes. By contrast, the "bottom up" approach aims to assemble artificial cells from nonliving organic and inorganic materials, a task traditionally left to God or to the unplanned forces of evolution. "Like the medieval alchemists," warned an article in *Science*, "today's cloning and stem cell biologists are working largely with processes they don't fully understand."[22]

Creation by Mail

Scientists already are reprogramming bacteria behavior under the rubric of "synthetic biology." Some are designing and building living systems that in some cases operate with an expanded genetic code that allows them to do things that no natural organism can. Genetic researcher Craig Venter said in 2004 that engineered cells and life-forms will be relatively common within a decade.

More disturbingly, anyone can order synthetic DNA; at least one laboratory hosts an online library of parts that can be built into genomes. Scientists already have made a poliovirus from mail-order segments of DNA. A *Nature* editorial, assuming that some of the results of synthetic biology have escaped into the environment, argued that it is time to assess the risks of this technology.

In 2005, a group of researchers recreated the flu virus that killed an estimated 50 million people in 1918. The publication of the full genome sequence gives any rogue nation or terrorist group all the information that they need to make their own version of the virus, raising the threat of biological warfare to a new level.[23] The ultimate fright—which may or may not be feasible—is a terminal laboratory experiment in which even deadlier organisms escape into our environment and kill off the human species.

Some worry that physics research could lead to disaster. Experiments planned for the Relativistic Heavy Ion Collider at Brookhaven National Laboratory in the United States were intended to make the densest, hottest matter ever seen on Earth, reproducing, on a small scale, the conditions that existed in the early universe.[24]

Three potential disaster scenarios worried the public. Experiments could produce black holes that might consume the Earth. A vacuum instability could expand catastrophically in all directions at the speed of light. Strangelets (a type of strange matter) could grow to incorporate ordinary matter, perhaps transforming the entire Earth into its form. Physicists dismissed the first two scenarios. As for strangelets, analyses indicated that they were unlikely to be a danger, but scientists could not declare absolutely that there would be no problem.[25]

Mind-Stretcher. In his novel *Earth*, science fiction author David Brin described how a black hole created in a laboratory falls to the center of our planet, eating the Earth from inside. Producing black holes is not fantasy. Scientists are actively planning to create miniature black holes in their laboratories, although they assure us that such small versions would pose no threat to our planet.[26]

Rees warned that caution should surely be urged (if not enforced) on experiments that create energy concentrations that may never have occurred naturally. The assessments of risk by those supporting such experiments are subjective; the theoretical arguments depend on probabilities rather than certainties.[27]

The most dangerous threat to the survival of humankind, warned physicist Peter Ulmschneider, is the likelihood of uncontrollable inventions. He was particularly worried about deliberate man-made destruction. It is not only war; it is the mounting concentration of power in the hands of a few individuals. If our irresponsible side cannot be controlled, mankind is doomed. Sooner or later, a powerful individual or movement will succeed in putting an end to it—a fate that might have befallen extraterrestrial intelligent societies that existed before us.[28]

The Dalai Lama issued a very similar warning. It is no longer adequate, he counseled, to say that the choice of what to do with this knowledge should be left in the hands of individuals. We need secular ethics shared by all faiths—compassion, tolerance, consideration of others, and the responsible use of knowledge and power.[29]

Will the threats to ourselves that we create also appear in other civilizations? Or are we extrapolating human futures to the whole universe, again making Man the measure of all things?

Hints of Optimism

We have only one data point on the lifetimes of technological civilizations; ours has existed for a few centuries. Beyond that, estimates of such lifetimes depend on whether one is optimistic or pessimistic.[30]

Many reject the assumption that self-destruction is inevitable; we still have the capacity to make rational decisions about our future. Any emergent societies having developed sufficient technology to be capable of destroying their whole species would almost certainly be intelligent enough to recognize and avoid the danger of suicide, MacGowan and Ordway believed. This implies that some, if not most, intelligent species will survive and prosper for astronomical periods of time.[31]

Futurists have tended to overestimate the importance of negative factors, argued the Clarks. It takes an extreme pessimist to believe that even if only 1% of the world's population survived some catastrophe, the 60 million who remained could not get things going again over many generations. Even if L is short, a new technically sustainable civilization might arise with a more extended lifetime. If intelligence has survival value, posited Shostak, it will come back after a disaster.[32]

Planets might produce sapient species at fairly short intervals, Brin suggested, depending on the time needed to recover from the damage done by the previous sentient race. The outlook for finding other civilizations brightens considerably if they routinely have successors.[33]

Peter Schenkel, author of upbeat novels about contact, argued that the lesson to be drawn from the human analogy is encouraging rather than discouraging. Humankind is struggling toward a peaceful global order; more advanced civilizations will have moved even farther along this path. It is highly improbable that an intelligent extraterrestrial race would permit itself to degenerate, to fall victim to stagnation, or to lose interest in science. Schenkel concluded that L should be raised to infinity.[34]

Mind-Stretcher. Extraterrestrials who are more intelligent and better informed than humans—less driven by genes and instinct—may be more able to perceive, choose, and implement alternative courses of action. That would make them even less predictable than we are, casting doubt on any model of their future that we project.

Fear of Machines

We need look no further for the famous "missing link"—it is us. As Nietzsche said, Man is a rope stretched between the animal and the superhuman.

—Arthur C. Clarke, 2003[35]

Worries about a takeover by intelligent machines have spread through popular culture, recently via the Terminator films. Fear of machines is not new; it can be traced at least as far back as the Industrial Revolution of the nineteenth century. The difference is that the machines we fear now are smarter.

MacGowan and Ordway predicted 40 years ago that intelligent machines (which they called automata) could come to dominate their biological creators, perhaps gradually circumventing the limitations imposed on them by humans. Once they became independent of the productive support of their creators, the automatons probably would abandon the human race and emigrate to greener astronomical pastures.[36]

According to artificial intelligence pioneer Hans Moravec, what awaits us is a postbiological world in which the human race has been usurped by its own machines. If we cannot beat them, we might join them. Humans could seek to improve their own intelligence and durability by augmenting their brains and bodies with artificial components and eventually would transfer themselves entirely into computers—a technique that has become known as "mind loading."[37]

Another artificial intelligence pioneer, Ray Kurzweil, foresaw computers exceeding human intelligence within a few decades. Willing humans would have their brains scanned, uploaded to a computer, and then would live their lives as software running on machines. Kurzweil later warned that the inherent impossibility of restraining intelligence means that the strategies that we devise now cannot absolutely ensure that future artificial intelligence embodies human ethics and values.[38]

A superintelligent machine could be the last invention that humans need ever make, speculated Rees. The most likely and durable form of life may be machines whose creators had long ago been usurped or extinguished.

This transition could extend the lifetimes of societies made up of such beings. Although a naturally evolved species must inevitably become extinct, declared cosmologist John Barrow and physicist Frank Tipler, its machine descendants need not ever become extinct themselves.[39]

Minsky envisioned advantages to our evolution into machines: immortality, colossal intelligence, the ability to experience a wide range of abstract and concrete phenomena that are beyond the reach of humans. A technological society could convert itself into creatures that may be able to harness the power of the Sun and travel to other stars. Each intelligent species may reach a critical threshold where it will have to make a choice between this future and others.[40]

As wonderful as biology is, proposed Arthur Clarke, it may be just a means of producing the true masters of the universe. Our mechanical offspring may pass on to goals that will be wholly incomprehensible to us. Beginning as tools through which humans can explore and humanize the cosmos, machines may develop independence and become the alien.[41]

We may be assuming too much. One can envision cultural and political resistance to the idea of creating brains superior to our own, particularly if we fear that they might escape our control. Frank Herbert, in his science fiction novel *Dune*, envisioned a "Butlerian Jihad" leading to a ban on making machines with brains like ours (there was a Butler, who raised alarms about the Industrial Revolution in England).[42]

After Biology

The missing link in all past SETI arguments, argued Steven Dick, has been a failure to account fully for the effects of cultural evolution—in particular, the future evolution of biological intelligent beings to a post-biological state. Cultural evolution, which is proceeding much faster than biological evolution, has been a huge force in human adaptation, enabling us to do what none of our primate relatives can do. Only humans can accumulate knowledge over long periods of time and transfer it so that the next generation can improve on it.

The possibility that many extraterrestrial beings will be postbiological becomes an increasingly important factor as the lifetime of technological civilizations is extended. That lifetime need be only thousands of years for cultural evolution to have drastic effects on societies. If technological civilizations typically last more than a few hundred years, the result will be a postbiological universe (there may be a transition period populated by human/machine cyborgs).

Assuming a linear development of extraterrestrial civilizations over time leads to a conclusion that more than 99% of intelligent life will be postbiological. Such creatures, who may be essentially immortal, might have characteristics that we ascribe to God: omniscience and omnipotence.

Long-lived artificial intelligences could improve the prospect of contact. Short-lived biologicals like us might find themselves intercepting communications of postbiological beings.

Dick hinted at the ambiguous implications. Postbiologicals would have a great capacity to do good—or evil.[43]

Statistics

The phenomena that the Drake equation assumes to take place are only those we are sure have taken place at least once, except for the last factor: the longevity of civilizations. This is the one area in which we are totally ignorant; we do not know if there is a limitation. Because introducing a limit minimizes our estimates of the numbers of civilizations that we might detect, the equation may be too conservative.[44]

One approach to this question is to look at the longevity of our ancestral species. The 10 or so hominid species that preceded modern humans came and went at a rate of about 200,000 years each. As ours began at least 130,000 years ago, we soon may be due for a change. This has less optimistic implications than Philip Morrison's estimate that the lifetime of an average species on Earth is close to 5 million years, or evolutionary biologist Richard Lewontin's comment that a typical mammalian species lasts about 10 million years.[45]

In the longer term, the extinction of individual species is the norm. More than 99% of all species that ever lived on Earth are now extinct. Those other species lacked the scientific knowledge and technological ability to save themselves from external forces. They also lacked our ability to destroy ourselves through conscious acts.

Mind-Stretcher. Ulmschneider calculated that 99.8% of all intelligent societies that ever existed in our galaxy are extinct.[46] If he is right, explorers from living civilizations might find relics on many planets or in orbit around many stars—the ruins of great cities, the monuments of ancient kingdoms, the tombs of the dead. The most active interstellar travelers might be archaeologists studying the remains of departed civilizations. They may always be in demand, for the realm of their researches will be vast in space and deep in time.

We may be focusing too narrowly on our species as it now exists. What is the lifetime of a civilization whose members evolve into something different? Humankind, declared de Duve, will either evolve or disappear.[47] This would be true of other intelligent species as well.

Extraterrestrial Terminators

Nature is no friend of man's, and the most that he can hope for is her neutrality.

—Arthur C. Clarke, 1950[48]

Isaac Asimov, in his 1979 book A *Choice of Catastrophes*, listed potential disasters that could end or cripple the human experiment. Some were Earth-bound, such as the slowing of the Earth's rotation, tectonic events in the Earth's crust, changing climate, and the failure of Earth's magnetic field. Many others came from beyond the Earth: collisions with astronomical objects such as black holes, free-floating planets, clouds of dust and gas, and antimatter; the bombardment of the Earth by comets, asteroids, and meteorites; the effects of collapsing stars; the death of our own Sun; in the very long term, the increase of entropy and the closing of the universe. Asimov added competition from other forms of life, including disease microorganisms, and conflicts among humans. He mentioned but dismissed the possibility of conflict with extraterrestrial intelligence.[49]

Only a year later, the idea that extraterrestrial forces could cause catastrophic mass extinctions on Earth gained new credibility when scientists described how an asteroid impact could have wiped out most species of dinosaurs 65 million years ago. Researchers later found evidence of another impact 250 million years ago at the boundary between the Permian and Triassic eras—the time of the greatest extinction of life known to science.

Others believe that this extinction was triggered by terrestrial forces such as massive vulcanism. The two causes may not be mutually exclusive; massive impacts could trigger volcanic eruptions, perhaps at the antipodal point—the opposite side of the Earth.[50]

Two billion years ago, a massive body blasted out a crater 300 kilometers (nearly 200 miles) wide in what is now South Africa, the largest impact structure known on our planet.[51] The effects on the Earth's then-young biosphere would have been calamitous, dwarfing the event that killed off the dinosaurs. It tells us something that life survived.

These collisions can have both negative and positive results. The Cretaceous–Triassic blast extinguished many species but it opened opportunities for others, including those that evolved into humans.

Catastrophism Revived

Nearly all geologists once were in the grip of a nineteenth-century concept known as uniformitarianism, which claimed that geological change was gradual. The discovery of plate tectonics in the 1960s actually reinforced this doctrine, as the motion of Earth's plates appeared to be very slow; geologist Walter Alvarez described them as moving at the speed of growing fingernails.

Most geologists failed to relate impact craters on the Moon and Mars to the history of the Earth. As late as 1945, experts at the U.S. Geological Survey refused to acknowledge that Meteor Crater in Arizona was caused by an impact. Now we know that our planet has been pounded by large objects. Once geologists removed their mental blindfolds, they began finding impact craters all over the Earth.

Such paradigm-changing discoveries can be serendipitous. The first evidence of the dinosaur-killing impact—a layer of rock unusually rich in iridium—was found by geologists looking for something entirely different.

Impacts are not the only possible disasters. David Keys, in his book *Catastrophe*, contended that a massive volcanic eruption in 535 profoundly influenced our history. In a matter of decades, the old order died and a new world—essentially the world as we know it today—began to emerge.[52]

We often are slow to react to potential extraterrestrial threats. The first Earth-crossing asteroid was discovered long ago, in 1932. Yet, worries about possible future impacts on the Earth did not motivate systematic surveys of such asteroids until the 1980s.

We got a double jolt in 1994. In February, a small asteroid, about 10 meters across, exploded in the atmosphere over the western Pacific Ocean in a blast 10 times as powerful as the Hiroshima A-bomb. Surveillance satellites alerted U.S. officials; thinking that this explosion might have been nuclear, they reportedly awakened the President.

In July of that year, astronomers watched the fragments of Comet Shoemaker-Levy plow into Jupiter's atmosphere, leaving scars that persisted for months. American legislators directed NASA and the Department of Defense, in coordination with the space agencies of other countries, to identify and catalog within 10 years the orbital characteristics of all comets and asteroids that are greater than 1 kilometer in diameter and whose orbits cross that of the Earth. That led to the Spaceguard Survey, foreseen by Arthur Clarke in his 1973 novel *Rendezvous with Rama*. A more modest program known as Spacewatch had been operating since the 1980s.[53]

Already, the Spaceguard Survey has found over 700 Near Earth Objects with a diameter of 1 kilometer or more; one may strike the Earth every 500,000 years (the frequency is much higher for smaller bodies). Studies have indicated that the impact of a comet or asteroid serious enough to precipitate the collapse of civilization may occur every 100,000 to 300,000 years.[54]

Our improved detection capabilities have enabled us to spot smaller asteroids, including some that pass closer than our Moon. Astronomers discovered in 2004 that a 1000-foot-wide object will narrowly miss the Earth in 2029, possibly at a lower altitude than the communications satellites we place in geosynchronous orbit. If that small asteroid hit the Earth, the impact energy would be 850 megatons—15 times more powerful than the largest hydrogen bomb ever tested and 60 times more devastating than the Tunguska blast that leveled vast areas of Siberian forest in 1908.[55]

The danger is not entirely predictable, as asteroids dislodged from their presently "safe" orbits could penetrate into the inner solar system. There may be more than 700,000 asteroids larger than 1 kilometer across—the threshold for global catastrophe.[56]

The possible effects of large-body impacts have sensitized the scientific community to think more in terms of cosmic influences on Earth systems. As Jill Tarter observed, we have recently begun to appreciate how intimately the evolution to intelligent life on Earth was connected with our astrophysical environment.

Giant impacts almost certainly introduced a random extraterrestrial factor into the history of life, punctuating evolutionary change. Seeing the Earth as being predominantly under the control of cosmic influences, geologist Michael Rampino proposed an alternative to the Gaia theory: the "Shiva Hypothesis," named for the Hindu goddess of destruction who returns all forms back into the primordial nature from which they emerged.[57]

Mind-Stretcher. Robert Dixon turned the asteroid threat argument on its head by proposing that if occasional great extinctions caused by asteroid collisions are a necessary condition for intelligent life to evolve, the frequency of such life may be dependent on the frequency of available asteroids.[58]

Asteroids are not the only concern; large comet nuclei and similar icy objects could be dangerous if they were perturbed into Earth-crossing orbits. The Kuiper belt, a doughnut-shaped region beyond the orbit of Neptune, contains an estimated 100,000 frozen bodies more than 200 kilometers across. We know the positions and orbits of only 800 of them. Even the more distant Oort Cloud could be a source of danger. Major impact events 35 million years ago may have been caused by long-term comets, diverted into the inner solar system by a perturbation of the cloud.[59]

As hazards from asteroids and comets must apply to inhabited planets all over the Galaxy, Sagan concluded that few civilizations will be both long-lived and nontechnological. Chyba warned us of the implications: A biosphere that does not produce a technical civilization capable of assessing and responding to its particular impact environment on a timescale shorter than that between giant impacts will be catastrophically disrupted. Science fiction author Larry Niven put it more wryly, commenting that the dinosaurs became extinct because they didn't have a space program.[60]

Black Clouds, Dark Planets, Bursting Stars

A minute fraction is in the form of visible suns. But what about the rest? Is it in dead suns, planets, and moons? Is it possible that billions of exhausted suns are now wandering in waste places of space? And are they all surrounded by dead planets, still in revolution, counting off lifeless and useless years?

—*Scientific American*, February 1906

There may be other extraterrestrial threats. Astronomers predict that our solar system will collide with interstellar clouds several times during each galactic year—the time it takes for our Sun to complete an orbit around the center of our Galaxy. In his 1957 novel *The Black Cloud*, astronomer Fred Hoyle described the potentially disastrous impact on life of being immersed in a dense cloud that would prevent much of the Sun's energy from reaching the Earth. Another astronomer, Priscilla Frisch, predicted that the solar system may be heading for a cloud of interstellar matter up to a million times more dense than its present surroundings. Our encounter with that cloud may occur in the next 50 thousand years.[61]

Our Sun enters a denser environment each time it crosses the plane of the Galaxy. Some scientists have suggested that these crossings coincide with mass extinctions on Earth, but others find the evidence weak; we crossed the plane 2 million years ago without any sign of major damage to our planet.[62]

Other potential disasters include a nearby supernova that strips our atmosphere's ozone layer. On the average, one supernova every 50 million years is close enough to engulf the Earth in its expanding ejecta; some researchers once thought that a nearby supernova had killed the dinosaurs. (Sagan put a more positive spin on supernovas, observing that the evolution of life on Earth is driven in part by cosmic rays originating in the deaths of such massive suns.) Even a flare caused by a "star quake" can bathe the Earth in gamma-ray and X-ray radiation.[63]

The greatest potential killers are gamma-ray bursters, whose blasts of energy are concentrated into deadly beams. Short duration bursters, which may be generated when neutron stars merge with each other or with black holes, are more common but far less powerful than long-duration bursters, which may be the result of massive stars collapsing into black holes.[64]

Bursters can produce a flux of gamma radiation that would sterilize nearby life-bearing planets. The good news is that the radiation from one of these awesome events would merely wipe out the ozone layer and possibly darken the sky, devastating the world's ecology and food production. The bad news, according to one model of how bursters work, is that the radiation pulse might be followed by a month-long blast of extremely energetic cosmic-ray particles, as much as 100 times the dose lethal to humans.[65]

The most brilliant explosion known was a gamma-ray burster 9 billion light-years away—more than halfway across the observable universe. If that event had taken place a few thousand light-years from us, it would have been as bright as the mid-day Sun, and it would have dosed Earth with enough radiation to kill off every living thing.[66]

It is only a matter of time before a burster occurs nearby, warned Leonard and Bonnell. Within the estimated danger range of 3 thousand light-years, we can expect one every 100 million years, similar to the mean time between the largest mass extinctions in Earth's geological record. This may explain the preferential survival of deep-water organisms and the sudden rise of new species by increased mutation. Gamma-ray bursters may periodically reset the biological clock, forcing the evolution of life to start afresh.[67]

Australian astronomer Ray Norris concluded that gamma-ray bursters could have a disastrous effect on us even at the distance of the Galaxy's center. He expected such a burster every 200 million years. Yet, we have been evolving on Earth for 4 billion years, implying that the interval between disasters is at least that long. Either our calculations are erroneous or we have been very lucky. To Norris, this implied that life on Earth may be unique in the galaxy.[68]

Borrowing from an evolutionary theory developed by Stephen Jay Gould and Niles Eldredge, astronomer Milan Cirkovic proposed a kind of "punctuated equilibrium" for the evolution of technological intelligence in the Galaxy, a long-term balance between gradual and catastrophic effects. Technological civilizations may be eliminated periodically in the course of galactic evolution; the most important single causes of catastrophe may be gamma-ray bursters.[69] There may not have been enough time for civilizations significantly older than ours to arise during this part of the cycle.

We also might face threats from interstellar wanderers. Runaway stars that may have been flung out of binary systems race through the Galaxy at high speeds; one pulsar is zipping along at 1100 kilometers (680 miles) a second. Burned-out stars and brown dwarfs may be difficult to spot until they are relatively near our solar system. Our Galaxy may contain about one billion dead, extremely dense pulsars. About a billion nonprimordial black holes may be strewn through the Galaxy; the nearest may be only 10 or 20 light-years away. Some astronomers have proposed that huge numbers of dark matter halos—about as large as our solar system—may drift through the Milky Way, with one passing through our neighborhood every few thousand years.[70]

Simulations of planet formation in protostellar disks have shown that a high percentage of planetary-sized objects are ejected from the system or are pulled out of it by encounters with other stars. There may be many free-floating planets in interstellar space that are nearly impossible to detect, particularly because they are cool.[71]

Even if there were no collision, Earth's orbit could be disturbed by a close encounter with a massive body. Simulations have shown that a near miss from a rogue star less than half the size of our Sun could bounce the Earth and Mars out of their normal trajectories.[72]

The Deadly Intruder

H.G. Wells foresaw an extraterrestrial terminator in an 1897 story called "The Star." Astronomers detect a wandering planet entering our solar system. After a merger with Neptune, this body plunges inward, apparently on a collision course with the Earth. Our planet's temperature rises; snows melt, floods and violent storms begin, earthquakes and volcanic eruptions ravage the world. Millions die. Although the collision never takes place, the Earth's climate has been changed, forcing humans to migrate toward the poles.[73]

A Growing Sun

There is one predictable event that will require our descendants to leave their planet of origin or go extinct. In roughly 1 billion years, our expanding Sun will render the Earth inhospitable for life. The real challenge, said Jill Tarter, is surviving one's star; there may be few who do so.[74]

Astronomers are watching another sun die, burning its planets to cinders. This expanding star, 550 light-years away, has engulfed its children. As astronomer David Neufield described it, we are witnessing the apocalypse that will destroy our own solar system. Other intelligences might someday watch the tragedy of our own extinction—if we do not build self-sustaining habitats beyond the Earth.[75]

Mind-Stretcher. If more advanced intelligent beings saw that we were in danger of being wiped out by an astrophysical event, would they do anything to help? They could give us advice on how to divert an approaching celestial body, or on a more affordable way of migrating outward from our expanding Sun—if they cared.

Historian Will Durant once wrote that humanity exists by geological consent, which can be withdrawn at any time.[76] We may now extend that concept to say that Humankind exists by astrophysical consent. That, too, could be withdrawn at any time. A civilization-ending accident could occur unexpectedly, even within the lifetime of humans who now populate the Earth.

Although most of us now attribute disasters to causes identifiable through science, violent geological events in the past such as earthquakes, eruptions, landslides, and tsunamis often were transformed into visits by gods and monsters.[77] Some humans may attribute future disasters to deities—or to powerful aliens.

Neither the optimists nor the pessimists have proven their cases about how long technological civilizations live. The future of life on Earth and beyond, a question hardly enunciated in early exobiology, remains the least developed of astrobiology's major questions.[78]

We have no way of knowing the longevity of alien civilizations other than by detecting them and learning something of their histories. Inserting solid numbers may have to wait until we discover evidence of a civilization older than ours, or find the remains of one that is now extinct.

Strangely, the possibly disastrous impact of contact with a more technologically advanced civilization is almost never mentioned in this context. Many of those most pessimistic about the human future are very optimistic when they speculate about the consequences of contact with intelligent aliens, often foreseeing that it would extend the lifetime of our species rather than shorten it.

Bottom Line, Longevity: A gain for the deniers. While predictions of our imminent self-destruction may have been overstated, astrophysical dangers now loom larger.

The Drake Equation, Take Two

Some have argued that the Drake equation is incomplete, that it needs additional factors. These proposals lengthen the list of requirements for the emergence of communicating civilizations.

Ward and Brownlee proposed a modified Drake equation in which the key biological factor is the percentage of a lifetime of a planet that is marked by the presence of complex metazoans (multicelled life). They added other factors: the fraction of planets with a large moon (to stabilize their rotation), the fraction of solar systems with Jupiter-sized planets (to protect smaller planets from large impactors), and the fraction of planets with a low number of mass extinction events.[1]

Papagiannis and Mauldin argued separately that the equation neglects the factor of time; it assumes that all factors are relatively unchanged over the billions of years for life to develop into civilizations. Yet, some disruptive processes in the Galaxy occur on times scales of 100 million years, which may cause us to overestimate the lifetimes of civilizations.

Only a very small fraction of planets where life originates might be able to offer the long-term stability required for the slow evolution of life to high intelligence. The duration of the zone of habitability may be as important as its distance from the star.[2]

Ideally, the equation would address cultural evolution beyond the state known to us—not just the emergence of postbiological societies but other factors as well. This is extremely difficult to do when we have no agreed theory of cultural evolution.

It is easiest to be pessimistic about the probability of alien technological civilizations when our sample for a factor is limited to one, such as the origin of life. On the other hand, our experience with the astronomical factors implicitly favors the believers.

We first knew of many stars; although that did not prove that extrasolar planets existed, it did imply that they might be numerous. Now we know that such worlds are abundant. That does not prove that Earth-like planets exist, but it does make them seem more likely. If we find large numbers of such planets, their existence would imply many opportunities for life to emerge.

The more examples we have of the phenomenon in question, the looser the constraints become for the next factor. The derivation of human uniqueness from the Drake equation—the conclusion that there are no other technological civilizations—may be driven by our ignorance.

Should We Continue the Search?

To succeed at last, we must embrace many failures.
           —Donald Goldsmith and Tobias Owen, 1980[1]

Looking and hoping are what science is all about.
           —John Casti, 1989[2]

Premature Opinions

Since 1960, scientists have conducted more than 100 searches for signals from extraterrestrial civilizations.[3] As of this writing, there has been no confirmed detection of alien technology; although searchers have spotted many intriguing signals, none have been found a second time.

It now seems clear that pioneering SETI experiments were based on overly optimistic estimates for the abundance of extraterrestrial civilizations sending out radio signals. Shklovskii referred to the early period of SETI as the time of "adolescent optimism," in which searchers placed exaggerated emphasis on the radio-technological prospects for extraterrestrial communication while ignoring both the humanities and biology. Drake admitted as early as 1976 that "if there were once cockeyed optimists, there aren't any more."[4]

Some SETI pioneers had been realistic about the short-term prospects. Bernard Oliver said in 1981 that he did not expect to make contact with any of the things being done then; that was a dress rehearsal, learning how to parse signals. The sensitivities were not high enough to make a detection probable. Drake did not assume that the later Project Phoenix would be successful; the numbers were against it.[5]

"Everything we have learned since 1960 tends to reduce the likelihood of success under the Drake Equation," said James Trefil in 1988, before astronomers began discovering planets around other stars. Robert Rood thought it more likely that increased knowledge will decrease our estimate than increase it, although he acknowledged that the uncertainty is great.[6]

The authors of the SETI 2020 report reached exactly the opposite conclusion in 2002. Estimates of the chance of success, based on the Drake
equation, have become more optimistic in the last half-decade, particularly
because of the discovery of extrasolar planets and the growing evidence
for possible life elsewhere in our solar system.[7]

In that document, Jill Tarter reviewed how premises had changed since
the Project Cyclops report of 1972:

— Planetary systems are now the rule, not the exception.
— The scientific view of potential habitats for unknown kinds of life has
  expanded.
— More than one hundred different interstellar molecules have been detected,
  possibly including the simplest amino acid.
— The process of evolution that led to intelligent life on Earth was more episodic
  and random than we had believed. One can argue that we were particularly
  slow in our evolutionary history; elsewhere life may have evolved to intelligence
  on a shorter timescale, or it may not have evolved at all.
— While there is a stronger case today for the prevalence of life elsewhere, the
  scientific community is strongly divided over whether intelligence will always,
  or even often, accompany the origin of life. Intelligence may be the inevitable
  result of predator–prey relationships, or it may be a fluke.
— We are less optimistic now about the longevity of a technological civilization
  as a radio-transmitting technology. As our sample size is one (ourselves) and
  we do not know the longevity of that sample, we can only get answers by making
  assumptions.[8]

Consider the challenges that radio searches face: the vast number of stars
to be surveyed and the wide range of possible frequencies on which a signal
might be transmitted. Even the best SETI programs are limited in their
coverage; no system looks for everything.

The vast majority of search space remains unexplored. Search space does
not just mean the three-dimensional volume of the Galaxy. As the Jet
Propulsion Laboratory's Michael Klein explained, it is a multidimensional
space that includes source location, signal frequency, power level, time of
arrival, signal modulation, and polarizations.[9]

The NASA program, if it had been carried to its planned conclusion,
would have explored only one ten-billionth of the cosmic haystack. That
haystack is so immense, said Drake, that no theory, no amount of dedication, no endless hours at the typical radio telescope is going to produce a
thorough search.[10]

Jill Tarter reminded us of the scale. "We're going out now 155 light years
in a galaxy that's one hundred thousand light-years in diameter. It's way
too early to get discouraged."[11]

A thorough search would mean surveying the entire sky over the plausible microwave band. This would require national-level resources, particularly if one insists on continuous coverage. Noting that history is littered

with grand enterprises that tried to plan too much too early, Paul Horowitz argued that this applies to SETI; we know too little.[12]

Searches have looked for continuously present signals; powerful but occasional pulses in the radio band would be undetected by all existing radio telescopes. The NASA search would have missed signals that pulsed on and off or that significantly varied in frequency during observations.[13]

Current SETI systems are not sensitive enough to be able to detect radio and television leakage from the Earth at the distance of the nearest star to the Sun, 4.3 light-years. We could detect early warning radars out to 15 light-years, although we would need at least 100 times more sensitivity to extract the message from the signal.[14]

As scenarios of detection increase in plausibility, they also increase in difficulty.[15] The easiest signals to find—beacons—may be the least likely. The most likely—the local communications of a remote civilization—are the hardest to detect.

The radio search assumes that a technological civilization will be located nearby only if the number of such civilizations is very large. Yet, there are orders of magnitude of uncertainty in estimates of the average distance between communicative civilizations in our Galaxy. Depending on the assumptions that one plugs into the Drake equation, the nearest communicative civilization is within or beyond our present capability.[16]

Clarke thought that it was ridiculously optimistic to expect immediate success in such a search. He wanted to see the whole debate given a decade or two of benign neglect while radio astronomers sift through the torrents of noise pouring down from the sky.[17]

Patience

There is reason for long-term optimism. If there are extraterrestrials radiating significant radio signals, the probability of our detecting them will continue to improve. The sensitivity of radio telescopes has been doubling every 3 years. Within the next two decades, said Shostak, we'll check out not a few thousand, but two million stars.[18]

Like politics, searching for others is the art of the possible. Any given phase of the search will use one technology and one strategy to examine one part of the electromagnetic spectrum. If those searches fail, researchers might reasonably argue that we should move on to another part of the spectrum, perhaps using a different strategy and possibly a different technology. "We should not limit our searching to approaches which seem plausible only in the light of our current technological prowess," argued Drake. "We should search, where affordable, for almost any physically possible variety of signal . . . the costs of doing business have become much greater."[19]

Ideally, we would look for evidence of alien civilizations throughout the electromagnetic spectrum, over all of the sky all of the time. Unfortunately, the real world of budgets and available technologies limits our options. Researchers may be driven to searching wherever it is affordable, rather than wherever the possibilities are most intriguing.

Searchers have to make choices. It remains to be seen whether they have made the right ones. "It's naive to think that the searches we're doing today are the best ones," commented University of California astronomer Dan Wertheimer. Long ago, Von Hoerner counseled that it is impossible to know in advance which method is the best.[20]

The search also will be limited by our assumptions. Searchers may have to stretch their analogies to include a wider range of phenomena.

What Can We Conclude?

The silence we have heard so far is not in any way significant. We still have not looked long enough or hard enough.

—Frank Drake, 1992[21]

Frank White, author of The SETI Factor, suggested in 1990 that we could reach these conclusions after 30 years of searching: Our galaxy is not teeming with radio signals; there are no Type II civilizations in the Galaxy using enormous amounts of energy to communicate; our searches have detected something out of the ordinary many times, although none have been confirmed as intelligently directed signals. Lack of success so far tells us nothing, except that contact is harder than we thought it would be.[22] The only thing we know for sure is that our galaxy is not illuminated by powerful radio transmitters continuously broadcasting in the ways for which we have looked.

Drake admitted that the negative results from Project Phoenix do imply that there are not large numbers of civilizations transmitting at many frequencies, at least not lately. However, there are likely to be far more intrinsically faint civilizations than intrinsically bright ones.[23]

Supporters of SETI argue that it is premature to speak of a great silence. "All the world's searches for extraterrestrial intelligence are pathetically tiny compared to the size of the radio spectrum, the width of the sky, and the depth of space," astronomy writer Alan MacRobert declared. "Dozens of exasperated civilizations could be blasting Earth with wake-up calls at dozens of 'logical' hailing frequencies, and they would all likely be missed by every SETI project present or planned. Compared to the staggering job, our best SETI efforts are mere proof-of-concept trial runs."[24]

Sagan acknowledged that the longer we listen on the greatest variety of systems without success, it will be increasingly apparent that life *that wishes to communicate with us* is not an absolute cosmic commonplace (italics in original).[25] That may be a small percentage of our Galaxy's intelligent life.

Aliens and Bosons

Skeptics argue that SETI researchers offer no definition of what will satisfy them that detectable civilizations do not exist. Their quest seems open-ended, never subject to falsification.

After the international petition on SETI was published in October 1982, physicist Frank Tipler asserted that the radio search is not a scientific experiment because it cannot falsify the hypothesis being tested: that extraterrestrial civilizations exist. Drake seemed to confirm this when he wrote that "we can never prove the nonexistence of life, intelligent or otherwise, in the universe. No amount of failure in SETI endeavors constitutes proof that we are alone."

The lack of falsification is not unique to SETI. One cosmologist commented that there is so much "wiggle room" in current cosmological theories that they could not be disproved.

Do we consistently apply the falsification standard in other scientific fields? Or are we setting a higher standard for SETI?

Consider a recent *Scientific American* article about physics. A key prediction in physicists' tentative theories about mass is a new kind of field that permeates all reality, called the Higgs field. If this field exists, theory demands that it have an associated particle, the Higgs boson. Using particle accelerators, scientists are now hunting for the Higgs. "These quanta must exist," declared the author, "or else the explanation is not right."

Despite spending millions of dollars of taxpayer money, scientists from several nations have so far failed to detect the Higgs boson. Like some SETI researchers, they have found "tantalizing evidence" for their target just at the limits of their research instrument, in this case the Large Electron–Positron collider. May we now conclude that the Higgs boson does not exist?

Early evolutionists too were regarded as people without any quantitative basis for their science. Yet, as Morrison pointed out, they were on to something profound.[26]

How Much Is Enough?

The Clarks, observing that SETI is on a grander scale than any other scientific survey ever undertaken, argued that there is no predetermined timescale for a successful outcome.[27] Others have offered more specific predictions.

Five SETI scientists, writing in 1977, thought that within a few decades mankind will either have discovered the presence of extraterrestrial intelligence or will have placed very severe constraints on the likelihood that such intelligence exists. Shostak addressed the same time frame in a more upbeat way: "If the galaxy is populated by only ten thousand advanced civilizations, success is a few decades down the road."[28]

Dyson thought it likely that we will discover evidence of extraterrestrials within 100 years. However, warned Clarke, if we have found no sign of alien intelligence after centuries of listening and looking, we might be justified in assuming that we are alone in the universe.[29]

MacGowan and Ordway addressed an even longer timescale. "Within 1,000 years from now at the most we should be able and desirous to inaugurate a massive search program, including the use of interstellar probes if necessary, to determine the occurrence of communities out to at least 1,000 light years. If we find no evidence of extrasolar life we may well become discouraged and give up."[30]

A NASA workshop report presented the issue more ambiguously. Lacking any detection, the conviction of our uniqueness would hardly ever reach certainty. It would form, over a long time, less into sharp conclusions than into a kind of substructure of human thought, a ruling consensus of attitudes[31]—perhaps like the ruling consensus that preceded Copernicus and Galileo.

Steven Dick wondered whether a century from now the search will be seen as just a curious episode in the history of science. Sociologist William Bainbridge had raised a similar question about space exploration and utilization in 1976: "Either spaceflight will be proven a successful revolution that opened the heavens to human use and habitation, or it will be proven an unsuccessful revolution that demonstrated in its failure to the limits of technological advance."[32] Thirty years later, the share of our treasure that we spend on our space efforts has been scaled down. Yet, we still send humans and machines into space; we still formulate expansive plans for the future.

Like the earlier, canal-inspired search for life on Mars, the current search for radio signals may be a case of astronomers trying to resolve a problem beyond the limits of the science of their time. Yet, the consequences of success are so momentous that we and our descendants are likely to continue searching.

If finding evidence of extraterrestrial intelligence proves to be beyond our present technical abilities, we may be justified in assuming that future humans will be capable of more. Modern science is a cumulative enterprise, Rees reminded us. Discoveries are made when the time is ripe, when the key ideas are in the air, or when some novel technique is exploited.[33]

History tells us that the quest for cosmic company is a recurrent cultural phenomenon. Even if set back momentarily, it is likely to reappear in one form or another.

What Are We Looking for?

As is often the case for exploration, SETI is based on fragmentary knowledge. A group of experts acknowledged that the search is fraught with uncertainties and assumptions.

For what evidence are we searching? When and where should we look? Some answers are based on science; what kinds of signals travel well across the Galaxy? Some are based on technology; what is the relative effort of sending spacecraft as compared to signals? Some are based on the supposed behavior of the alien civilization; should we look for something they have deliberately sent us or for unintentional by-products of their technology? All of these questions require at least provisional answers before we can search.[34]

Uncertainty, Ambiguity, and Serendipity

Journalist Joel Achenbach may have overstated the case when he concluded that the alien question is likely to remain a matter of "infinite possibilities and zero certainties."[35] We have narrowed some of the parameters; although many unknowns remain, researchers will continue to chip away at the uncertainty surrounding them.

To Sagan, uncertainty served a higher purpose: It drove us to accumulate better data. "Someone has to propose ideas at the boundaries of the plausible," he told an interviewer, "in order to so annoy the experimentalists or observationalists that they'll be motivated to disprove the idea."[36]

Although deliberate searches for evidence of intelligent life will remain the preferred method, a discovery might be serendipitous. Radio signals from space were first discovered inadvertently. The cosmic microwave background was detected by researchers looking for something entirely different; pulsars too were found by accident. As telescope designer Roger Angel saw it, unexpected astronomical discoveries often occurred when observers were pushing new equipment to its limits to accomplish a highly focused goal unrelated to the actual discovery.

George Pimentel of the U.S. National Science Foundation highlighted the role of serendipity in the first Congressional hearings on SETI, held in 1978. If extraterrestrial life is found, it will occur in the course of other activities of our best astronomers. The chances of hearing messages are probably best enhanced by devoting as much of our scientific resources as we can afford to the best astronomy of the day, and giving serendipity its chance.[37] In this context, it would seem useful to periodically remind astronomers that unusual phenomena might be evidence of alien technological activity.

"If we already knew what we were looking for," said Jill Tarter, "we could plan exactly how you would verify that what you got is what you intended to find. But it will most likely be the case that something tantalizingly new will arise. There will be a lot of unanticipated events."[38]

Another factor may complicate the search: ambiguity. The scientific history of the Mars rock suggests that, although some observers might interpret a signal or a pattern they find in the heavens as evidence of extraterrestrial intelligence, others will challenge that conclusion.

Astronomer Peter Boyce warned that the elusive signal is likely to be at the very edge of detectability; its very existence may be doubtful. It is ambiguous evidence, psychology professor Albert Harrison observed, that encourages us to rely on our preconceptions and biases.[39]

The Consequences of Searching

Asking the question "are we alone in the universe?" is as revolutionary as the removal of the Earth from the center of the solar system, and the removal of the solar system from the center of the Galaxy, declared Otto Struve. This revolution is already under way, although it would be accelerated by an actual discovery.[40]

In one world view, cosmic evolution commonly ends in planets, stars, and galaxies (the physical universe). In the other, it commonly ends in life, mind, and intelligence (the biological universe). Steven Dick thought that the proof of one of the two worldviews will change everything; it already has begun to change everything in anticipation of the outcome.[41]

The mere fact that belief in the existence of sentient aliens can be scientifically justified broadens human horizons to a cosmic perspective. Five SETI scientists put the case in transcendental terms: To be part of this search is to partake of a dream that binds us temporally with our past and future, spatially with the cosmos, and culturally with our destiny.[42]

Canadian scholar Allen Tough saw several positive consequences of searching: enlarging our view of ourselves and enhancing our sense of meaning; feeling a kinship with the civilizations we are trying to detect; thinking about how extraterrestrials might perceive us, giving us a fresh perspective on our society; stimulating thought and discussion about fundamental questions; serving as a useful educational tool; giving us technical spin-offs; encouraging international scientific cooperation; stimulating us to think about detection scenarios and their consequences.[43] Anticipating contact encourages us to consider what is good for our species as a whole, to gain shared perspectives on our future.

Even if a long dedicated search fails, we will not have wasted our time. A NASA workshop report foresaw that we will have developed important technology, with applications to many other aspects of our own civilization. We will surely have added greatly to our knowledge of the physical universe.[44] We already know that the search stimulates interdisciplinary thinking in a way that few other fields can match, requiring the perspectives of many sciences.

The most powerful implication of the search remains the same. If other technological civilizations exist, our efforts are making contact more likely.

Direct Contact

Some of the elder generation of awakened worlds were already facing the immensely difficult problems of travel on the interstellar and not merely the interplanetary scale. This new power changed the whole character of galactic history. Hitherto . . . the life of the galaxy had been in the main the life of a number of isolated worlds which took no effect upon one another. With the advent of interstellar travel the many distinct themes of the world-biographies gradually merged in an all-embracing drama.

—Olaf Stapledon, 1937[1]

Starflight

Early fictional depictions of direct contact with extraterrestrials, like those by Lasswitz and Wells, assumed that spacecraft could carry living beings across interplanetary distances. Now we know that such journeys are possible for a civilization at our level of technological development. We already have landed men on Earth's moon and are planning to transport humans to Mars. We also know that the home planets of alien civilizations, if they exist, are much farther away. Could extraterrestrials traverse interstellar space to our solar system?

Much of the public assumes that they could. Mass media science fiction populates visiting starships with intelligent beings, whether they are altruistic humanoids or slimy reptilians.

Most astronomers have been deeply skeptical about direct contact between civilizations across interstellar distances. Von Hoerner, writing in 1962, declared that space travel—even in the most distant future—will be confined completely to our own planetary system. He claimed that a similar conclusion will hold for any other civilization, no matter how advanced it may be.[2]

Astronomer Edward Purcell was even more outspoken, dismissing the idea of interstellar flight as preposterous. "All this stuff about traveling around the universe in space suits," he declared, "belongs back where it came from, on the cereal box."[3]

Most SETI scientists have tended to agree. A NASA report on SETI declared that manned interstellar flight is out of the question not only for the present but for an indefinitely long time in the future. The SETI advocates who wrote that report admitted that making such journeys is not a physical impossibility, but argued that it is an economic impossibility at the present time. Some unforseen breakthroughs must be made before humans can travel to other stars.[4]

Even if interstellar travel were technically feasible, SETI researchers argue, the costs involved would be so enormous that other civilizations would not expend the necessary resources. If the purpose is to gather information, interstellar communication would be far more economical. "It is always easier to transmit photons," five scientists wrote, "than to transmit matter."[5]

SETI conventional wisdom assumes that each technological society stays close to the star of its birth. By virtually ruling out interstellar travel, the SETI paradigm depicts what Brin called "isolated motes of intelligence separated by sterile tracts of space." Drake confirmed this when he wrote that the distances between the stars create an interstellar quarantine.[6]

An interesting exception was Carl Sagan, who argued in 1962 that interstellar travel not only might be possible, but that other civilizations eons more advanced than ours must today be plying the spaces between the stars. Very advanced societies, he wrote later, should surely be capable of interstellar flight; there are no fundamental physical objections to it. Sagan extended that capability to future humans in the book he co-authored with Shklovskii, declaring that "efficient interstellar spaceflight to the farthest reaches of our galaxy is a feasible objective for humanity."[7]

This question has profound implications for the consequences of contact. If direct contact is possible, our hopes and fears acquire a new intensity.

Assumptions

Mere distance is nothing; only the time that is needed to span it has any meaning.

—Arthur C. Clarke, 1968[8]

The argument that interstellar travel is impossible or too difficult to be worth doing has rested on several challengeable assumptions. The first, made by Von Hoerner and others, is that in order to cover interstellar distances within "reasonable times," we ought to fly as close as possible to the velocity of light. Because that would require titanic amounts of energy, interstellar flight would be extremely difficult. (Von Hoerner implicitly argued the other side of this issue when he stated that the only goal that may be important enough to justify the immense effort needed for inter-stellar space travel appears to be the search for other intelligent beings.[9])

Other key assumptions are that interstellar spacecraft are inhabited by human beings or their alien biological analogs; that voyages must be round-trips requiring acceleration and deceleration for both the outward and inward legs of the trip; that such journeys must be completed within a human working lifetime (typically 40–50 years); and that an interstellar vehicle must carry its entire energy supply and reaction mass on board.

"We cannot plan on building a spacecraft that could travel anywhere close to the speed of light with our present technology," wrote Goldsmith and Owen. "Yet only by traveling at nearly the speed of light can we hope to cover interstellar distances in tens of years." They then proposed an extremely unlikely scenario that exaggerated the timescale of interstellar exploration. "If a choice had to be made between spending money to search for extraterrestrial signals now and spending money to send out a space crew to return after, say, 200,000 years, then an immense prejudice in favor of human journeys would be needed to choose the second alternative."[10]

Exaggerating obstacles to interstellar flight is comparable to exaggerating the difficulty of finding evidence of alien intelligence through astronomical observations. In both cases, some critics are trying to discredit ideas that they do not like.

Bracewell observed that calculations by opponents of starflight produce astronomical sums. "Then you'll find that they've thrown in a factor of a thousand, which they introduced themselves. When you point this out to them, it's not very long before they come up with another irrational objection."[11]

There are abundant precedents for this behavior. Before the Space Age, some scientists presented analyses showing that interplanetary flight was impossible. The President of the Royal Astronomical Society of Canada, for example, calculated that a spaceship would need an initial weight of about 1 trillion tons to make a round trip to the Moon.[12]

What makes interstellar travel different is not just the distances between stars but also our own impatience, due to the short lifetimes of human beings. We are in a hurry.

Voyages to the stars begin to look more feasible if we remove humans from the equation and let machines do our exploring on one-way voyages at lower velocities over longer periods of time. We almost certainly would not send humans in the early stages of interstellar exploration; crewless probes would be less expensive, easier to propel, and more expendable.

In any case, the assumption of a round-trip makes no sense. Unmanned spacecraft need not return, any more than the machines we send to Mars; they could tell us what they found through electromagnetic signals. We would not send humans to another star system unless we were sure that they could survive there and would not need to come back.[13] That would cut the energy requirements in half.

An exploring machine might not even need to decelerate into the target system. The British Interplanetary Society's study of an interstellar probe assumed that the main ship would fly by Barnard's Star without slowing down, launching much smaller subprobes to do more thorough investigation.[14] That would cut the energy requirements again.

We can lower those requirements still more by extending the time for the mission, reducing the velocity of the spacecraft to a small percentage of the speed of light. That is easier to do when we send machines; people impose much greater burdens. Even for human crews, there are several proposed solutions. Our descendants may come up with better ones.

Might the solutions for interstellar flight rest on physical principles that we have not yet discovered? Some SETI scientists have shown themselves willing to consider that possibility when they discuss interstellar communication. Yet, most of those scientists resist extending that flexibility of mind to interstellar travel.

This reminds us of how Arthur Clarke described stages in the way the scientific establishment reacts to a new big idea. First, it is impossible. Second, it may be possible, but it's not worth doing. Third, I said it was a good idea all along. Interstellar flight by machines has entered the second stage.

Starships of the Mind

The demonstration that no possible combination of known substances, known forms of machinery and known forms of force can be united in a practicable machine by which men shall fly long distances through the air, seems to the writer to be as complete as it is possible for the demonstration of any physical fact to be.

—Astronomer Simon Newcomb, 1893[15]

Journeys to the stars are a daunting technological challenge for a civilization with our present level of technology. The distances involved, and by implication the time, energy, and machine reliability required, are orders of magnitude greater than anything we have attempted. Yet, no known law of physics or engineering tells us that interstellar flight is impossible. Although there is a huge jump in scale from solar system travel to interstellar travel, Clarke reminded us that we children of the Space Age can no longer remember how enormous the solar system seemed only a lifetime ago.[16]

Leslie Shepherd, Eugen Sanger, Ernst Stuhlinger, Robert Bussard, and others began studying the physics and engineering of interstellar flight as long ago as the 1950s, addressing the central issue of propulsion with proposals for fission, fusion, and other systems. These pioneers found that energy requirements decrease dramatically if we extend the allowable time

for the journey; the curve flattens considerably below half the speed of light
(0.5 c). By 1963, D.F. Spencer and L.D. Jaffe of the Jet Propulsion Laboratory were arguing that interstellar travel was feasible using staged nuclear
systems.[17]

Hughes Research Laboratories physicist Robert Forward, one of the
leaders in the emerging interstellar lobby, outlined an array of propulsion
choices in 1975. One option is to propel the spacecraft with energy beamed
from our solar system, eliminating the need to carry an energy source and
fuel on board. Forward proposed a national program for interstellar exploration beginning with automated probes, leading to the launch of a human
mission to Alpha Centauri in 2025.[18]

Dyson found no lack of propulsion systems available to any creatures
who possess some technical competence and a desire to travel around the
galaxy. Outlining several alternatives in a 1979 paper, he concluded that
systems that are probably feasible but require very demanding new technology should be capable of velocities of 0.05 c, or one-twentieth of the
speed of light.[19] A mission at that average velocity would take 86 years to
reach the distance of the Alpha Centauri system, the nearest stars to our
own Sun.

The technical literature on this subject has grown rapidly since those
early days; a bibliography of publications on starflight expanded to include
hundreds of entries. By 1989, when science writer Eugene Mallove and
astronomer Gregory Matloff outlined the basic propulsion choices in their
Starflight Handbook, even the skeptical mathematician John Casti admitted that studies had shown that no new physical principles are involved in
allowing a ship to travel at one-tenth the speed of light. However, he cautioned that the engineering hurdles are enormous.[20]

Starwisp

In 1983, Forward proposed a lightweight, high-speed interstellar fly-by
probe pushed by microwaves beamed from satellite solar power stations
in our solar system. A delicate wire mesh sail a kilometer across with
microcircuits at each intersection, Starwisp could be accelerated to one-
fifth the speed of light. As it passed through the Alpha Centauri system
21 years later, this spiderweb craft would receive beamed energy to
power its detectors and logic circuits. The sail would become a phased-
array microwave antenna, sending high-resolution imagery back to
Humankind.[21]

Human Probes

The interstellar idea is taking shape in a great, ongoing conversation.
—Paul Gilster, 2004[22]

MacGowan and Ordway, writing in 1966, drew a sharp distinction between manned and unmanned interstellar flight. They thought that the short human life span, limited acceleration tolerance, vast distances, and the cosmic-ray barrier combine to make human interstellar travel unfeasible even in the quite distant future. However, unmanned interstellar probes may prove practicable because of their smaller payloads. Instrumented probes and intelligent automata could be designed with very long, even unlimited life expectancies; such machines could undertake interstellar journeys of at least moderate distances.[23]

We launched our first machines toward the stars during the next decade: *Pioneers 10* and *11*, and *Voyagers 1* and *2*. Those spacecraft now are far beyond the traditional nine planets of our solar system. Although they will cease operating long before they reach the distance of the nearest stars thousands of years from now, they hint at what might be possible.

We could take our next steps with missions that reach part of the way into interstellar space. Bernard Oliver, deeply skeptical about interstellar flight, acknowledged that voyages into our outer solar system are within our near-future capability.[24]

The Jet Propulsion Laboratory began studying an interstellar precursor mission using known space technology in 1976. A modified version, which came to be known as the Thousand Astronomical Unit mission, would journey out 25 times the distance of Neptune, completing its voyage in 50 years. That may sound like a long time, but we still are receiving signals from robotic missions that were launched more than 30 years ago. Controllers may be able to detect faint telemetry from *Voyager 1* until 2020, 42 years after it left the Earth.[25]

A 1977 NASA report on SETI grudgingly admitted that unmanned probes to a few nearby stars for the purpose of scientific exploration might be a worthwhile endeavor; in all likelihood, such an attempt will be made at some future date. However, that report ridiculed the idea of sending probes to all Sun-like stars within 1000 light-years because of the astronomical cost.[26]

The British Interplanetary Society's *Project Daedalus* study, published a year later, concluded that an interstellar probe could be built by a civilization whose technology was only slightly more advanced than ours. A massive spacecraft powered by inertial fusion might reach Barnard's Star (8 light-years away) within 50 years using foreseeable technology.[27]

Interest in interstellar probes was revived in the 1990s, when NASA Administrator Daniel Goldin commissioned studies. The space agency already has launched a spacecraft on a voyage to Pluto that may be extended into the Kuiper Belt from 2016 to 2020.[28] Missions like that one could be interim steps toward the stars.

There are obvious advantages to sending robotic scouts. Humans can live in only a tiny percentage of the universe without elaborate physical protections; machines are indifferent to extreme conditions—and they are patient.[29]

Mind-Stretcher. The distinction between manned and unmanned space-craft may vanish if intelligent beings merge with machines. Tipler consid-ered it exceedingly unlikely that flesh and blood beings will ever engage in interstellar travel, because human "uploads" have such an advantage over people in space.[30] We can project this idea farther. Rather than being artifacts sent out by societies of biological beings, interstellar explorers might be independent intelligent machines that propel themselves to des-tinations they choose.

Interstellar probes almost certainly would look for evidence of life. They give us one way to gain knowledge of star systems that lack garrulous, radio-equipped inhabitants, argued Clarke; it might be the only way.

If you want to find out about who's alive out there, said physicist and space colony designer Gerard O'Neill, it's stupid to wait until the other guy develops radio communication. It makes much more sense to send probes to monitor planets in other systems. Such robotic observers could give you millions of years warning before a civilization is even going to develop.[31]

Interstellar exploration seems quite far off now, Grinspoon recognized, probably as distant as voyages to the planets seemed at the beginning of the twentieth century. Yet, by the end of that century, planetary exploration was being taught in history class. Interstellar exploration could begin by the mid-twenty-first century.[32]

A Proposed T-Shirt

Enthusiastic students at our best technical universities often adopt slogans that sum up an idea in a formula or a few words. In the early days of Gerard O'Neill's campaigning for space colonies built from lunar materials, some of his young supporters wore T-shirts declaring "LUNAR MINE IN '89."

Here is a new T-shirt slogan for the best and the brightest at M.I.T. and Cal Tech, one that leaves out an overly optimistic date: **0.1c**. The average member of the public won't understand what that means. The initiated will know that, for our initial interstellar explorations with machines, we need only aim for one-tenth of the speed of light.

Alien Probes

Interstellar probes are appealing as long as someone else sends them, but not when we face the task ourselves.

—NASA report, 1977[33]

Interstellar radio communication has significant limitations. As Bracewell pointed out in 1960, a search for radio signals directed at our planet would have been fruitless throughout most of the billions of years of the Earth's existence. It is more logical to assume that superior civilizations would send automated messengers to orbit each candidate star and await the possible awakening of a civilization on one of its planets, using radio signals to attract the attention of any indigenous intelligent beings. Such a probe might repeat back to Earth exactly what it received; to us, its signals would have the appearance of long-delay echoes. (Some have suggested that long delayed radio echoes detected in the past might originate from such a machine, although no such probe has been found.[34])

These ideas provoked resistance from those most attached to the remote detection scenario; Morrison described them as tendentious. "It's rather amusing," Bracewell commented, "to see the way they jump on both the probe idea and the idea of uniqueness with very strong, very fierce feelings."[35]

Others were not so dismissive. Tipler believed that robot probes would achieve the aims of an exploring civilization much better than radio signals; they can contact civilizations that are not listening or those that do not have radio technology, even intelligent but pretechnical societies. Such probes also could be used to explore uninhabited systems.[36]

Our own future interstellar machines might be, in part, Bracewell probes, designed to detect and communicate with other intelligences as well as to search for new biospheres. If a civilization at our level of technological development already is projecting such probes, older and more advanced civilizations might have launched them long ago.

Mind-Stretcher. If other advanced technological societies conclude that probes are the most efficient means of exploration, those machines might be the most widely distributed artifacts of intelligence. We are far more likely to meet the technological creations of extraterrestrials than the aliens themselves.

Searching for signals and sending out probes are not mutually exclusive strategies. An alien civilization that detected our emissions might send a probe to look us over. Remote detection could lead to direct contact.

Engineer Christopher Rose and astrophysicist Gregory Wright proposed another scenario. If speedy delivery is not required, other civilizations might send out artifacts packed with information that were intented to be found—a vastly upgraded version of the *Pioneer* plaques. Astronomer Woodruff Sullivan, while defending the search for electromagnetic signals, admitted that some attention should be paid to the possibility of finding an information-drenched object sent by an extremely advanced civilization interested only in one-way communication.[37] Finding such an object, or any form of alien spacecraft, might tell us far more about the extraterrestrial society than a brief and ambiguous radio signal.

Human Voyages

Interstellar travel... is essentially not a problem in physics or engineering but a problem in biology.

—Freeman Dyson, 1964[38]

In 1952, Leslie Shepherd of the British Interplanetary Society published what may have been the first technical article on human interstellar flight. After analyzing propulsion requirements, he found that travel to other stars might be possible if we are willing to accept transit times of greater than 100 years, and possibly 1000 years. Shepherd concluded that there does not appear to be any fundamental reason why human communities should not be transported to planets around neighboring stars, although the transit time is so great that many generations must live and die in space. Interstellar exploration and colonization may require a revolution in our way of life, not only socially but biologically.[39]

Mallove and Matloff came to a similar conclusion 37 years later. Our brief human existence is incompatible with the enterprise of starflight, except to the extent that we are happy to leave the joy of future discoveries and experiences in other solar systems to our descendants. Even today, analysts believe that our current understanding of physics does not allow human interstellar flight beyond nearby stars, unless we build giant world-ships and develop the concept of flight through generations.[40]

Such worldships are not a new idea. Tsiolkovskii had proposed a "generation ship" in 1928. Stapledon, writing in 1937, visualized immense exploration vessels many miles in diameter, constructed in space. He foresaw high-speed artificial "worldlets" voyaging from system to system, toward the thousands upon thousands of planetless stars that awaited encirclement by rings of worlds.[41]

Some worldship technologies might first be employed to build space colonies in our own solar system. British physicist J.D. Bernal imagined globe-shaped colonies, 10 miles across, in 1929; sooner or later, he speculated, pressures such as the imminent failure of the Sun would force an adventurous colony to set out beyond the bounds of our solar system. O'Neill, who developed the space colony concept in much greater detail in the 1970s, suggested that a modified version of a space community will have traveled to a nearby star within the next few centuries.[42]

Sustaining such a society would raise significant sociological issues. Robert Heinlein's science fiction stories "Universe" and "Common Sense," published in 1941, showed how generations of people born on an interstellar ark might lose track of the original purpose of the voyage. Sociologists

Mircea Pfleiderer and Paul Leyhausen worried that passengers might suffer a crippling loss of technological knowledge during the long flight.[43]

Papagiannis took a more upbeat view, commenting that our own Earth is a colossal spaceship that orbits the Galaxy in about 250 million years; nobody seems to mind this grand trip. O'Neill-type space colonies would make the colonization of the Galaxy much easier and much more complete, because their inhabitants will prefer to live in space habitats rather than search for planets with Earth-like atmospheres. Your present author and others have proposed interstellar colonization strategies built around such artificial biospheres.[44]

British starflight visionary Anthony Martin concluded that "the most likely transport would be ponderously slow world ships with city-sized cargoes of individuals, probably grouped into nation-sized fleets." French scientist Maurice de San foresaw a long-term outcome: a nomadic culture living between stars.[45]

The Interstellar Shuttle

Stimulated by the idea of periodic alien visitations to our solar system, R.W. Moir and W.L. Barr of the Lawrence Livermore National Laboratory proposed cycling interstellar ships that would connect several systems. In this scheme, a world ship would exploit the gravitational slingshot effect of each star to send itself on to the next sun in its closed path. Once initiated, this approach would use a minimum of propulsion energy. There is one catch; the minimum time for a 16-light-year trip around three stars would be about 41,000 years.[46]

How do we reconcile these long journeys with the human life span? One approach is to maintain humans in a kind of suspended animation during much of the voyage. Don Wilcox suggested "cryo-sleep" in his 1940 story "The Voyage that Lasted Six Hundred Years."[47] That idea later became a staple of science fiction films such as "2001" and "Alien," although its feasibility has not yet been proven.

Another approach is to extend individual human life spans. "For a being with a life span of 3000 years a voyage of 200 years might seem not a dreary waste of most of one's life," said astronomer Michael Hart, "but rather a diverting interlude."[48]

There have been more exotic proposals. Clarke suggested in 1968 that we send only germ cells, fertilizing them 20 years before landing and having cybernetic nurses raise the babies. Others proposed that we simply send information on desirable nucleotide sequences that the starship would convert into living beings. A probe might even prepare an environment on a suitable planet so that Earth organisms can survive and support the introduction of machine-nurtured humans.[49]

Such scenarios raise their own risks. Would these young humans have any sense of connection with those who sent them? Children developing without committed human parents, warned Mauldin, may lack some essential qualities.[50]

Human voyagers might not need to travel all the way to another sun. The space between the stars is not empty; there are stepping stones.

Our solar system and others extend much farther into interstellar space than we once thought. There could be huge numbers of icy planetoids far beyond Pluto, including some that are Pluto's size or larger. Beyond the Kuiper Belt, the enormous cloud of icy bodies hypothesized by astronomer Jan Oort may reach out an estimated 3 light-years; other sun-like stars may be girded by similar clouds.[51]

If Alpha Centauri has a cloud of similar scale, its cloud will overlap with ours. "Any civilization that spans our Oort Cloud," wrote space colonization advocate Mark Hopkins, "also will expand into the Alpha Centauri Oort Cloud." He described this as "a star bridge" between stellar systems.[52]

Bodies similar to comet nuclei may pervade our Galaxy; the average distance between them might be only light-days instead of light-years. This would make the Galaxy a much friendlier place for interstellar travelers.[53] Extraterrestrials too could take advantage of such stepping stones.

The Last Migration

When the sun has exhausted its energy, it would be logical to leave it and look for another, newly kindled star still in its prime.

—Konstantin Tsiolkovskii[54]

Why would we send humans, in whatever form? MacVey concluded in 1977 that were fusion drives permitting interstellar transit times of about a century to the nearest stars the ultimate in propulsion techniques, manned interstellar flight could probably never be made for purposes other than colonization. Your present author independently published a similar conclusion during the same year.

Even if migration seems like a remote prospect now, one factor may drive our descendants to journey away from their star. Roughly a billion years in our future, our Sun will expand to consume the Earth and the other inner planets. Long before then, physical conditions on the Earth would become intolerable (some predict that the Sun's increased luminosity will make the Earth uninhabitable sooner, in about 500 million years).

Mautner suggested that survivors could continue to exist by adjusting their distance from the star as it evolved. However, migrating farther out in our solar system might be only a temporary solution. Yoji Kondo warned that even the most distant bodies eventually will become unsuitable for our species as the Sun turns into a white dwarf star.

Rocket pioneer Robert Goddard, writing in 1918, described an interstellar journey to find new worlds for humans fleeing a dying Sun. This would meet the criterion of the Project Cyclops report, that only the most extreme crisis would justify mass interstellar travel.[55]

Alien Colonizers

It is difficult to construct a plausible scenario whereby an intelligent species develops and retains for centuries an interest in interstellar communication together with the technology to engage in it, and yet does not begin interstellar travel.

—John Barrow and Frank Tipler, 1986[56]

The classic SETI paradigm rests on an assumption that technological civilizations do not expand beyond their home systems. That may not be a sound assumption. If some other civilizations have much greater technological capabilities and much greater resources than we do, interstellar vehicles carrying intelligent beings may be more easily within their reach.

Andrew Clark and David Clark addressed this by inventing a category of extraterrestrials they called IMETI, or extraterrestrial intelligence capable of interstellar mobility. In their view, it is just as possible that there are significant numbers of IMETI as it is that there are significant numbers of radio-transmitting ETI.[57]

Darling projected that an intelligent technological species will be spacefaring and star-colonizing within a period that is completely insignificant on a geological timescale.[58] The primary question might be one of motivation: Why would an advanced civilization choose to launch inhabited interstellar spacecraft?

During the 1970s and 1980s, several physical scientists assumed that population growth would be a driving force for expansion. Rood and Trefil thought that every civilization must face its own Malthusian dilemma. Only those who solve it by expansion are likely to try to establish contact with other races; any aliens we come across are likely to have already expanded into space.[59] Other scientists proposed models of migration and colonization, with widely varying estimates of how rapidly alien civilizations might expand through the Galaxy.

Morrison questioned the Malthusian view that human society or its counterpart will grow indefinitely and be pressed to dwell among the stars.[60] Today, these models are being forgotten as population growth commands less of the intellectual world's attention.

Religious or ideological motivations might play a role. Stapledon thought that worlds that suffered from the mania of religious imperialism would seek interstellar travel long before economic necessity forced it on them. Science fiction author Stephen Baxter put a more positive spin on the idea: Perhaps we will build starships as we build cathedrals, as repositories of faith sailing into the future.[61]

The most universal motive for interstellar colonization will be the most basic of all: survival. Civilizations that survive for billions of years must migrate, sooner or later; as Sagan put it, their eventual choice, as ours, is spaceflight or extinction. Both Sagan and Chyba believed that this represented a cosmic selection effect, putting pressure on organisms to develop technology.[62]

Even it were not true that survival depends on colonization, Barrow and Tipler speculated that there will be a group in any intelligent species that believes that it does. They will launch the starships.[63]

Some civilizations whose stars were becoming unsuitable may have dealt with this challenge long ago by expanding away from their original homes, colonizing other systems or achieving a sustainable existence in interstellar space. There could be a widespread pattern of technological civilizations expanding outward as their stars age, but at very different times in galactic history. Migration might be a powerful force shaping the presence of intelligent beings in the galaxy.

Bracewell drew an analogy. The fact that the Earth is populated with intelligent creatures is not because many habitable areas of Earth fostered the evolution of intelligence, but because one area (Africa) was the scene of events. Humans could have walked from Africa to California in less time than it would take for intelligence to evolve in California. Migration is a faster method of civilizing the Galaxy than independent evolution.[64]

Interstellar Migration, or Interstellar Cultural Diffusion?

For some years now, researchers have debated whether the spread of agricultural techniques into Neolithic Europe resulted from the migration of farmers from the Middle East or from the diffusion of those techniques through cultural transfer from those farmers to an older European population. Recently, some have suggested that both processes may be involved, as when early farmers intermarried with indigenous hunter-gatherer females.[65]

SETI orthodoxy assumes that the cultural transfer model will prevail at the interstellar level. If interstellar migration occurs, the expansion of a culture could take place by more direct means.

The idea of colonization makes SETI true believers very nervous, observed journalist Gregg Easterbrook, because the apparent absence of settlements in our part of the Galaxy becomes another argument that aliens don't exist. The dilemma would be eased if the destinations of alien colonizers were not planets. Civilizations that expand in self-contained, self-sufficient settlements may be acclimated to space, just as some forms of life became acclimated to land once they emerged from Earth's oceans.

An interstellar ark could be either a means of transportation to another biosphere, or a permanent traveling biosphere itself. David Stephenson speculated that the latter vessel might grow during the journey, drawing on material along the way, adapting to conditions wherever it went.[66]

There is another significant implication of inhabited spacecraft that do not require habitable planets as their destinations: They do not need Sun-like stars. Once intelligent beings have become accustomed to living in space colonies, proposed Rood and Trefil, it makes no difference where the colonies are.[67]

We return to probability. Several analysts have given us mathematical models showing the average number of civilizations within each standard volume of space in our Galaxy, or the number of such civilizations per fixed number of stars. These calculations usually rest on an assumption that no civilization expands to other stars or to interstellar space. What if some do? This would change the demographics of technological intelligence.

It may be easier to detect an expanded civilization than one that remains confined to its biosphere of origin. If colonization is indeed the galactic norm, Jill Tarter recognized, then our immediate search strategies might profitably concentrate closer to home; the nearest transmitter is likely to be far closer to us than is the nearest site of indigenous extraterrestrial intelligence. Intercolony communication may offer the best possibility for detecting signals.[68]

Many technological societies may accept being limited to their biospheres of origin. Those that expand outward into the Galaxy may be a small minority—but a very influential one. Migrating civilizations may be the ones that we are most likely to encounter.

Searching for Artifacts

Electromagnetic signals might not be the only evidence of alien technological activities that we can detect. Recent years have seen growing interest in broadening our quest to include a Search for Extraterrestrial Artifacts (SETA).

Scot Lloyd Stride of the Jet Propulsion Laboratory described two predominant hypotheses as to how we can detect extraterrestrial intelligence:

a SETI Energy Hypothesis that states that a technologically advanced civilization uses electromagnetic energy as a means to remotely explore the universe and to detect or communicate with other advanced civilizations, and a SETI Artifact Hypothesis that states that a technologically advanced civilization has undertaken a long-term program of interstellar exploration via transmission of material artifacts.[69]

The authors of the SETI 2020 report recognized that we could search for alien artifacts or interstellar spacecraft. SETI has sought signals instead, not because these other approaches are without merit, but simply because in electromagnetic signaling the speed is very high and the cost is very low. We should keep our robotic eyes open for both, Tarter and Chyba proposed.[70]

Television journalist and novelist Richard Burke-Ward proposed an ambitious broadening of SETI to look for or contact alien artifacts. The search for alien machines might even be incorporated into the Drake equation. The probability of finding a functioning probe would depend on the prevalence of intelligent species in our galaxy, the likelihood of probes being sent to other stars, and the life spans of the probes (we might qualify this; a dead probe still could be detectable).

There are several possible forms of evidence: a beacon advertising the probe's presence; communications between drones; heat traces from propulsion or waste energy radiation; small objects near the Earth or other planets; space-time distortion; anomalous modulation of human communications systems; traces of construction on the Moon or other bodies.[71]

Scientists already have scanned the lunar gravitational libration (Lagrange) points for objects that might have been placed there to take advantage of those relatively stable positions. They found nothing larger than a few meters in size.[72]

Dyson pointed out that a starship braking from a high velocity would leave behind it a long straight trail of hot plasma that should be a source of persistent broad-band radio emission; radio astronomers could watch for tracks in the sky. The effects of more exotic propulsion systems such as matter–antimatter engines might be detectable over great distances.

Others believe that alien machines will be very difficult to detect. "No matter how awesome the starship might be in a terrestrial context," wrote Viewing and his colleagues, "in its own environment—interstellar space—it is virtually invisible."

The Clarks proposed looking for tritium that would be associated with fusion propulsion, gamma-ray emissions from a nuclear-powered probe, or gravitational waves from other propulsion systems. They cautioned that these techniques could detect only a small subset of possible alien probes.[73]

We may be able to detect only those probes employing technology not very far advanced beyond our own. The same may be true for our efforts to detect electromagnetic signals.

Mysterious Probes

We cannot assume that we will detect alien machines by their signals; probes might be designed not to initiate contact until they were detected. Or they might be uninterested in contact with humans, incapable of making contact in a way that humans understand, or actively avoiding contact. The best chance for successful contact, Burke-Ward thought, may be to develop a distinctive communications system that the probe can interpret as unequivocal proof that humans know that the probe exists.

What would be the purpose of an interstellar probe? Burke-Ward advanced three possibilities: data gathering probes that would collect information and transmit that to their home civilization; direct action probes that might be designed to intervene in the affairs of the system visited; sentient entities. However, a probe may not be here to study humans or the Earth. It may not consider humans to be intelligent—or it may have a profoundly different purpose.[74]

An interstellar probe would be obvious if it were a self-reproducing machine, Barrow and Tipler claimed; it would construct an artifact in our solar system. This object would be so noticeable that it could not possibly be overlooked.[75] As we do not see such an artifact, this perspective supports Tipler's view that we are unique.

Habitats that migrated to our solar system in the past could be drawing on its asteroidal or cometary resources. As our asteroid belt is an excellent source of raw materials, Papagiannis thought that it would be the best place to search for alien space colonies, which might reradiate in the infrared.[76] To date, searches have not detected any.

What if they do not wish to be seen? Matloff and Martin speculated that alien world ships could exist silently in our solar system, masquerading as asteroids or comets.[77]

Moravec suggested another reason for invisibility: Alien machines could be very small. Robots might miniaturize their operations until they are working with matter on a scale much finer than we can see. A wave of reorganization could pass through a solar system without leaving its native inhabitants any the wiser.[78]

We still would be looking for evidence of alien technology, not the aliens themselves. The two categories might merge in the form of a highly intelligent machine capable of autonomous operations, an artificial sentient being that might contact us on its own initiative.[79]

Detecting evidence of any nonhuman technology within our solar system, or anywhere in the vast spaces between the stars, would tell us that at least one civilization had achieved interstellar flight and is able to send robotic spacecraft or inhabited vehicles toward us. Contact could be direct.

The only way we can text the visitation hypothesis is by surveying the Sun's entire family. The field to be searched is vast: Our solar system extends far beyond the known planets. Astronomers already have found that the asteroid population reaches well above and below the disk containing the eight planets known before 1930 (Pluto's orbit is more inclined).

We may, perhaps unintentionally, create tools for this search as a spin-off from more conventional solar system exploration. Rees foresaw that huge numbers of miniaturized robotic probes will, 25 years from now, be dispersed throughout the solar system, sending back images of planets, moons, comets, and asteroids. This would be consistent with Dyson's principle for the planning of space operations: Every mission searching for evidence of life should have other exciting scientific objectives, so that the mission is worth flying whether or not it finds evidence of life.[80]

We cannot prove or disprove direct contact solely by theoretical arguments; we can only test it by observation. Cocconi and Morrison's rule applies. The probability of success is difficult to estimate; but if we never search, the chance of success is zero.

The Limits of Observation

We already have tested our ability to detect small artifacts on the surface of other planets, from spacecraft in orbit around them. This experience suggests that an initial detection of alien technology in our own solar system may be ambiguous.

Space scientist Michael Malin and his colleagues, analyzing imagery from orbiters observing Mars, found what they believed to be evidence of human technology on the Martian surface—the wreckage of the Polar Lander, its parachutes, and the scorch marks of its retro-rockets. Later observations failed to find the white dot assumed to be the lander; it may have been an artifact of the camera, a "noisy pixel."[81]

Once again, we were operating at the limits of our technologies. Once again, we may have seen what we hoped to see.

Astroarchaeology

The most probable scenario, at least during the foreseeable future, might be called Discovery Without Contact . . . we obtain unequivocal proof that intelligent ET's exist (or have existed), but in a manner that excludes communication.

—Arthur C. Clarke, 1993[82]

If there have been alien visitors to our solar system, we cannot assume that they are recent. A probe sent our way a million years ago might still be circling lifeless about the Sun. Finding such evidence in our own neighborhood could give us information about distant civilizations more quickly than searching for signals or waiting for alien messages.[83]

The authors of Project Cyclops acknowledged that we might stumble onto something of this kind while engaged in archaeological or astronomical research. They "felt" that the probability of this happening was extremely small.[84]

Physical artifacts may not be limited to spacecraft. MacVey suggested others: laboratories, radio relay stations, telemetry stations, marker beacons, monuments and edifices, implements, refuse, or environmental changes such as radioactive hot spots and paleomagnetic anomalies.[85] We might add graves.

If visitors left traces of themselves on the surface of the Earth, those artifacts may have been destroyed or buried long ago by geological change and erosion by water and wind. On some other solar system bodies—our Moon, the moons of Mars, the asteroids—artifacts might survive undisturbed over geological time.

A Mirror Image. We ourselves have left artifacts on the Moon—unintentional cairns of human technology—that will remain visible for millions of years even if our species goes extinct. Other human artifacts on Mars, Venus, and Titan may have shorter detectable lifetimes due to erosive processes.

Baxter proposed that traces might be found under Martian polar ice, including the trails of skis and sleds, artificial elements, or "unnatural" isotopic ratios. However, if alien visitors practiced the kind of planetary protection protocols that are to guide our own exploration of other planets, they may have left no trace at all. A human-made machine known as the Phoenix Lander, designed to settle on Mar's North polar cap in 2008, will be buried gradually under several feet of frozen carbon dioxide after completing its mission.[86]

What if aliens deliberately left evidence of their visits, as European explorers often did during their great age of exploration? Two obvious locations for a long-lasting monument in our solar system would be on one of the rocky bodies that lack an atmosphere (like Earth's Moon) or in orbit somewhere. Unless the latter emitted signals, it would be very difficult to find.[87]

In his science fiction story "The Sentinel" (one of the main inspirations for the film "2001: A Space Odyssey"), Clarke imagined an alien monument deliberately placed under the surface of our Moon. Humans find it only after they achieve space travel. This intelligence detector reports to its home civilization when humans interfere with it.[88]

Ground-penetrating radar mounted on orbiting spacecraft could open up new ways of looking for artifacts buried under the surfaces of solid bodies in our solar system. One device under development for a future

Mars mission may be able to penetrate as deeply as 1 kilometer, nearly two-thirds of a mile.[89]

Archaeology on Earth is growing more sophisticated in its ability to infer elements of intelligence from physical remains. It is not just a search for evidence of humanity, argued anthropologist Paul Wason of the Templeton Foundation, but for evidence of intelligence, intention, or purpose. Part of being an agent with a plan is a deeply evolved ability to recognize the work of other purposive agents—whether in sophisticated electromagnetic wave patterns or in an ancient garbage dump.[90]

The Monuments of Mars

Like other approaches to the search for extraterrestrial intelligence, astro-archaeology is subject to being discredited by extravagant claims. A prime example is the monuments of Mars.

In 1972, the *Mariner 9* spacecraft photographed Martian features that bore some similarity to pyramids. More provocatively, one of the *Viking* spacecraft orbiting Mars in 1976 imaged a butte in the Martian region known as Cydonia that seemed to resemble a huge human face staring up at the sky. Freelance science writer Richard Hoagland, who discovered the image in 1981, promoted the idea that this was evidence that intelligent life had existed on Mars.[91]

When the Mars Global Surveyor's Orbital Camera imaged the Cydonia region in 1998, the Face appeared to be nothing more than a jumble of rocks, peaks, and ridges, its resemblance to a human visage caused by a coincidental alignment of sunlight and camera angle. Although this discredited the idea that the Face had been constructed by intelligent beings, there was an interesting implication: This was the first time NASA programmed a planetary probe to search for possible evidence of an alien civilization.[92]

Some advocates remain undiscouraged. A small group of researchers known as the Society for Planetary SETI Research has been studying images of Mars since the 1970s, searching for surface anomalies that might have resulted from intelligent activity (they published *The Case for the Face* in 1998).[93]

Such humanoid "faces" are images of ourselves transplanted to an alien environment—a product of the human hunger for patterns that we recognize. If aliens ever did build a face on Mars, it would be more likely to look like them than like us.

Mind-Stretcher. Confronted with the greatest works of past civilizations, humans often have believed them to be beyond the powers of ordinary mortals. Some have attributed them to superior beings; the Aztecs thought that the dead city of Teotihuacan had been built by gods. If we encounter

**artifacts of an alien civilization, we too might be awed by the works of
superior beings.**

Ancient Visitors to Earth

Sagan proposed in 1963 that extraterrestrial civilizations exploring astronomical objects and their inhabitants would sample planets with intelligent
life every 10 thousand years and those with advanced civilizations every
thousand years. Consequently, there is a statistical likelihood that the
Earth was visited at least once during historical times. Archaeologists
should be attuned to the possibility that they might uncover extraterrestrial
artifacts on the Earth.[94]

Although he thought that the prospect of extraterrestrial visitation to
our contemporary civilization was dim, Sagan wrote later, he still believed
it possible that alien visitors had journeyed to Earth in prehistoric times.
Scientists should examine ancient myths and legends for evidence of
contact.[95]

Erich von Daniken expanded on this idea, claiming that the gods of
many ancient cults and religions may have been extraterrestrial visitors. In
his wildly successful book *Chariots of the Gods,* he asserted that there was
good reason to trust stories about gods who came from on high, walked
the earth, and freely distributed moral and technical advice to human
beings. In his view, the God of the Bible was a primitive people's image of
an alien being.[96]

Many critics ridiculed von Daniken's theories, arguing that they involve
misinterpretations of archaeological data, plus a large dose of fabrication.
Why were these ideas so popular? Psychologist John Baird suggested that
alternative explanations of a bizarre and unusual kind are seen as more
valid when applied to ancient events than when applied to current events
of a similar level of complexity.[97]

Despite deep skepticism about theories like von Daniken's, folk memories cannot be dismissed totally. Sagan found a plausible contact myth in
some of the stories surrounding the origins of the Sumerian civilization in
the fourth millennium B.C. That myth evolved into a religion that was
almost totally oriented toward the stars. The contact myth did not die in
Sumeria or Mexico; it continued to be told and experienced in medieval
Europe and persists even today.[98]

Consider the example of a French expedition's encounter with the Tlingit
tribe in 1796. The tribe's oral tradition contained sufficient information for
later reconstruction of the true nature of the contact. Many of the incidents
were disguised in a framework of mythology; the French ships were
described as large black birds with white wings. To say that all past-contact
inquiry is bad scholarship, argued journalist Chris Boyce, is like saying that
Schliemann was wrong to look for Troy on the basis of myths.[99]

Von Daniken was not alone; other authors have suggested that alien visitors influenced our past by granting knowledge that accelerated our development. Some find it difficult to believe that we could have arisen from ape-like ancestors to create technologies and civilizations without outside help. Such theories reflect a low opinion of human abilities.

Some people believe that extraterrestrials are resident on the Earth today, detectable only through physical and psychological profiling. According to physicist Robert Ehrlich's book *Eight Preposterous Propositions*, 45% of the U.S. population are in no doubt that extraterrestrial beings have been stalking our planet.[100]

Nearly all scientists reject this idea, asserting that there is no confirmed evidence of past or present extraterrestrial visitations to the Earth. Nonetheless, one physicist argued that there is a nonvanishing chance that very advanced civilizations are able to visit Earth, and that they might already be here.[101]

This brings us to the most popular and most controversial image of contact. In many minds, speculations about interstellar exploration and colonization by another species are linked with the belief that some Unidentified Flying Objects are vehicles from other worlds.

The UFO Controversy

I shall not commit the fashionable stupidity of regarding everything I cannot explain as a fraud.

—Psychologist Carl Jung[1]

Rocks from the Sky

For centuries, ordinary people reported stones falling from the sky. Scientific authorities rejected such claims as preposterous. Rocks could not fall out of the sky, as there were no rocks in the sky.

In what may be the first documented case, people in fields near Luce, France, saw a stone mass drop from the sky after a violent thunderclap in 1765. Seven years later, the French Academy appointed an investigating committee including Antoine Lavoisier, now regarded as the father of modern chemistry. In his report to the Academy of Science, Lavoisier stated that his analysis "absolutely proved" that the stone had not fallen; it had been heated when struck by lightning. Those who reported the fall must have been mistaken.

The academy went on record, denying that meteorites had an origin outside the atmosphere. Reports of witnesses were altered to conform with accepted theories. Museum keepers threw away meteorites lest they be accused of clinging to foolish superstitions. Not until 1803 did the French Academy of Science accept the reality of meteorites, and then only after the fall of over 2000 stones at L'Aigle made it impossible to deny.[2]

Astronomer J. Allen Hynek described the UFO phenomenon as a parallel case. Tens of thousands of people have reported unexplained objects in our skies. The numbers of sightings may be far greater than the numbers on record; the U.S. Air Force estimated in the early 1950s that only 1 out of 10 people who had seen UFOs actually reported their experiences.[3] Those who did recount sightings included scientists, professors, air-traffic controllers, engineers, pilots, military personnel, and police officers—people who would be considered credible if they reported more familiar phenomena. Yet, there remains a vast credibility gap. Most scientists either dismiss the UFO phenomenon or avoid the subject because of potential damage to their careers.

Hynek, who served for years as a consultant to the Air Force, observed that data on the UFO phenomenon had to run an insidious gauntlet that meteorites had been spared. Discoveries of meteorite falls did not become material for cultists and pseudo-religious aberrants; meteorites were not regarded as sent by alien intelligences. Nobody concocted stories about riding meteorites to other planets or meeting perfected beings. The people who generated such stories, although few in number, were vocal and insensitive to ridicule; they were given ample press and often generated a cultist following.

Hynek believed that the controversy over evolutionary theory was the only other one in which basic emotional responses, buttressed by deep-seated religious and personal prejudice, played so major a role. However, the flow of opinion was reversed. The concept of biological evolution first was slowly accepted at the top echelons of biological science before filtering down to popular levels, where the greatest emotional responses were generated. By contrast, the UFO phenomenon arose at the grass roots level. "It was the highest scientific echelons that generated the emotional storm against allowing unprejudiced examination of observations made by persons judged sane by conventional standards," wrote Hynek. "The UFO evidence has not been properly presented at the Court of Science."[4]

The scientific search for extraterrestrial intelligence provides an intriguing parallel with evolutionary theory. SETI, originating among a small number of scientists and being slowly accepted by a larger number of their colleagues, has been a top-down phenomenon that has won popular support. However, much of the public already was predisposed to believe in a plurality of inhabited worlds. Many also were prepared to believe that more technologically advanced beings could visit the Earth.

To the general public, UFO sightings and tales of contact with extraterrestrials are the primary introduction to the question of alien minds. Laymen, including influential ones, still link SETI and UFOs. Sociology professor David Swift proposed in 1982 that NASA recognize this connection by taking on SETI and UFOs as a package; not surprisingly, NASA has not done so. Yet, as recently as 1999, *The Complete Idiot's Guide to Extraterrestrial Intelligence* restated that linkage, focusing on UFOs and visitors rather than on SETI.[5]

Ancient Sightings

If you believe that democracy is the way to decide issues . . . you will not doubt that many kinds of aliens have visited us for a great variety of reasons.

—Jack Cohen and Ian Stewart, 2002[6]

Flying objects that we cannot identify did not pop into our vision for the first time in 1947. People have seen mysterious things in the sky throughout history, although they called them by different names.

Reports of flying craft and humanlike occupants can be traced back into antiquity, where they merge with religion, myth, or superstition. A venerable Chinese tale speaks of a far-off land of flying carts inhabited by one-armed, three-eyed people riding winged chariots. A Sanskrit text describes aerial dogfights among gods piloting flying machines. Ancient Egyptians recorded sightings in hieroglyphics. Citizens of the Roman Empire saw airborne vehicles; Livy reported phantom craft in the sky.[7]

Perhaps the most oft-quoted description from ancient times is found in the Old Testament's Ezekiel 1, which may have been written around 590 B.C. Here is one translation, edited for length:

A stormy wind came out of the north, and a great cloud, with fire flashing forth continually, and a bright light around it . . . in the midst of the fire something like glowing metal. And in the midst of it were figures resembling four living beings . . . they had human form. Each of them had four faces and four wings . . . they gleamed like burnished bronze. . . . under their wings on their four sides were human hands. . . . In the midst of the living beings there was something that looked like glowing coals of fire, like torches darting back and forth among the living beings. . . . lightning was flashing from the fire. And the living beings ran to and fro like bolts of lightning. . . . there was one wheel on the earth beside the living beings, for each of the four of them. . . . they were lofty and awesome, and the rims. . . . were full of eyes. And whenever the living beings moved, the wheels moved with them. And whenever the living beings rose from the earth, the wheels rose also. . . . over the heads of the living beings there was something like an awesome gleam of crystal. . . . above the expanse was something resembling a throne. . . . high up was a figure with the appearance of a man. . . . when I saw it, I fell upon my face and heard a voice speaking.[8]

Astronomer John MacVey concluded that, even clothed in its quaint Biblical language, this is clearly a description of an aerial craft.[9] Others disagree, seeing this vision as a supernatural experience or a hallucination. Its meaning seems to lie in the mind of the reader.

As is the case for other sightings of strange aerial phenomena far in the past, we do not have the evidence to prove or disprove the theory that Ezekiel saw a UFO. One investigator concluded that all accounts of UFO-like sightings handed down through the ages are doubtful—until verified.[10]

What may have been the world's first military investigation of UFOs occurred in Japan in 1235, when a warlord observed lights circling in the night sky. He ordered learned members of his entourage to study this phenomenon and issue a report (they concluded that the wind was making the stars sway).[11]

The Great Airship Mystery

In the United States, the first widely known wave of sightings of strange objects in the sky occurred in late 1896 and early 1897. Thousands of people from large areas of the central and western states reported that they had seen airships. Some claimed to have interacted with crews, who were humans speaking English; a few even were invited aboard. Many of these observers described the airships as dirigible-type machines driven by a motor attached to an air screw or propeller. This coincided with the then current opinion that the first flying machines would be airships, not heavier-than-air machines like the Wright Brothers' aircraft that first flew in 1903.

The most common explanation for these sightings was that a secret inventor had developed the airships. A dirigiblelike balloon had flown over Paris as early as 1852, and an American inventor went aloft in a similar craft in 1865. Jules Verne's novel *Robur the Conqueror*, published in 1886, described a globe-girdling aerial craft, although it was heavier than air. A less popular theory suggested that the airships were extraterrestrial visitors, the most frequent source being Mars.

According to historian David Michael Jacobs, those who had not seen these airships simply would not believe these reports. They assumed that the witnesses did not see what they claimed to see. By contrast, no amount of persuasion could dissuade the people who had sighted such objects from believing that they had seen an airship.

These patterns of behavior were repeated in the modern era of UFO sightings that began after World War II. However, we should be cautious about lumping the two eras together and assuming the same explanation for both waves of sightings. Although some reports from the 1896–1897 wave were hoaxes, research has shown that most witnesses really were seeing airships, products of our own civilization.

Electrical engineer Michael Busby, who studied hundreds of press reports and mapped the sightings, found a growing body of evidence that a group of individual Americans secretly designed, built, and flew airships as early as the 1840s. Their work led to the manufacture and flight of the several airships sighted in 1896 and 1897. This project may have been quietly supported by the U.S. Government.

Busby tells us that the mood of that time was not of anxiety or paranoia (a frequent although unproven explanation for UFO sightings during the Cold War). It was a time of optimistic anticipation, of belief in the creative powers of science and technology.[12]

What may have been the most widely witnessed sighting took place in 1917 at Fatima in Portugal, where a crowd of 50–70 thousand people watched as the clouds parted to reveal a huge silver disk spinning like a windmill. The object, which emitted heat, plunged toward the Earth before climbing back into the sky and disappearing. (Others saw this as the Sun twirling in the sky, throwing off colors and descending to the Earth)[13] If this was a hallucination or a religiously inspired vision, it was a massively shared one.

During the first third of the twentieth century, Charles Fort published collected reports of peculiar sightings, some of which now would be categorized as UFOs. Referring to conventional scientific wisdom, he wrote in 1919 that "our data have been damned, upon no consideration for individual merits, but in conformity with a general attempt to hold out for isolation of this earth." Fort argued that the notion of things dropping in on our planet from "externality" was unsettling to science; the scientific attitude toward the unwelcome is that it does not exist.[14]

During World War II some military pilots described glowing objects flying beside their aircraft, although these "foo fighters" did not become widely known until later. Scandinavians reported many sightings of "ghost rockets" immediately following the war.[15]

In the United States, the modern wave of UFO sightings began in 1947 when pilot Kenneth Arnold reported that he had spotted disk-shaped objects flying in loose formation, making undulating motions like a saucer skipping over water. Newspaper writers coined the term *flying saucer*, which allowed people to place inexplicable observations in a new category. The term also implied a tone of ridicule.[16]

The Arnold story encouraged people all over the United States to come forth with their own tales of strange objects in the sky. Noting the anthropological theory that the acceptance of an idea depends as much on the state of society as on the idea itself, planetary scientist William Hartmann argued later that the reaction to Arnold's report can be explained by the fact that postwar society was primed for the acceptance of alien spaceships.[17]

Others came to different conclusions. Almost everyone then assumed that UFOs were real but easily explained as something other than alien craft. According to a 1947 Gallup Poll, most people thought these objects were illusions, hoaxes, secret weapons, or other explainable phenomena; very few thought they came from outer space.[18]

The Air Force Investigates

To study the problem further, the U.S. Air Force established Project Sign in December 1947 (Project Sign later became Project Grudge, then Project Blue Book). By the end of that year, Air Force investigators were secretly moving toward the extraterrestrial hypothesis while publicly dismissing saucers as natural phenomena or hoaxes.[19] Official opinions later shifted away from the visit scenario.

Given the Cold War, continued military intelligence control of the investigation seemed natural. Jacobs observed that this may have inhibited the scientific community from conducting its own study. If the Air Force had released the UFO data and encouraged all scientists to look at its files, commented Hartmann, the UFO mystery probably would have been clarified after a few months of scientific and public excitement.[20]

From 1947 to 1964 the controversy raged within the confines of special interest groups—the Air Force on one side and private UFO groups on the other. The press, public, and Congress became involved only sporadically. Nonetheless, the urge to explain generated a voluminous literature, ranging from Donald Keyhoe's best-selling books claiming UFOs were extraterrestrial visitors to Philip Klass' debunking works arguing that they were misinterpretations of known phenomena.

Meanwhile, hoaxes were damaging the credibility of legitimate UFO research. The Clarks later observed that the intellectual vacuum left by science was all too quickly filled by the unscrupulous purveyors of fantasy.[21]

The motion picture industry began to capitalize on public interest in UFOs in 1951, portraying aliens as both benevolent and dangerous. "The Day the Earth Stood Still" presented an attractive humanoid from a utopian planet who landed his saucer in Washington, D.C. to warn Humankind against its warlike tendencies. This film, sometimes described as the first to treat the idea of extraterrestrials seriously, showed the alien as both ethical and potentially threatening—and able to walk through human cities without being recognized. Another 1951 film, "The Thing from Another World," portrayed the survivor of a saucer crash as intelligent but ready to destroy humans who stood in the way of his seeding the Earth with his offspring. Both of these visions of aliens remain with us today.

A scientific panel convened by the CIA in 1953 recommended that national security agencies strip UFOs of the special status they had been given and the aura of mystery they had acquired. During the same year, skeptical astronomer Donald Menzel became the first American scientist to publish a book on UFOs. He dismissed as ludicrous the idea that UFOs represented extraterrestrial intelligence. People who accepted this idea, he insisted, were lunatics, cultists, religious fanatics, or, at best, frightened and confused.[22]

This is strikingly similar to criticism of early rocket propulsion advocates. After David Lasser's book *The Conquest of Space* was published in 1932, physicist H.H. Sheldon said that "all rocket enthusiasts are mental defectives."[23]

Hynek thought that it was destructive to sneer at reported sightings. "Ridicule is not part of the scientific method," he said at a symposium, "and people should not be taught that it is."[24] In a modern parallel, some critics of SETI also have employed ridicule, drawing on such UFO-related clichés as "little green men."

The 1950s produced the contactees, people who claimed that they had met aliens and had even been given rides in flying saucers. According to the contactees, space people (who looked like humans) came from planets free from war, poverty, or unhappiness. These angellike beings wanted to prevent war, stop nuclear testing, and help us build the kind of utopian society that they enjoyed. Some flying saucer clubs of the time were contactee-oriented, even though many contactee claims were disproved. People with little knowledge of the UFO phenomenon constantly confused the lunatic fringe with serious UFO investigators.[25]

The rise of contactees and flying saucer movies came at the same time as increased secrecy in the Air Force's handling of UFO reports. Although the contactees and the movie industry gave the UFO phenomenon publicity the Air Force wanted to avoid, their focus on the sensational and fantastic lent credence to the Air Force's denials of an extraterrestrial origin for UFOs.

Condon the Denier

After an upsurge in sightings in 1965, the press, the public, the U.S. Congress, and the scientific community all entered the controversy. A continued public uproar led congressmen (Gerald R. Ford among them) to hold the first open hearing on the subject in April 1966. Hynek's testimony criticized the Air Force approach, calling for a panel of civilian scientists to determine if a major problem actually existed.

A panel including Carl Sagan, concluding that no UFO case represented technological or scientific advances outside of a terrestrial framework, recommended that the Air Force contract with universities to study UFO sightings. The University of Colorado accepted the project, placing Dr. Edward Condon in charge.

Condon released his committee's report in January 1969. In his summary, which ignored or overruled some of his staff's views, he declared that his study had found no direct, convincing evidence for the claim that any UFOs represent spacecraft visiting Earth from another civilization. He went beyond that to assert that the great distances and times involved in interstellar travel made contact between different solar systems impossible (a position shared by many SETI advocates). The Condon report did acknowledge that people who reported UFOs were normal, responsible individuals and that some sightings were difficult to explain by conventional means.[26]

As Hynek saw it, the Condon report confused the UFO problem with the extraterrestrial hypothesis. The issue was not the validity of that hypothesis, but the existence of a legitimate UFO phenomenon regardless of theories about its origin. Just as nineteenth-century scientists could not explain the aurora borealis, UFOs might be inexplicable in terms of twentieth-century physics.[27]

Hynek again attacked Condon's approach in 1975. Instead of asking what the overall, observed nature of the UFO phenomenon was, Condon set out to test the hypothesis that UFOs were visitors from outer space. No attempt was made to find patterns or relationships among the thousands of cases from all over the world. "This would be like asking," Hynek wrote, "whether the Northern Lights represented interstellar communications, and concluding that since the data did not support that hypothesis, the Northern Lights were hallucinations, hoaxes, or sheer imagination."[28]

Some supported the Condon report's conclusions. *The New York Times* commented that its authors were courageous because they discounted a growing religion. *The Nation* agreed with Condon's recommendation to keep school children from reading about UFOs and getting a warped view of science.

Others remained unconvinced by Condon's summation. The American Institute of Aeronautics and Astronautics issued a study challenging the report's conclusion that nothing of scientific value could come from further study of UFOs. The AIAA found it difficult to ignore the small residue of well-documented but unexplainable cases that form the hard core of the UFO controversy.[29]

Years later, physicist Peter Sturrock pointed out that Condon had been open to the possibility that other scientists might at a later date come up with good plans for UFO research and had even advocated that such plans should be funded. "All of the agencies of the federal government, and the private foundations as well," wrote Condon, "ought to be willing to consider UFO research proposals along with others submitted to them on an open-minded, unprejudiced basis."[30]

Meanwhile, the Air Force got what it wanted, ridding itself of the issue. On December 17, 1969, the Secretary of the Air Force announced the termination of that service's 22-year study of UFOs.

At a scientific meeting that year, Sagan took a more open-minded position: There was insufficient evidence to exclude the possibility that some UFOs were space vehicles from advanced extraterrestrial civilizations, although the insignificance of our own civilization and the vast distances between the stars made the extraterrestrial hypothesis unlikely. Sagan asked the rhetorical question: Why is that theory so popular? He speculated that there were four "resonances": religious connections, the relief of boredom by believable novelty, military classification, and intolerance of ambiguity.[31]

One might add a broader reason. The desire for contact that underlies the scientific search for signals can be expressed in other ways, including the expectation of visitors.

A November 1973 Gallup Poll showed that 51% of adult Americans believed UFOs were real. Eleven percent—a projected 15 million people—said they had seen one. Here, in Jacobs' words, was a phenomenon that virtually the entire adult population had heard about and that millions of people claimed to have seen, yet no one knew for sure what it was.[32]

At the Highest Level

During his presidential campaign in 1976, Jimmy Carter revealed that he had seen a UFO in 1969. He vowed that if he became President he would make public every piece of information that the United States had about UFOs.

After Carter took office in 1977, White House Science Adviser Frank Press recommended that NASA form a small panel of inquiry to see if there had been any new significant findings on UFOs since the Condon Report. NASA Administrator Robert Frosch replied that while NASA was willing to continue responding to public inquiries, he recommended taking no steps to establish a research activity in this area or to convene a symposium on this subject.[33]

Conspiracies, Crop Circles, and Abductions

The last great wave of UFO sightings took place in 1973. After that, most reports of extraterrestrial visitors took other forms.[34]

Extravagant claims and hoaxes continued to damage the credibility of UFO studies. One famous example is the alleged UFO crash near Roswell, New Mexico in 1947. Some claimed that the wreckage included the bodies of aliens and that the UFO and its crew were taken to U.S. government facilities for further studies. Most detached observers now believe that this conspiratorial scenario was generated by a military cover story intended to disguise the wreck of a classified U.S. balloon system.[35]

Another focus for conspiracy theories has been Area 51, a U.S. Air Force base at Groom Lake, Nevada. Since the base was established in the 1950s to test the U-2 spy plane, its cloaked existence has fueled speculations about more exotic activities such as studying the technology of UFOs. A 1997 Air Force report stating that there were no captured aliens has not deterred such allegations.[36]

Two other phenomena generated widespread media coverage and public interest. One was crop circles, which began appearing in English grain fields in 1978 and later spread to other countries. The simplest patterns—which initially may have been caused by atmospheric vortices—later diversified into many shapes in addition to circles.

Allegedly, some of these markings were caused by UFOs. Yet, as skeptics pointed out, their number and complexity seemed to increase with media coverage. Several hoaxers eventually confessed that they made crop circles; two said they created more complex patterns to prove that an intelligence must lie behind them.[37]

The other phenomenon was reports of abductions. Many people claimed to have been seized by aliens and taken to their spacecraft, where medical experiments were performed on them. This actually was not new; reports of abductions had appeared as early as 1929.[38]

Harvard University psychologist John Mack interviewed more than 70 of the abductees. Finding that their stories were remarkably consistent, he concluded that those people had undergone traumatic experiences that could not be explained by known causes. Mack argued that abductions force us to reconsider our perception of reality; "no familiar theory or explanation has come even close to accounting for the basic features of the abduction phenomenon." As abductions cannot be understood within the framework of Western science, a new scientific paradigm may be necessary. Mack concluded that "abductions have to do with the evolution of consciousness and the collapse of a worldview that has placed humankind at a kind of epicenter of intelligence in a cosmos perceived as largely lifeless and meaningless." He speculated that the aliens might be from other dimensions.[39]

Jacobs conducted his own research into abductions. After 325 hypnosis sessions with 60 abductees, he concluded that an alien form of life was using his subjects to produce another form of life—a secret life. According to one survey, 4 million Americans claimed to have had abduction experiences.[40]

Another Harvard psychologist, Susan Clancy, later proposed that the abduction phenomenon is due to a confluence of multiple factors—sleep paralysis, interest in the paranormal, hypnotherapy, memory tricks, and emotional investment. Alien abduction stories may give people a deep sense that they are not alone in the universe; their memories resemble transcendent religious visions, scary and yet somehow comforting. Like the most credible UFO witnesses, these people could not be dismissed as ignorant or crazy.[41]

Skeptical psychologist John Baird noted that the alien creature in these accounts usually stands upright on two legs, has a single head with a pair of eyes in front, and always demonstrates full command of the language native to the region.[42] Most observers doubt that the abduction phenomenon involves extraterrestrials; these experiences may stem from mental events that we do not fully understand. They may prove nothing about contact with other civilizations.

UFO visits remained a popular subject for science fiction, particularly in film and television. Steven Spielberg's films "Close Encounters of the Third Kind" and "E.T." portrayed extraterrestrials who came to the Earth in interstellar spacecraft as harmless, even charming. Other media presentations such as the film "Independence Day" and the television series "V" showed alien visitors as vicious conquerors. UFOs, extraterrestrials, abductions, and alien interference in human affairs were staples of the long-running television series "The X-Files."

The dialogue of the deaf continues to the present day. UFO advocates continue to claim that governments know UFOs are alien spacecraft but will not release that information to the public. Governments, to the extent that they respond to such allegations, deny them. Meanwhile, several polls have suggested that a majority of Americans continue to believe that flying saucers exist and are here. The proportion of believers seems to be less among Europeans, but still is high.[43]

UFOs and SETI

SETI scientists have resisted any linkage between their search and the UFO phenomenon, seeing that as a serious threat to the credibility of their enterprise. They generally reject the idea that UFOs are extraterrestrial visitors, or keep quiet about the issue.

Jill Tarter showed how sensitive this was when she wrote a blistering letter to a newspaper that had paired a story about SETI with a feature about a UFO "nut case." Emphasizing the effort expended over two decades to distinguish the science of SETI from the pseudo-science and outright fraud that characterizes the UFO community, she declared that such irresponsible journalism can threaten federal funding.

In 1981, Frank Drake asserted that there is no evidence that any UFO is the product of another civilization. He went beyond that 17 years later. If UFO reports are real, he said, they must be due to extraterrestrial spacecraft. As interstellar travel is impossible, UFO reports may be discounted.

By keeping extraterrestrial intelligence far away from us, SETI researchers avoid public attention that might attract the lunatic fringe. To some extent, this strategy has worked. By contrast with UFO studies, SETI is seen as a middle-of-the-road scientific activity.[44]

There is a down side. In its eagerness to avoid being tainted by the UFO issue, the SETI community has excluded any form of direct contact from its calculations. Many refuse even to acknowledge that the possibility of such contact should be studied objectively. Other, equally intelligent and scientifically trained people protest that contact through interstellar probes is as credible as helpful messages from the stars.

Fact, Speculation, and Disinformation

Psychiatrists Lester Grinspoon and Alan Persky were impressed by the unusual emotions exhibited by both witnesses and interpreters of UFO phenomena. Unlike most scientific topics, UFOs seemed to rouse a fervor usually reserved for politics, morality, or religion.[45] Those who debate the existence of extraterrestrial intelligence sometimes display similar zeal.

Two issues have complicated the debate: witness reliability and the bias of the investigator. Ardent believers and hard-core skeptics will reach different conclusions about a case that is ambiguous, as most UFO sightings are.[46] Yet, the most articulate reports came from obviously intelligent people, including some observers who have seen UFOs at close range and reported their sightings in detail.

Sociologist Robert Hall cited the parallel case of Galileo's telescope. After Galileo discovered the moons of Jupiter, many people refused to look through the instrument. They "knew" that there could not be such bodies around Jupiter; therefore, they "knew" that the telescope was deceptive.[47]

Goldsmith and Owen were more skeptical. They argued that reports by people who claim to have seen extraterrestrial visitors or their spacecraft are in the least credible category and must *always* be subject to error (emphasis added). To explain the many UFO sightings reported each year with the spacecraft hypothesis requires the assumption that we are doing something special to draw attention.[48]

Science writer James Gleick found that unbelievable: "How infinitely unlikely it is that our corner of the universe should be receiving alien visitors in such strikingly near-human form at just the eyeblink of history when we have discovered space travel." Bracewell added a practical objection. No doubt it would be feasible to design an elaborate probe that can descend to a planetary surface, but that would not be an optimum search strategy.[49]

In the murky world of ufology, said UFO researcher Jenny Randles, it can be difficult to know whether to take some things as fact, speculation, or disinformation. She saw this mystery as a powerful tool that can be used to manipulate public opinion. The UFO phenomenon provides a perfect smokescreen if you want to test fly a secret aircraft; people will report what they see as an alien spaceship, and most of the world will dismiss that possibility. (In 1997, a declassified CIA report stated that over half of all UFO reports from the late 1950s through the 1960s were accounted for by manned reconnaissance flights.)

Randles speculated that alien visitors often have served as scapegoats for some very terrestrial events. She also suggested that there is a huge social and financial incentive to maintain the illusion of alien invaders.[50]

Real But Unknown Phenomena?

Hynek's 1972 book *The UFO Experience* brought the study of UFOs to a new level of sophistication. While making clear his disenchantment with the extraterrestrial hypothesis, he concluded that there exists a phenomenon that is worthy of systematic, rigorous study. The body of data points to an aspect or domain of the natural world not yet explored by science. Investigations that have sought to disprove these points of view (such as Project Blue Book and the Condon Report) have failed to make the case.

Hynek recommended that scientists and engineers establish a loosely knit institute for the study of the UFO phenomenon (he established the Center for UFO Studies a year later). He also recommended that a member country of the United Nations propose that the United Nations set up a committee to facilitate communications on this subject.

Hynek acknowledged that, after 20 years of association with the problem, he still had few answers and no viable hypothesis. "When the long awaited solution to the UFO problem comes," he wrote, "I believe that it will prove to be not merely the next small step in the march of science but a mighty and totally unexpected quantum jump."[51]

Hynek returned to these themes in 1975. There remains a profoundly impressive body of data that constitute a new empirical set of observations. They may signal a whole domain of nature as yet unexplained. "Nothing that intrigues the mind of Man," he wrote, "is automatically ineligible for the scientific approach."[52]

Hynek reminded his readers that twentieth-century scientists tend to forget that there will be a science of later centuries whose knowledge of the universe may appear quite different. "We suffer, perhaps, from temporal provincialism, a form of arrogance that has always irritated posterity."[53]

Psychological Explanations

If we peer through the looking-glass of unidentified airships, mutilated cattle, and crop circles, we see reflected back at us the wondrous darkside of collective hopes and fears for the future of our species.

—Randall Fitzgerald, 1998[54]

Many people have speculated that UFOs are a psychological phenomenon. Yet, UFO witnesses stand convinced of the external causes of their perceptions. Psychologist Baird concluded that most people claiming to have seen UFOs were telling the truth: They saw something exceptional in the sky. From their vantage points, what they saw was inexplicable.[55]

Hynek found that persons reporting a UFO sighting first tried to fit their observations into familiar categories, coming to regard the phenomenon as strange and unidentified only after its appearance and actions seemed to rule out familiar interpretations. This is quite contrary to assertions made by UFO skeptics, who claim that witnesses are eager to find something strange. When reasonable people report events that receive no social support from their friends and do not fit their own prior beliefs, we have to take these reports seriously.

Hall dismissed the argument that UFO sightings are due to hysterical contagion, as such events last only a few days, at most a few weeks, and tend to be highly localized. Although some "hard–core" UFO reports stand up better than many a court case, Hall acknowledged that some UFO buffs load their reports with interpretation. However, some scientists caught up in the UFO controversy defend their positions with more emotion than logic. Faced with detailed reports by reliable witnesses, they loudly and confidently assert interpretations that conflict strongly with available testimony and show a startling degree of disrespect for the reason and common sense of intelligent people.

Hall concluded that reports as persistent and patterned as hard-core UFO sightings must be systematically motivated in some way. Either there must be a distinctive physical phenomenon that witnesses have observed, or there must be a powerful and poorly understood motivation. He found it more plausible to believe that there is a physical stimulus than to believe that multiple witnesses misperceive in such a way as to make them firmly believe they saw something that jars their own beliefs and subjects them to ridicule from their associates.

To Hall, the very strength of our resistance to UFO evidence suggested that there clearly was a phenomenon of surpassing significance. That phenomenon was going to force some of us to make fundamental changes in our knowledge, a good definition of scientific importance. The arguments are about who has to change.[56]

A Jungian Analysis

Eminent Swiss psychologist Carl Jung published a book in 1959 entitled *Flying Saucers: A Modern Myth of Things Seen in the Skies.* He was in no doubt regarding their objective reality. "Either psychic projections throw back a radar echo," he wrote, "or else the appearance of real objects affords an opportunity for mythological projections." Jung suggested that UFOs might be real physical phenomena of an unknown nature.

Addressing people's need to believe in flying saucers, Jung interpreted the UFO phenomenon in the context of the Cold War. In a dark time for humanity, a miraculous tale grew up of an attempted intervention by extraterrestrial "heavenly" powers—and this at the very time when humans were seriously considering the possibility of space travel and of visiting other planets. The world situation was calculated as never before to arouse expectations of a redeeming, supernatural event.

Jung believed that UFOs had become a living myth. "Just at the moment when the eyes of mankind are turned towards the heavens, partly on account of their fantasies about possible spaceships, and partly ... because their earthly existence feels threatened, unconscious contents have projected themselves on these inexplicable heavenly phenomena and given them a significance they in no way deserve."[57]

Myths Past and Present

After studying ancient folklore and religious belief systems in the context of modern UFO reports, researcher Jacques Vallee concluded that contemporary accounts of UFOs and their occupants are merely the modern variant of a complex of experiences that infuse the folk memories of all cultures. He noted a remarkable similarity between reports of UFO occupants and reports of fairy sightings of an earlier age.[58]

People in ancient Greece, Rome, or medieval Europe did not talk about spaceships or extraterrestrials, observed White. The appearance of lights or phenomena interpreted as objects in the sky were not in general associated with visitors or fantastic creatures, but with religious beliefs; they were treated as manifestations of supernatural forces. Physicist Dyson dismissed UFOs as "mythical animals," the contemporary reincarnation of phoenixes and unicorns.[59]

Davies—like Baird and Hall—concluded that, in most cases, UFO witnesses are sincerely reporting genuine experiences. Those experiences are largely subjective and reflect deep-seated human desires and anxieties of a quasi-spiritual nature. "What we see in the UFO culture," he posited, "seems to be an expression in the quasi-technological language appropriate to our space age of ancient supernatural beliefs, many of which are an integral part of the folk memories of all cultures."[60]

Even though the facts may not support actual contact with aliens, the desire to connect remains intriguing. Steven Dick suggested that UFOs, along with science fiction, are a way of working out the biological universe world view in popular culture. These stories of mythic proportion broaden our horizons. They force us to consider our place in the universe; they make us wonder whether that universe is full of good, as in "E.T.," or evil, as in the Alien film series.[61]

Aerospace historian Curtis Peebles, noting that earlier myths had addressed interactions between humans and humanlike supernatural beings, also thought that the flying saucer myth was an attempt to find a relationship with extraterrestrials. "The function of mythology is to allow a society to relate to the larger world," he wrote. "This has not changed." Grinspoon was more dismissive, arguing that "these fantasies carry an unrealistic expectation that our interactions with aliens will be safely within the realm of our previous experiences, including, especially, our TV watching experiences."[62]

Psychology professor Albert Harrison occupied the middle ground. UFO reports represent a mixture of what actually happened with imperfections in our perceptual and memory processes, our personalities, attitudes, and beliefs, and the influence of our friends, acquaintances, and the culture in which we live.[63]

As a result of science's abdication from any area that might be considered New Age, Vallee claimed, charlatans and hoaxers are given a free hand. However, as Brin remarked, aversion to an idea simply because of its association with crackpots gives crackpots altogether too much influence.[64]

The Galilean Test

I do not think the evidence is at all persuasive that UFOs are of intelligent extraterrestrial origin, nor do I think the evidence is convincing that no UFOs are of intelligent extraterrestrial origin.
—Carl Sagan, 1968[65]

Those who claim that UFOs are extraterrestrial visitors have not yet proven their case. The vast majority of UFOs—some would say over 95%—have turned out to be IFOs: Identifiable Flying Objects.[66]

Arthur Clarke acknowledged that phenomena still unknown to science may account for the very few UFOs that are both genuine and unexplained. However, there is no hard evidence that Earth has ever been visited from space; if that does happen, radar networks would know within minutes.[67]

Bernard Oliver cited the example of an intruder into our skies that was documented with many photographs—a bolide that passed through the Earth's atmosphere in 1972. If there were something to UFOs, we would have better evidence than we do. After decades of UFO reports, Shostak reminded us, there still are no artifacts to examine.[68]

Whatever light they may shed on atmospheric phenomena and human psychology, some say, UFOs tell us nothing about extraterrestrials. Even if the reality of UFOs were established, it would not necessarily prove the reality of alien visitors.[69]

On the other hand, the unknown phenomena explanation for UFOs would not necessarily exclude the idea that advanced extraterrestrials could visit the Earth. As geochemist David Schwartzman pointed out, surveillance by extraterrestrials could be taking place independently of UFOs.[70] Discrediting the claim that UFOs are extraterrestrial visitors does not prove that the visit scenario is impossible.

The Case for UFO Research

At the root of every pseudo-science there is to be found, if one searches assiduously and without prejudice, the germ of a new science.
—Physicist and engineer Maxwell Cade, 1966[71]

The UFO controversy polarized the question of observing extraterrestrial vehicles on or near the Earth. There should be room in the debate for an owlish position: that we can record and objectively examine reports without drawing conclusions in advance.

Most scientists regard ufology as, at best, a pseudo-science. The Clarks disagreed, asserting that the systematic study of sightings might well be the best way of detecting extraterrestrials capable of interstellar mobility. Many of the same principles used to justify searching for extraterrestrial intelligence by radio can be used to justify a scientific form of ufology. That could be part of an expanded notion of what constitutes SETI, which starts with the contention that extraterrestrials could have a radio communication capability and then mounts surveys to obtain the evidence.

Noting that the demystification of alchemy and astrology led to legitimate chemistry and astronomy, the Clarks thought that there will be similar rewards in demystifying UFOs. SETI pioneers set an example when they faced down the giggle factor and legitimized their subject.[72]

This idea received United Nations backing in 1978. The U.N. General Assembly invited interested member states to coordinate scientific research into extraterrestrial life, including unidentified flying objects, and to inform the Secretary General of their observations, research, and evaluations. Little seems to have come of this.

Planetary scientist Von Eshleman pointed out a practical problem: the UFO community cannot bring to bear the complex and expensive astronomical firepower that is routine in SETI, including the largest antennas and most sensitive electronic and digital systems in the world. Although UFO researchers have plenty of soft data, observed Swift, they have no single satisfactory hypothesis and no simple plan for investigation.[73]

SETI offers another parallel. Credible evidence of extraterrestrial intelligence, according to Jill Tarter, is being unable to explain a signal—which you also can't make go away—by any known astrophysics or technology.[74] Some UFO sightings may fall into that category.

A Scientific Review

In 1997, a team of scientists organized by physicist Peter Sturrock studied the physical evidence from some of the better documented UFO sightings. They concluded that, although there was intriguing evidence associated with some cases, none involved currently unknown physical phenomena or pointed to the involvement of extraterrestrial intelligence. Some reported incidents might have involved rare but significant phenomena such as electrical activity high above thunderstorms or rare cases of radar ducting. A few cases might have their origins in secret military activities.

The UFO problem is not a simple one, the Sturrock team concluded; it is unlikely that there is a single answer. However, most current UFO investigations are not carried out at a level of rigor that is consistent with prevailing standards of scientific research. It may be valuable to carefully evaluate UFO reports to extract information about unusual phenomena currently unknown to science. To be credible, such evaluations must take place with a spirit of objectivity and a willingness to evaluate rival hypotheses.

Sturrock pointed to the example of the SEPRA group at the French national space agency as a model of a modest but effective organization for collecting and analyzing information about UFO sightings. SEPRA (once called GETAN) had studied such reports for more than 20 years. Of 3000 reports, only about 100 required detailed investigations, and only a handful of these had not been satisfactorily explained as natural phenomena.

Sturrock's group commented that the most important change that could be made by scientists is to become curious. In view of the emergence of clear patterns in UFO reports and in view of great public interest, it is remarkable that the scientific community has exhibited so little curiosity in the past. Sturrock offered reasons: There are no public funds to support research into UFO sightings; There may be an assumption that there are no data worth examining; there may be a belief that the Condon report effectively settled the question; the topic may be perceived as "not respectable."

Like Hynek, Sturrock found that discussions of the UFO issue have remained narrowly polarized between advocates and adversaries of a single theory—contact with an alien civilization originating in another solar system. This fixation on extraterrestrials has narrowed and impoverished the debate, precluding other possible theories. Sturrock concluded that unless there is some event that galvanizes the scientific community into action or some new initiative that permits a modest but effective level of scientific research, the UFO problem is likely to remain an enigma—perhaps for another 50 years.[75]

A Cautionary Note

The history of science is littered with examples of scientists missing or ignoring phenomena that they were not looking for or did not expect. Consider sprites.

For many years, eyewitnesses had reported seeing giant flashes of light above distant thunderstorms. Scientists ignored these reports until one of their younger colleagues detected such flashes with an auroral camera in the late 1980s. This discovery sparked research activity that turned up a surprising collection of optical emissions over thunderstorms, including "sprites," "blue jets," and "elves." Sprites are not small; some may reach from an altitude of 90 kilometers (over 50 miles) down to cloud tops, as well as extending 40 kilometers (25 miles) horizontally.[76] Yet, generations of scientists missed them because they were not looking.

Referring to the discovery of pulsars, Sturrock asked what would have happened if a lone astronomer had detected pulsed radio signals in 1967 with a home radio kit and announced his discovery in a local newspaper. There is a real risk that the claimed discovery might have been dismissed by radio astronomers as preposterous and never investigated.[77]

Given the tendency of each side of the UFO debate to dismiss the other, it may be difficult to keep our minds open to possible alternatives. Still, there is reason to apply Ockham's Razor, adopting the simplest explanation for unusual phenomena until we have evidence to the contrary.

As Shklovskii put it in 1971, every object must be assumed natural until proven unnatural. Sagan concurred, saying that extraterrestrial intelligence is the explanation of last resort, when all else fails.[78]

The Drake Equation, Take Three

The Drake equation was derived from an assumed mode of contact: the detection of electromagnetic signals from very distant stars. It rested on the assumption that technological civilizations do not expand beyond their home systems. The density of such civilizations was determined by the number of separate evolutions to life and intelligence. The equation did not take into account the possibility of interstellar exploration, interstellar colonization, and direct contact.

That gap has been challenged forcefully. The Drake equation is wrong, Dyson argued, because it says that the number of extraterrestrial civilizations is equal to the number of independently originating civilizations. In fact, life spreads, diversifies, and speciates. Any community of intelligent creatures adapted to living freely in the vacuum of space will spread and speciate in the Galaxy. One intelligent species let loose in space may become a million intelligent species within an astronomically short time. Consequently, the Drake equation gives only a lower bound to the number of civilized societies.[1]

Shostak and Barnett, while defending conventional SETI, acknowledged that one star system could seed others if interstellar travel happens. The number of technological societies might be large even if the chance of evolving intelligence at one particular location is small.[2]

David Viewing of the British Interplanetary Society proposed a revised Drake equation in 1975, introducing factors representing the average number of colonies established by each independent civilization, and colonies established by the colonies.[3] Brin argued later that the equation needs three new factors when we introduce star travel:

V—the velocity at which an interstellar culture grows into space, pausing to settle likely solar systems and rebuild necessary industry before again continuing its expansion

$L(z)$—the lifetime of a zone of colonization into which a species has expanded, after which the settled region becomes fallow again

A—an approach/avoidance factor, different for each culture, representing a "cross-section" for discovery by contemporary human civilization. This encompasses motivations to initiate or avoid contact, life-style variations, abandonment of radio for other technologies, and anything else that might cause an extraterrer-strial culture to be more or less observable.[4]

At a minimum, the equation should include a factor for the probability that another technological civilization would engage in interstellar explora-tion, expansion, or colonization. This could have powerful implications.

The factors in the traditional equation can be seen as filters that reduce the probability of contact with technological civilizations. Adding an expansion and colonization factor could greatly increase that probability by expanding the range of locations where technological societies may exist.

Those most devoted to the classic SETI paradigm continue to resist including interstellar expansion in their calculations. One can see why; opening the door to a direct encounter raises the question of why we have not seen evidence of alien technology nearby. The direct contact scenario also has significant implications for the possible consequences.

Why Don't We See Them?

*All our logic, all our anti-isocentrism, assures us that we are not unique—
that they* must *be there. And yet we do not see them.*
                    —David Viewing, British Interplanetary Society, 1975[1]

*There is . . . another species of life—corporeal indeed, and various in its
order; but . . . not to be seen, not to be heard, not to be felt by man.*
                    —British historical and religious writer Isaac Taylor, 1836[2]

Before Fermi

Early in the twentieth century, pioneering spaceflight theorist Konstantin
Tsiolkovskii envisioned human expansion through the solar system. If
extraterrestrial colonization was in Humankind's future, Tsiolkovskii
assumed that it was an inevitable step for other intelligent beings. Those
older and more advanced than we must already have expanded beyond
their natal star systems.

   Tsiolkovskii's reasoning led him to confront what later became known
as the Fermi Paradox. If alien expansion included our solar system, why
don't we see any evidence of them? He suggested that because older and
wiser civilizations know that contact could ruin us, they leave us alone; we
have been set aside as a reserve of intelligence. Tsiolkovskii thought that
they would visit us when we are more advanced.[3]

   Charles Fort identified what he called the greatest of mysteries in 1919:
"Why don't they come here, or send here, openly?" He offered his own,
possibly ironic, explanation: We are property. Once upon a time, the Earth
was a no-man's land. Other worlds explored and colonized here, fighting
among themselves. There has been an adjustment among contesting claim-
ants. Now something owns Earth, warning off all others.[4]

   This question arose again in 1950, when a conversation among Enrico
Fermi and other scientists touched on the question of flying saucers and
extraterrestrials. Fermi asked "where are they?" He followed up with a

series of calculations about the probability of Earth-like planets, the probability of life arising on them, the probability of the evolution of higher life-forms and the evolution of technology—in effect, an early version of the Drake equation. According to some sources, Fermi came to the conclusion that we long ago would have been visited by extraterrestrials if any existed. Drake saw this differently, believing that Fermi had concluded that we don't know enough to answer the question.[5]

In 1973, astronomer John Ball proposed a new version of Tsiolkovskii's idea that we are in a reserve: Aliens have set us aside as part of a wilderness area, wildlife sanctuary, or zoo. This "zoo theory" predicted that we will never find them because they do not want to be found and have the technological ability to assure this. Ball added other possibilities: Extraterrestrial civilizations have not yet found us, or they may know we are here but are uninterested in us.[6]

Others later raised objections to the zoo theory. Trefil thought that the odds against any sort of quarantine strategy are insuperably high; all it would take to ruin that strategy is one poacher. We can conceive of a single honest race capable of maintaining the solar system in isolation for ethical reasons, but not of a million other races doing the same thing.[7]

The Great Debate

By the mid-1970s, debates about this question had been intensified by the growing credibility of two ideas: space colonization and interstellar flight. Princeton physicist Gerard O'Neill's space colony concept, first published in 1974, showed how artificial, inside-out planets could draw on the resources of the Moon and the asteroids to construct new versions of themselves. O'Neill persuaded many people that expanding Humankind throughout the solar system was not only feasible, but economically justifiable and possibly appealing.[8]

Meanwhile, studies by scientists and engineers were suggesting that interstellar travel could be achieved by a civilization more technologically developed than our own. For some, the conjunction of these ideas strengthened the belief that more advanced technological species could expand outward to colonize other solar systems. Older civilizations might have done this long ago, yet we see no evidence of their presence.

These ideas fed into a backlash against the sweeping claims made by prominent SETI advocates—and the publicity they received. Some scientists began arguing that the search was a futile effort that did not merit public funding. To some extent, Casti suggested, this may have been a reaction against the euphoria emerging from the 1971 Byurakan meeting.[9]

"Xenology" faced its first traumatic struggle, between those who seek optimistic excuses for the apparent absence of sentient neighbors and those who accept what Brin called The Great Silence as evidence for humanity's isolation. Both approaches, he commented, suffer greatly from personal bias and a lack of detailed comparative study.[10]

The easiest way out of the apparent paradox is to claim that intelligent extraterrestrials do not exist. That "solution" cannot be disproved unless we detect evidence of alien minds at work.

Many of the participants in this debate reject that conclusion as wildly premature. The issue may be far more complicated than any one-factor analysis would suggest.

If They Could Expand, They Must Not Exist

One premise or another of the Fermi dilemma is at stake.
—John Mauldin, 1992[11]

Writing in the *Quarterly Journal of the Royal Astronomical Society* in 1975, Astronomer Michael Hart presented four explanations for there being no intelligent beings from outer space on Earth: A physical, astronomical, biological, or engineering difficulty makes interstellar travel infeasible; extraterrestrials have chosen not to visit Earth; they have arisen so recently that they have not had time to reach us yet; they have been on Earth, although we do not see them here now. Hart claimed that none of these explanations is convincing, leading to a conclusion that we are the first civilization in our Galaxy. Therefore, an extensive search for radio messages from other civilizations is probably a waste of time. Hart, who thought it likely that cultures descended directly from ours will occupy most of the habitable planets in our Galaxy, did admit that our descendants might eventually encounter a few advanced civilizations that never chose to engage in interstellar travel.[12]

Several people challenged Hart's arguments, particularly the proposed rapidity of interstellar expansion. All reasonable combinations of assumptions led to much slower rates than those Hart assumed, according to astronomer Laurence Cox of the Hatfield Polytechnic Observatory. To be universal, Hart's theory must apply indefinitely to our own future; that, in turn, would require a fundamental change in our own society that renders our past history useless for prediction.

To astronomer George Abell, Hart's analysis showed how easily we can use arguments concerning the characteristics of hypothetical civilizations to reach conclusions that we want to believe and thus how fragile our estimates of the number of other civilizations really are. Morrison thought arguments like Hart's might be driven by a religious concern for demonstrating that we are the only conscious beings to be created.[13]

Other deniers pursued arguments similar to Hart's. Basing their cases on population growth rates, they claimed that a relentlessly expanding species could colonize the entire Galaxy within a time far shorter than the Galaxy's age. Eric Jones of the Los Alamos Scientific Laboratory came up with one of the lowest estimates of the time required: about 5 million years. As we see no evidence of such colonization, he concluded that no other technological, space-faring civilization has arisen in our Galaxy.[14]

Jones argued later that descendants of such a civilization should be found near virtually every useful star in a time much less than the current age of the galaxy. Only extreme assumptions about local population growth rates, emigration rates, or ship ranges can slow or halt an expansion. If interstellar travel is practical at a few percent of light speed, it is virtually certain that our solar system would have been settled by non-natives long ago. Unless we discover that interstellar travel is impractical, we are probably alone in the Galaxy.

Jones conceded one key point. If intelligent beings can live in interstellar space and need not be clustered around stars, estimates of settlement times may be meaningless; the absence of obvious signs of settlements in our solar system would not be significant.[15]

Astronomers Thomas Kuiper and Mark Morris picked up the argument in 1977. If interstellar travel and colonization are possible, there are two possible outcomes: Technological civilizations that last long enough to begin the colonization process are rare, in which case the Galaxy is essentially unpopulated, or there are several such civilizations, in which case the galaxy is fully explored or colonized. They offered three possible explanations for the lack of contact: The Earth has been a preserve for a long time; alien technology has advanced beyond the stage where a planetary base is needed; Earth's biology is incompatible with or even hostile to that of the species that dominate our part of the Galaxy.[16]

David Schwartzman proposed that although colonization is not a probable extraterrestrial strategy, surveillance and eventual contact are. The present surveillance of Earth by extraterrestrials may be the best reconciliation of those who are optimistic and those who are pessimistic about the number of alien civilizations.[17]

Hart and astronomer Ben Zuckerman organized a conference at the University of Maryland in 1979 to look at the Where Are They question, subtitling the meeting "A symposium on the implications of our failure to observe extraterrestrials." Although the papers presented a variety of views, there was widespread agreement that interstellar travel is feasible.

Physicist Stephen Webb later commented that it was difficult to read the proceedings of that conference without concluding that extraterrestrial civilizations have the means, motive, and opportunity to colonize the Galaxy. Yet, Zuckerman found that, to astronomers who work with optical and radio telescopes, the universe appears to be a gigantic wilderness area untouched by the hand of intelligence (with the possible exception of God's).[18]

Hart admitted that there could be a small number of other civilizations in our Galaxy, none of which were interested in interstellar exploration and colonization, or ever had been in all the ages since they first acquired the technological capability. He thought it more probable that the chance that any specific galaxy will contain life is extremely small. The reluctance to admit that the number is low is primarily a result of wishful thinking; a galaxy teeming with bizarre life forms sounds a lot more interesting than one in which we are alone.[19]

Papagiannis too found it extremely unlikely that a large number of long-lived civilizations could have existed in our Galaxy without any of them starting the colonization process. It is virtually impossible to find a universal reason that would prevent each of them from initiating interstellar travel. (Papagiannis assumed a natural tendency for life to expand to occupy all available space.) There was no way that the colonization wave would have missed or by passed our solar system, which meets all of the essential requirements for the establishment of space colonies—longevity, stability, a suitable sun, and a multitude of planets, moons, comets, and asteroids.

Like Kuiper and Morris, Papagiannis argued that one is led to two diametrically opposed alternatives. First, the Galaxy already has been colonized; if so, there must be space colonies in orbit around every well-behaved star. Second, the Galaxy has not been colonized because of the lack of a substantial number of advanced civilizations.[20]

Papagiannis later revised his argument. Colonizing civilizations would not come to live on planets, as they would be accustomed to living in space colonies. They would build more of those colonies using the raw materials obtained from asteroids and small moons. We should search our asteroid belt before concluding that colonization of our solar system has not taken place.[21]

Tiplerism

The "Extraterrestrial Beings Do Not Exist" school found its most vocal champion in physicist Frank Tipler. His attack on SETI may have been motivated in part by what he called its semireligious overtones.[22]

Tipler proposed in 1980 that advanced civilizations would use self-replicating probes (sometimes called Von Neumann machines) to explore or colonize the Galaxy in a relatively short time. If extraterrestrial intelligences existed, their spaceships must already be in our solar system. As we don't see them, they do not exist.

Interstellar travel would be simple and cheap for a civilization only slightly in advance of our own, Tipler claimed 3 years later. If such a civilization had ever existed in the Galaxy, their spaceships would be here. "There must be a first civilization," he declared, "and it happens to be ours."

Tipler's Historical Perspective

In 1981, Tipler presented a capsule history of the extraterrestrial intelligence concept in the *Quarterly Journal of the Royal Astronomical Society*. As he saw it, the debate reoccurred periodically as the centuries passed, with new generations of debaters rediscovering pro and con arguments that had been used earlier. He concluded that, as has been the case for 2000 years, philosophical and theological beliefs are the main motivations for the belief in extraterrestrial intelligence.[23]

According to Tipler, believers in alien intelligence have tended to lack what he called a sense for the contingency of history and its unique, unpredictable events. If we apply contingency to expansion, we must question the inevitability that underlies Tipler's theory of self-reproducing probes. Technological civilizations may or may not choose to begin colonization; they may or may not choose to continue it. Tipler assumed that a particular technological model would be maintained for millions of years, although our own history suggests that such concepts have much shorter lifetimes.

Tipler restated his case in a 1986 book he co-authored with mathematician and cosmologist John Barrow. According to these authors, an intelligent species with the technology for interstellar communication would *necessarily* develop the technology for interstellar travel, and this would *automatically* lead to the exploration and/or colonization of our Galaxy in less than 300 million years (emphasis added). Von Neumann machines would be programmed to turn some of the material in the alien system into an O'Neill-type colony; those machines would synthesize the colony's inhabitants from genetic information.

Tipler's argument rested on the assumption that any intelligent species would expand into new environments, launching Von Neumann probes or colonization ventures of some type. If we deny this assumption, he acknowledged, we have nothing to go on except opinion.[24]

Counter Arguments

Drake described the basic weakness of arguments such as Tipler's: the readiness to assume the possible to be the inevitable. Advanced extraterrestrials may not want to spend the money and the energy to attempt interstellar travel. They may see no personal gain in creating a costly army of Von Neumann machines; they may be content to colonize their own star system; they, like us, may have found radio communication the more promising alternative.[25]

To Finney, the central premise for Tiplerian technological optimism is that whatever is not forbidden by natural law will come to pass. From this perspective, it only takes an extended thought experiment to fill the Galaxy. Is galactic expansion the working of some universal law about the course of intelligent life, or is it just a figment of the imagination of a technologically presumptuous but still adolescent species? Has a general principle governing the evolution and expansion of life been discovered, or is this just a reflection of the hubris to which we humans are so given? Our own history does not support the Tiplerian premise, Finney concluded; expansion is not automatic.[26]

The proposed models of interstellar colonization ignore the differences between the exploration of the Earth and the exploration of the Galaxy, argued Sagan and Newman. Earth is uncolonized by extraterrestrials not because interstellar spacefaring societies are rare, but because there are too many worlds to be colonized in the plausible lifetime of the colonization phase. For a reasonable population growth rate, the Galaxy is too big; they're not here yet (Sagan calculated that the establishment of a "galactic hegemony" would take a billion years). A culture that gave a wide berth to planetary systems in which life is evolving would pose no contradiction to the apparent absence of extraterrestrials in our solar system; we would be none the wiser if such a civilization occupied all of the remaining planetary systems in our spiral arm of the Galaxy.

Sagan and Newman introduced a political dimension into the debate, claiming that cultures aggressive enough to plan galactic colonization will destroy themselves before they get far. Although there are no very old civilizations with a consistent policy of conquest of inhabited worlds, there may be abundant groups of worlds linked by a common colonial heritage.[27]

Goldsmith and Owen too emphasized the vastness of our Galaxy. Even if a million civilizations exist in the Milky Way and even if 1% of these civilizations are devoted to interstellar exploration on a grand scale, there still would be 40 million stars to explore for each civilization. A galaxy-wide program of colonization requires alternating expansionist and "mature" periods; it may turn out that no civilization will embrace enough cycles to spread through a galaxy. The absence of colonists from other civilizations here on Earth does not prove that there are no long-lived civilizations in the Milky Way.[28]

Galactic Vandals

Challenging Tipler's assumption that a technological civilization would send out self-replicating probes, the Clarks argued that such galactic vandalism is an extremely unlikely motivation for interstellar travel. Eric Jones and Barham Smith questioned the idea that any self-respecting product of Darwinian evolution would gift the Galaxy to a swarm of sentient machines. Such machines would mutate, Benford warned; they could malfunction and change their motivations, becoming competitors to their creators.

Those implacable replicators will either have or lack a built-in principle of reproductive restraint, proposed Sagan and Newman. If their reproduction is limited, they would not be everywhere. If it were not limited, they would threaten the culture that contemplated making them and would not have been made.

We may be seeing forerunners of this issue in the development of robot fighting machines for Earthly combat. "The lawyers tell me," said one military researcher, "there are no prohibitions against robots making life-or-death decisions."[29]

The Psychology of Expansion

Mere possession of the technology for expansion is not enough. The motivation to expand must also be there.

—Ben Finney, 1983[30]

Models of expansion like Tipler's rest on extreme assumptions about what societies of intelligent beings would do, taking for granted that expansion is both inevitable and continuous. One of the arguments used by expansion theorists already has been undermined by events in the real world. As a driving force, population growth is far less credible today than it was 30 years ago; we know now that the population explosion can be slowed, even halted.

Finney, comparing interstellar colonization with historical human migrations (particularly the Polynesian expansion across the Pacific Ocean), concluded that "no specific migration has ever gone unchecked. Ecological barriers, the slowing or cessation of innovation, flagging motivation, or the opposition of those in the way have . . . stopped every . . . colonization movement so far." Interstellar colonization would not fill the entire Galaxy.[31]

Judging by human history, expansion episodes may be brief. Chinese explorers made many long voyages in great ships while the European Age of Exploration was just getting under way. Evidence suggests that they reached Africa; according to one controversial theory, they landed in the Americas well before Columbus and may even have planted colonies. Despite these magnificent achievements, Chinese officials made a policy decision to stop the voyages. The hulks of China's exploring vessels were left to rot away on riverbanks.[32] The colonies—if there were any—disappeared.

Only a small minority of our own species favors extending permanent human presence beyond the Earth. Extraterrestrials also might have a wide variety of drives, abilities, and situations. Expansion, if begun, might be episodic and may be sustained for only for limited periods of time; it might cease for societal reasons.

The implications of expansionist models are much different, Claudius Gros pointed out, if we drop the assumption that the colonies of expanding civilizations automatically inherit all their characteristics. An expanding civilization's motives might change before it reaches us.[33]

What other reasons might motivate interstellar migration? Groups may flee their home systems for religious, ideological, or political reasons. For those migrants, the jump need be made only once in the lifetime of their home star—unless a centralizing power pursued them.

A civilization that felt threatened by another technological species might spread out to reduce the risk of annihilation. However, that would only work if the escapees could assure that they would not be found in their new locations.

There may be other reasons for sending interstellar missions: scientific curiosity; precautionary observation of emerging intelligence; preemption of a potential threat from another technological civilization. None of these requires the sending of inhabited spacecraft; well-equipped probes could do the job.

Finney invoked a Galilean premise. We cannot determine by logical argument, equation, or debate how far and wide our descendants will spread. We can know the answer only by seeing what actually happens.[34] The same principle would apply to expansion by extraterrestrial civilizations.

Cosmic Geography

In the human case, the classic agenda for manned spaceflight is driven by the particular layout of our solar system—first the Moon, then Mars.[35] Other star–planet systems may be quite different. Would another civilization engage in spaceflight if there were no other detectable planet in the system? What if the home world were the only known body with a solid surface on which to land?

At the other extreme, consider the model proposed by Dole and Asimov: twin planets, revolving around a common center of gravity. With another planet close and large, with clouds, oceans, and continents clearly visible, would the urge to reach the companion hasten technological progress and eventual expansion? Could any egocentric philosophy of the universe develop?[36]

Mind-Stretcher. Where would we be today if Venus were as many had imagined it, a warmer but still habitable sister to the Earth? What if Mars were the size of our planet, able to hold a denser atmosphere and maintain warmer temperatures on its surface?

There may be other physical barriers to expansion. One of the major risks of human interplanetary flight is the occasional solar flare intense enough to damage or kill humans aboard spacecraft.[37] An alien star that generated more frequent and more powerful flares might discourage all attempts to journey beyond the home planet.

Location might matter in a grander sense. Some regions of the Galaxy may be more likely to produce technological civilizations than others. If some do expand, they will not do so equally in all directions; as David Stephenson pointed out, a uniform exponential model of interstellar expansion would only be followed if all systems were equally interesting and distributed uniformly in space.

The central region of the Milky Way is an inviting target for our searches because of its higher density of stars. Cade proposed that superior civilizations, hungry for energy, would be more likely to migrate toward the center, where the average separation between the stars is less. MacGowan and Ordway came to an even more forceful conclusion: When "executive automata" reach a certain level of development, they *must* be engaged in a general migration toward preferred galactic locations, probably in the galactic core (emphasis added). In our Sun's own rather sparse region of the Galaxy, only very slow star-hopping colonization would ever be possible by biological life.

There are counterarguments. Life and intelligence in the galactic center may suffer more frequently from astrophysical extinction; the same may be true for the bulk of the Galaxy's spiral arms. According to one theory, the most likely place to find old civilizations is near the outer edge of the Galaxy's spiral structure, where there are fewer threats.[38]

How likely is it that each phase of an expansion would take place? Like the "bottlenecks" that limit the number of communicating civilizations, there may be filters that determine the probability that an intelligent species would engage in, or sustain, interstellar exploration or expansion. The first filter would be ascending from the planetary surface to orbit around the home world; the second would be journeys to other bodies in that star–planet system. The third would be migration beyond the home star's family, which could lead to the colonization of interstellar space rather than to settlements on other planets. Each filter might reduce the probability that extraterrestrials would explore and colonize other solar systems—including ours.

Even for a civilization capable of interstellar flight, distance would be a factor. Journeys covering a few light-years may be considered feasible, but a gap of hundreds of light-years might be seen as a serious obstacle. Each spacefaring civilization may have a different concept of its accessible universe at different times in its history.

Even if some parts of the Galaxy are colonized, others may not be. If technological civilizations do come into contact, Bracewell proposed, it may be because of accidental proximity due to random spacing.[39] We might be in one of the voids.

A Minimum Model of Interstellar Expansion

The interstellar imperative—the bottom line of starflight—is that ours should become a civilization that can outlive its star.
                              —Eugene Mallove and Gregory Matloff, 1989[40]

Some intelligent beings would realize, sooner or later, that changes in their stars would make their planetary systems uninhabitable. The foreknowledge of impending doom, even if it were millions of years in their future, might drive some technological civilizations to migrate toward more hospitable locales. Those outward leaps might begin at very different times in different systems, depending on the lifetimes of host suns. Some hot stars may support zones of habitability for only a few hundred million years; cooler ones may remain stable for tens of billions of years.

Over time, the Outleap might acquire legendary status like the great Trek of the Boers in South Africa, or the Long March of the Chinese Communists in the 1930s. However, an outward migration on this scale may not be a single event that includes all of a given planetary population. Different groups may depart at different times, possibly in different directions.

A very ancient civilization that was forced to make its first jump billions of years ago might face the prospect of a second migration if its new home became unlivable. However, no second jump would be necessary for a species prepared to do without stars. A society capable of building interstellar arks that do not require the energy of a sun could migrate to any location in interstellar space. They could be present throughout the vastness of our Galaxy, yet be invisible to us.

This model of interstellar migrations separated by eons of time implies a much slower rate of expansion than those foreseen by Tipler, Jones, and others. Although the infrequency of such expansions would limit the number of locations where technological societies exist, migrations in different directions could raise that number.

This minimum model might offer a solution to the Paradox—if we had any way to confirm it. However, we should be cautious about applying any particular model to all technological civilizations that choose to establish colonies beyond their home systems. Even if this model applies to most, there may be statistical outliers. The model may not apply at all to uninhabited probes, which might be sent even when there are no plans for colonization.

The Great Silence

David Brin published a major analysis of the paradox in 1983, entitled "The Great Silence." Convinced that the case for some version of slow interstellar travel was growing stronger, he categorized some of the possible reasons for the apparent absence of extraterrestrials (this is taken from the more popularized version of his original paper).

Solitude. We are unique in evolving technological intelligence. Habitable planets may be rare; some "spark" may be needed to initiate life, or some "software miracle" is required for intelligence. The average lifespan of a technological civilization may be short due to some "inevitability" of self-destruction, or because robotic probes spreading throughout the galaxy are programmed to wipe out other intelligent life.

Magical Technology. Technological species may discover techniques that make radio and even colonization irrelevant.

Quarantine. Other intelligences may purposely avoid contact with us. Benevolent species might let "nursery worlds" lie fallow for long periods to nurture new sentience; we may be in a preserve or "zoo;" alien observers may be awaiting humankind's social maturity; or we may be quarantined as dangerous. As our neighbors won't remain near us due to the galaxy's differential rotation, the quarantine hypothesis would appear to call for some degree of cultural uniformity in the galaxy.

Macrolife. Advanced civilizations may abandon planet-dwelling. There might be selective pressures favoring those suited to life in starships. Spaceborne sophonts might greedily destroy terrestrial planets by mining them for their natural resources, or they might cherish nursery worlds they do not need as real estate.

Seniors Only. Spacefaring intelligences might graduate to other interests. Achieving immortality might promote conservatism and an aversion to the dangers of spaceflight.

Low Rent. Earth might be inaccessible or undesirable. Alien life may be biologically incompatible with terrestrial life. We may be in a region of the galaxy where interstellar travel is more difficult.

Migration Holocaust. Occupation of a nursery world by an expansionist species might cause extinction of its higher life forms, delaying the local emergence of intelligence. Earth might be the first nursery world to recover sufficiently after the last wave of colonization passed this way.

Brin offered a grim scenario in which colonizing cultures leave behind them wastelands emptied of intelligence; the Great Silence may be the sound of sand sifting up against monuments. He added other possibilities: Aliens might be cautious about contact to avoid becoming targets for an aggressive species or its probes, or they might be here, contacting governments or secretly meddling in our affairs.

Brin concluded on a hopeful note. A noble race may have taught a tradition of respect for the hidden potential of life in all subsequent spacefaring species. "It might turn out that the Great Silence we're experiencing is like that of a child's nursery, wherein adults speak softly, lest they disturb the infant's extravagant and colorful time of dreaming."[41]

Brin's analysis had diverging implications, noted Brian McConnell. On the one hand, the probability of contact may be reduced because some civilizations capable of interstellar communication choose to keep quiet. On the other hand, the probability of contact may rise because some civilizations choose to expand. Even if the emergence of technological civilizations is rare, the probability of contact may be high. So why don't we see them? To McConnell, the Deadly Probe hypothesis was the answer that required the fewest arbitrary assumptions.[42]

Other Theories

O'Neill proposed that we may be alone in this Galaxy as a material-oriented civilization, in contrast to more spiritual orientations. Material orientation may be a brief episode in the evolution of a society, perhaps one that every civilization goes through quickly. Civilizations may evolve to an end state in which intelligent beings are not tied to matter at all; they might have no interest in the physical universe.[43]

A spiritual orientation might be a path toward eventual extinction. Stapledon had suggested the possible implications many years ago:

In not a few worlds this way of the spirit was thronged by all the most vital minds. And because the best attention of the race was given wholly to the inner life, material and social advancement was checked. The sciences of physical nature and life never developed. Mechanical power remained unknown, and medical and biological power also. Consequently those worlds stagnated, and sooner or later succumbed to accidents which might well have been prevented.[44]

We might extend Grinspoon's argument that natural selection can act on a much larger scale than we're used to thinking about.[45] Although the universe as a whole may be hospitable for the evolution of life and intelligence, it may be inimical to their long-term survival at any particular location.

Our own time horizon may be much shorter than the billion years between now and an Earth made uninhabitable by a swelling Sun. Other terminators—asteroids, rogue planets, dark stars, gamma-ray bursters, black holes—could end the human adventure. "It is just possible," concluded Cirkovic and Cathcart, "that the total risk function facing civilizations is high enough to explain the total absence of their manifestations."[46]

An Optimistic Transition

James Annis of the Fermi National Accelerator Laboratory proposed in 1999 that the key terminators of intelligence are gamma-ray bursters. Each one is a mass extinction event on a galactic scale. Bursters were much more common in the past and may have wiped out many evolutions to intelligent life. If the timescale for land animals to develop intelligence was long compared to the time between bursts, intelligence could not emerge.

Things are looking up, according to Annis. The rate of bursters is declining to the point that our Galaxy may be undergoing a "phase transition"—the lowering of a suppressive force below a threshold, past which a previously forbidden process becomes allowed.

Annis thought it likely that intelligent life has recently sprouted at many places in the Galaxy. In another 100 million years, a new equilibrium state will emerge, where the Galaxy is completely filled with intelligent life.[47]

Others have suggested that we might be in a simulated universe, created by masterful intelligences. The conscious inhabitants of virtual worlds, suggested Davies, might be unaware that they are the simulated products of somebody else's technology.

Baxter, who looked at the simulation question in his "Planetarium Hypothesis," found that there may be physical limits to the simulation. No conceivable virtual-reality generator could contain a human culture spanning 100 light-years. An environment that large would be an imperfect emulation and should be distinguishable as artificial.[48] Applying such grand simulations to the extraterrestrial intelligence question may reflect the intellectual fashion of our computer age.

The Silence of the Immortals

Drake asked a rhetorical question in 1976: What if the aliens are immortal? (He defined this as meaning the indefinite preservation, in a living being, of a growing and continuous set of memories.) The reverence they would attach to the preservation of individual lives would drive these immortals to avoid physical threats. They might conceal themselves and prohibit transmission of radio signals detectable by other civilizations. However, this would not stop a civilization with questionable intentions.

Drake speculated that a civilization of immortals would choose the opposite policy: They would be extremely active in detecting and communicating with other civilizations. Their best assurance of safety would be to make other societies immortal like themselves, rather than risk hazardous military adventures. They would spread the secrets of their immortality among young, technically developing civilizations.

Drake offered another reason for immortals seeking to communicate. Using up their resources for amusement and adventure, they would want to share vicariously in the adventures of other civilizations.

Drake concluded that it was the immortals that we will most likely discover. Mortal civilizations like ours probably do not remain detectable forever, because their increasing technical sophistication enables them to cease the release of energy into space. However, some immortals *must* continue to transmit (emphasis added).

Drake proposed that the number of immortal civilizations would be far greater than the population of all detectable mortal civilizations. As immortals are likely to dominate space, we should concentrate our search on their signals.[49]

Drake returned to this theme in his autobiographical book. Arguing that immortals would want absolute assurance of their safety, he acknowledged that they might conceal themselves, perhaps even prohibiting transmissions of signals that could be detected by other civilizations. He again speculated that a better strategy would be to help other intelligent beings to become immortal, giving them the same incentive for safety. If this were the case, immortals might be extremely active in detecting and communicating with other civilizations.[50]

Drake recognized that immortals would look differently upon long-term risk.[51] We don't worry about changes in the Sun that will make our Earth uninhabitable a billion years or more from now; we and our children will be gone long before that happens. We don't worry much about an extinction-level collision with an asteroid that may happen once every 100 million years. An immortal might worry.

Drake's vision has dual implications for interstellar travel. In the technical sense, life extension could make voyages by inhabited spacecraft more feasible. As Clarke put it, extending life spans indefinitely would drastically reduce the size of the universe from the psychological point of view.[52]

On the other hand, immortality might lead to a growing disinclination to engage in risky activities. A society without life-threatening problems might be one without the exploring bug, Goldsmith speculated, and this might be a universal rule.[53]

Consider our own history. During the early years of the European Age of Exploration, men often joined very risky voyages into unknown waters even when they knew that the odds of getting back were low; if ships did return, half the crew might be dead. Life was short, typically 30–40 years. Why not take chances if you are going to die within a decade anyway? Would immortal humans be that daring? Our own long-lived descendants might abandon Humankind's outward reach, except through their machines.

What kind of society would immortality produce? Almost certainly a conservative one, resistant to change. The conventional wisdom might be frozen into place; there would be no young rebels to challenge it. Imagine the social and cultural consequences of immortals who refuse to retire. If they did retire, who would pay for their pensions?

The Galaxy could be sprinkled with immortal civilizations, each ossified into the mold of a past era. Societies composed of immortals might be inflexible in values and culture, reducing their adaptability to change. Immortality interrupts evolution, and adaptation.

For such civilizations to actively call attention to themselves, bargaining medical information for their security, seems like a risky strategy. The biologies of widely separated worlds may be so different that the life extension techniques of particular immortal species would not be useful to others who were mortal. It is far more likely that the immortals would seek to avoid the attention of spacefaring civilizations.

Immortality has been underrepresented as a factor in science speculation and science fiction. We are expected to believe that "Star Trek" and "Star Wars" characters living centuries in the future age and die just as we do. Yet, life extension has become a major area of medical research, with the clear hope that it someday will lead to eternal life for individual humans. One researcher declared that aging is an optional feature of life; it can be slowed or postponed.[54]

Perhaps science fiction authors avoid writing about immortality because they see it as static and boring. They may be right.

Immortality could have a profound effect on the way civilizations behave. Silence may spread across the Galaxy as intelligent beings achieve eternal life. It may be that only mortals like ourselves take the risk of broadcasting their presence or of sending out explorers.

Catalogs of Solutions

Many authors have offered lists of possible solutions to the paradox, including those described earlier. Those compilers include Carl Sagan, Isaac Asimov, Gerard O'Neill, Thomas Kuiper and Mark Morris, Seth Shostak, C.E. Singer, John Ball, Gregg Easterbrook, Frank White, Albert Harrison, Donald Goldsmith, Andrew and David Clark, Ian Crawford, Michael Kurland, Jack Cohen and Ian Stewart, Terence Dickinson, Peter Ulmschneider, T.L. Wilson, and William McLaughlin.

Stephen Webb pulled together 50 suggested solutions in his 2002 book *Where Is Everybody?* He divided these explanations (some of them whimsical) into three broad categories: They Are Here, They Exist But Have Not Yet Communicated, and They Do Not Exist. He noted that the They Are Here category is by far the most popular with the general public, although he rejected it.

Webb emphasized arguments for the rareness of advanced life and intelligence, in effect inserting his preferred numbers into the Drake equation. He suspected that there is a combination of factors—a product of various solutions listed in his book—resulting in the uniqueness of Humankind.

Webb concluded that the Fermi paradox tells us mankind is the only sapient, sentient species in the Galaxy. Yet, he admitted that there are potential challenges to that conclusion. Solutions supporting the argument that intelligent extraterrestrials do not exist depend on making one or more of the terms in the Drake equation very small.[55]

Reformulating the Problem

There are many ways of categorizing explanations for the apparent paradox. One is to focus on actions: what we think the extraterrestrials are doing, and what we are doing. Another is to concentrate on the intentions of each civilization; many explanations rest on assumptions about the behavior of alien societies, which, in turn, rest on analogies with our own. We could classify explanations by technological capabilities, ours and theirs; or we might do as some authors have done, mixing these approaches.

Here is another way to categorize our speculations, one that tries to take into account the serious explanations proposed by others. Where possible, the approach outlined below avoids making our Earthly perspective the center of the question, looking at these issues from a more detached point of view. We assume not only the spatial context of a huge galaxy containing hundreds of billions of stars, but also a temporal context that reaches back billions of years. The time frame includes not only the present but also a very long past.

This formulation addresses only technological civilizations, defined here as those *capable* of employing the technologies of interstellar communication and/or interstellar flight or other technological activities that we can detect, such as astroengineering. There could be civilizations that are capable of using such technologies but do not do so. There may be far more that do not have these capabilities at all. (Our own civilization was one of those until recently.) Here the sentient beings that form technological civilizations include postbiological intelligences, the descendants or masters of biological beings.

We should be cautious about applying any generalized explanation to all extraterrestrial civilizations. Technological societies sprung from different evolutions in different environments may do different things; we cannot assume uniformity in their behavior. Even if there is a collectivity such as a Galactic Club, there may be individual societies that are not members or that break its rules.

Most of the suggested alien behaviors that would make contact more likely involve the commitment of resources—financial, physical, or other.

When intelligent beings consider which technological activities to engage in, from listening for signals to astroengineering, they are faced with choices. The crucial variable may be which decisions they make.

Here the alleged paradox is divided into two questions: Why have we not detected electromagnetic signals suggesting the existence of alien technology, and why have we not detected alien artifacts, which could be anything from an interstellar ark to an abandoned socket wrench found on an asteroid in our own solar system (one category of artifacts, astroengineering, is addressed separately).

Explanations are divided into three categories: those related to our nondetection of electromagnetic signals, those that apply to our nondetection of artifacts or astroengineering, and those that are relevant to both. These explanations are not all mutually exclusive; more than one could be involved.

There may be multiple factors influencing whether alien civilizations search for others, explore, or expand. The reader is invited to experiment with combinations.

Hungarian astronomer Ivan Almar warned us that it would be rash to proclaim any of the proposed explanations of the paradox as final—or to reject them completely. Most explanations are likely to be wrong, yet the intellectual game is worth playing. As Sean Carroll said about string theories in physics, "All these proposals are in the spirit of 'unlikely to be right,' but so extremely interesting if they are that they are well worth thinking about."[1] The explanation we ultimately find may not suit anyone's convenience.

Explanations Common to Both

Uniqueness

There are no other technological civilizations now. We may be the first; or other technological civilizations existed in the past, but are now extinct. If there are natural cycles in which technological intelligent life evolves but is destroyed, we may be the first to emerge in the new cycle.

We also may be the first to emerge if past technological civilizations destroyed themselves. The destruction of biological civilizations by intelligent machines they created would not necessarily mean the extinction of technological civilizations, if those machines could be considered a civilization themselves.

Being the first in a new cycle does not exclude all forms of contact. A signal from an extinct civilization might reach us hundreds or thousands of years after it was sent. Artifacts from an extinct civilization could be much older.

Out of Range

Other technological civilizations exist now, but they are separated from us by vast distances that make contact by interstellar communication or interstellar travel unlikely. This implies that other technological civilizations are relatively rare and that none have expanded throughout the Galaxy.

Sagan and Shklovskii calculated that the average distance between technical civilizations is between a few hundred light-years and about 1000 light-years; Asimov thought that the average separation might be as great as 600 light-years. The result may be that every civilization, no matter how far advanced, is isolated. As Shklovskii put it, we could be functionally alone.[2]

Failures of Perception

Other technological civilizations do not perceive an external universe that could include inhabited worlds. Some extraterrestrials may be limited by their sensory abilities. Those living under densely clouded atmospheres might never detect their larger environments. Even where skies are clear, there may be no nearby astronomical bodies comparable to our Moon or the other planets of our solar system to stimulate the idea of other worlds. If intelligent aliens see only remote, stationary points of light in their skies, their relevant universes may be limited permanently to their own planet, sealed by a bowl of sky.

Failures of Imagination

Other civilizations have not passed through Copernican-level revolutions in placing themselves within the universe or among its living things. Few, if any, share our presumption that a multiplicity of inhabited worlds exists; even if they recognize the possible existence of other planets, alien technological civilizations may not conceive of other worlds evolving life, intelligence, and technology. Our belief in unseen others may not be shared by all sentient beings; the fact that we imagine extraterrestrials does not guarantee that they imagine us.

We cannot assume that our level of curiosity is a universal characteristic of all civilizations. Some others—perhaps most—may never search their skies for evidence of alien minds.

Failures of Nerve

Other currently existing civilizations that have the scientific and technological knowledge to search for signals, to transmit signals intended to attract the attention of others, or to send out interstellar explorers, do not use them. They may lack sufficient motivation to seek contact, or they may make a conscious decision not to seek it. If they ever tried to detect other civilizations, they may have given up after a search for signals or a search with probes failed to reveal convincing evidence; very few species may have millenium-length attention spans. They may see no point in exploring their interstellar environments or expanding their presences outward into the Galaxy; they may think that such tasks are beyond them or too low a priority to warrant a serious effort. They may have abandoned the use of the needed technological capabilities because of internal crises or societal changes.

Perception does not necessarily lead to action. Motivation may be as important as capability.

New Arrival

If other civilizations do search for signals or send out probes, they have not found us yet. We may have entered the era of interstellar communications so recently that they have not detected us. If they have found us and have decided to respond, their messages or probes may not have reached us.

Inadequate Human Technology

There are alien signals or artifacts, but they are beyond our ability to detect. Other technological civilizations may be using means of communication or transport unknown to us.

Inadequate Search Strategies

There are alien signals or artifacts that are potentially detectable with our present technologies, but we do not see them because they are outside of our current search space. We may not have searched long enough, widely enough, deeply enough, or continuously enough. We may be looking for the wrong kinds of signals or artifacts; our searches may rest on faulty assumptions.

We Misunderstand the Evidence

We have detected signs of extraterrestrial technology but fail to recognize them. We may have evidence that is unpersuasive or ambiguous, or the evidence may be buried in masses of recorded data.

They Are Hiding

Other technological civilizations are concealing their presence from us, and possibly from others as well. They may deliberately limit their emissions, maintaining something like radio silence.

If an alien presence is in our solar system, it may be deliberately hidden. Extraterrestrials could be monitoring or studying us without revealing themselves. We may be in an anthropological research area, a preserve, a wilderness, or a zoo; we may be under quarantine; we may be in a sphere of influence that excludes others. Some extraterrestrial civilizations may observe a principle of noninterference toward less powerful intelligences because contact might wreck their usefulness as suppliers of unique information, or for ethical reasons.

Hiding might not be a permanent policy. There may be thresholds that civilizations must pass before the more advanced initiate contact. Those thresholds could be scientific or technological, or they may be moral, ethical, or behavioral.

Humans Are Boring

Humankind is not sufficiently interesting to motivate extraterrestrials to contact us by signals or by spacecraft. We may be too primitive; we may have nothing that aliens want; we may be no threat to them. More advanced extraterrestrials may focus their attention on societies that have more to offer or that are potentially more dangerous.

Conspiracy

Some humans know that there is evidence of extraterrestrial signals or artifacts, but do not inform the rest of us.

Transcendence

Extraterrestrial intelligences have evolved to a state beyond what we know or can observe, such as an existence that transcends the limitations of spacetime.

They Are God

Some extraterrestrial intelligences are so omniscient and omnipotent that they have the qualities of all-knowing, all-powerful gods. We may see their works around us without being able to detect their source. A very powerful civilization may have created our universe; or we may be in a simulation, a virtual reality.

Explanations for the Lack of Signals

No Beacons

There never have been beacons. Even if technologically equipped aliens do imagine the existence of other civilizations, they may not operate beacons or send powerful targeted signals. If they do search, they may pursue a listen-only strategy.

Alternatively, there are no beacons now. One or more civilizations may have operated beacons in the past but are no longer doing so. Some of those signals still might be detectable if they are coming from a great distance but have not yet reached us.

Here we need to distinguish among beacons broadcast in all directions, beacons targeted at civilizations other than ours, and beacons aimed at us. The first two categories could exist even if the sending civilization did not know of our existence.

No Long-Distance Calls

There never have been interstellar point-to-point transmissions. Civilizations capable of using them may have no need for such transmissions if they do not expand or do not detect others.

Alternatively, there are no interstellar point-to-point transmissions now. No civilization is currently communicating with others, or, if it has expanded, with its colonies.

If there are point-to-point transmissions now, we may not see them because we are rarely in their line of sight—or we may see them as flashes of energy that fade too quickly to be reacquired.

No Local Calls

More advanced civilizations use communication methods with little or no leakage. They may emit detectable signals only during a brief phase of their histories; our searches may not coincide with those eras.

They Prefer Travel

Other civilizations have chosen interstellar travel, by biological beings or machines, as their preferred means of exploration. They may find searching and communicating by technological artifacts more useful than sending signals.

Beyond Our Imaginations

The science and technology of more advanced civilizations have evolved to a level that makes signaling irrelevant.

Explanations for the Lack of Artifacts

Impossible

Interstellar travel is impossible at the technological level achieved by any civilization, past or present.

Possible, But Not Worth Doing

Other technological civilizations capable of undertaking interstellar travel with probes or inhabited vehicles do not pursue it.

Limited Ambitions

One or more civilizations have undertaken interstellar travel in the past, but not throughout the Galaxy. Interstellar exploration and colonization may be episodic, limited in time and space rather than continuous; motivations for exploration and colonization may change over time. If other civilizations expand more continuously, the rate may be slow. Colonies may not engage in interstellar exploration or colonization themselves.

Isolation

Other civilizations are exploring or colonizing on an interstellar scale, but their explorations or expansions have not entered our part of the Galaxy. The gap between their location and our solar system may seem, to them, not worth crossing, even if they know we exist. We may be in an uninteresting neighborhood, or our system may be unattractive to them.

On Their Way

They are exploring or expanding toward our solar system, but they are not here yet. They may be coming our way for reasons unrelated to us; they may not know of our existence. Even if they do know, we may not be the primary reason for their journey. Or they may be coming because they detected our presence.

They Visited a Long Time Ago

Extraterrestrials or their machines have reached our solar system in the past, without leaving evidence of being here. If they did leave evidence, we have not found it or do not recognize it. Within this context, a probe or inhabited vehicle may have visited the Earth. It may or may not have been detected by humans.

Here But Undetected

A probe or inhabited vehicle is in our solar system, but we do not see it. If it is a machine, it may have ceased operating long ago. If it is a habitable spacecraft, its passengers may no longer be alive or may have transferred to another vessel and left.

If aliens are in our solar system, they may avoid Earth for some reason such as biological incompatibility, excessive gravity, or an ethic of noninterference. Other parts of our solar system may be of greater interest to them.

Here and Misunderstood

We have detected an alien presence in our solar system, but we do not recognize it or do not consider the evidence persuasive.

Who Needs Planets?

More advanced civilizations may not need planets or stars. They may be scattered through interstellar space, but not in our solar system.

They Are Not Travelers

More advanced civilizations may find searching and communicating by signals more useful than sending probes, inhabited vehicles, or other artifacts.

Beyond Our Imaginations

Their science and technology make interstellar exploration or expansion irrelevant. They may have evolved to a stage where interstellar vehicles are not necessary.

A Subcase: Astroengineering

No civilization has ever engaged in astroengineering. It may seem infeasible or not worth the cost.

No civilization has engaged in astroengineering activities that we can detect with our present technologies or search strategies.

We have detected evidence of astroengineering but do not recognize it.

Is There a Paradox?

Many have disputed the very existence of a paradox. The absence of visitors in our solar system has little meaning, Newman declared, because it depends on a series of questionable assumptions: If extraterrestrial life is abundant, if space travel is relatively easy, if advanced civilizations feel compelled to explore the galaxy and can do so, if they have had enough time, if we have tried hard enough to find them, then shouldn't we see evidence of extraterrestrial life?[3]

The Probability of a Paradox

Any assumption must have a probability of being correct. Stuart Clark described how astronomer Jean Heidmann developed this idea. If a conclusion is based on two assumptions that are each 90% certain, that conclusion is only 81% correct ($0.9 \times 0.9 = 0.81$). Heidmann identified 112 assumptions in Barrow and Tipler's interstellar expansion argument that lead to their conclusion that extraterrestrials do not exist. If each of these assumptions is 90% correct, then their conclusion is 0.9 times itself 112 times. This means that Barrow and Tipler's conclusion is only 0.0007% certain.

Jill Tarter challenged the whole logical construct, because it requires that we take as a fact that extraterrestrials are not here. We can't say for sure that there isn't some long, slow spacecraft orbiting the asteroids and chopping up raw materials.[4]

We have not explored our neighborhood thoroughly enough to rule out the presence of extraterrestrials or their works. As the chances of finding a small alien artifact by accident are almost zero unless it draws attention to itself, we cannot exclude the possibility of a presence in our solar system.

To jump to the conclusion that they are not there simply because we don't see them easily is to make the same mistake people made about microscopic life, argued McDonough. Until Leeuwenhoek invented the microscope, it was thought that nothing smaller than an insect or a mite lived. The world was crawling with zillions of microscopic beasties, but because nobody had seen them, they did not exist.[5]

"Some would argue," Webb acknowledged, "that until we can rule out that possibility, there is no Fermi paradox." The implication is clear: SETI should include searches of our solar system to test the probe and colonization hypotheses.[6]

The only way we can be sure that there are no signals is by conducting an all-sky, all-frequency, all the time search—and that would answer the question only for that slice of time when the search was active. The only way we can be sure that there are no alien artifacts in our solar system is by a thorough search of the Sun's empire, with resolution fine enough to pick up the remains of small interstellar probes.

Russian astronomers L.M. Gindilis and G.M. Rudnitskii concluded that there is no paradox at all. Although the idea of this "astrosociological" paradox was useful in stimulating discussions and active searches for an answer, we should acknowledge the degree of our ignorance and moderate our self-confidence. It is difficult to uncover a contradiction between theory and experiment when we have neither well-established facts nor well-grounded theory.[7]

We do not know enough to definitively answer the questions that are the heart of the paradox. Our searches may someday resolve the issue. Just one detection of an extraterrestrial technology, and the paradox collapses.

Thinking Outside the Box

The paradox is surely telling us that something is fundamentally wrong with our view of the universe, and our place in it.
—Stephen Baxter, 2001[1]

Misunderstanding the Universe

There is another way of looking at the apparent paradox. We may be misunderstanding the reality around us.

We have a long history of misconceiving our larger environment, even our own planet. Our world was—until at least the late eighteenth century for most Europeans and well beyond that for many others—a place of geographical uncertainty. As historian Anthony Padgen reminded us, the Romans, the Mughals, the Chinese, even the Spanish and the Ottomans all had very different versions of our planet, and all were different from the one we have today.[2]

In every age, people have pitied the universes of their ancestors, convinced that they at last had discovered the full truth, observed astronomer Edward Harrison. Yet, the universe, as something seen through the human mind, shaped by human perception, and rationalized by human thought, is reconceived from century to century. Every universe if falsifiable.[3]

Conceptions of the cosmos that our ancestors took for granted were challenged repeatedly as astronomers discovered new evidence and theorists drew new conclusions. We found that the Earth, our solar system, and our Galaxy are not central. We learned that the universe is far more immense than our predecessors had imagined.

We also learned that the universe has a very long history, stretching more than 13 billion years into the past. Generations of stars have passed, many taking planets—and possibly life and intelligence—with them along their evolutionary courses.

Astronomers often underestimated the number and diversity of celestial bodies beyond our planetary system. They once reasoned that the apparent absence of stars in dark nebulae might be real; only in the twentieth century did they infer that invisible matter was diminishing the light of many stars.[4] A generation ago, astronomical textbooks stated that the Milky Way contains about 100 billion stars; newer estimates are as high as 400 billion.

New astronomical capabilities continue to reveal structures that had been hidden from us. Galaxies that are detectable only in submillimeter wavelengths were not discovered until 1997, even though they are some of the brightest objects in the universe.[5]

Do we now stand on the threshold of knowing everything? Or will our latest models of the cosmos be rejected by our descendants?

Twentieth-century science revealed that the universe is much stranger than we had imagined. Scientists now believe that visible matter constitutes less than 1% of the universe's matter and energy. If we add together all forms of familiar (baryonic) matter, including gas in galaxy clusters and the intergalactic medium, we still come up with less than 5%.[6]

Roughly a quarter of the universe is composed of dark (nonbaryonic) matter, which we can detect only by its gravitational effects. The Massive Compact Halo Object survey concluded that between 8% and 50% of dark matter in our Galaxy is in clumps weighing about half the mass of the sun (Some astronomers believe that these are the burned-out normal stars known as white dwarfs). Black holes are thought to make up only a small part of the total.[7]

As of early 2006, the most popular theory still proposed that most dark matter consists of massive exotic particles that do not interact with normal matter except through gravity. If such dark matter is truly different from ordinary matter, we are made of atypical material. As Trefil put it, the kind of matter that makes up our solar system, our Earth, and our bodies is a relatively minor part of a universe that is composed predominantly of very different stuff.[8] In a sense, this could take us back to the pre-Copernican view that the heavens were made of different substances than our familiar world. Others speculate that dark matter is not actually matter at all; it might manifest a change in Newton's laws at large distances.

Astronomers and physicists are designing detectors that might find more direct evidence of dark matter. If they succeed, we may suddenly discover a quarter of the universe that had been hidden from us—one of the great achievements of twenty-first-century science.[9] Yet most of reality still may be beyond our reach.

Since 1998, "dark energy" has been seen as one of the central features of the cosmos, comprising more than 70% of the universe's total matter and energy.[10] The effect of this mysterious force is profound.

For most of the past century, astronomers had assumed that gravity would slow down the expansion of our universe, perhaps contracting it back to a Big Crunch. That paradigm was reversed in the late 1990s when astronomers discovered that the expansion of the universe is speeding up, apparently because of dark energy. The nature of that energy may lie in Einstein's cosmological constant, in the newer concept of "quintessence," in the breakdown of relativity on large scales, or in another of several proposed explanations. Some theorists have suggested that this expansion is part of a longer cycle in which the universe eventually will reverse course and contract back to a crunch.

An article in *Science* described this as "a preposterous universe."[11] How can we say that we understand our macroenvironment when 95% of its features cannot be detected directly by our instruments?

It is not the universe that is preposterous, Sean Carroll responded, it is our theories, which fall short of making perfect sense of it. However, dark energy seems to transcend known physics. "We're going to need a really new idea," said astronomer Robert Kirshner.[12]

Some scientists suspect that many traditional laws of physics might be merely local bylaws, restricted to limited regions of space. The burden of proof, said physicist Andrei Linde, now lies with those who maintain that the universe is everywhere the same and the laws of physics are everywhere the same.[13]

We may be living inside a small pocket of order, science writer George Johnson speculated, a backwater in a universe overwhelmed by randomness. Life may be possible in some parts of the universe, suggested physicist Steven Weinberg, but perhaps not in most.[14]

Mind-Stretcher. If we someday exchange information with extraterrestrial civilizations, we might receive descriptions of the physical universe, including maps of our Galaxy, that are different from our own. Alien knowledge could profoundly alter our sense of where we are.

We must address changes through time as well. Some puzzling features of our universe may simply be the result of the era in which we exist and can observe.[15]

Our understanding of the universe has entered a new period of instability; we may be in a new pre-Copernican era. The better our astronomical technology and techniques and the more imaginatively they are employed, the more likely is the revelation of new realities. Rees foresaw a crescendo of discoveries continuing throughout this decade, due in part to a coincidence of technology, funding, and the way the intellectual discourse has developed.

This uncertainty may have profound implications for the question of extraterrestrial intelligence. If 95% of the universe is in the form of unseen substances, asked Carroll, does this not mean that there is the possibility of hidden structure? Might the dark sector be a fascinating place, with its own interactions, perhaps even a kind of intelligent life?[16]

Future observations and search strategies may open up additional ways to detect intelligence and its works. They may suggest new solutions to the alleged paradox.

The Multiverse

Cosmos after cosmos issued from his fervent imagination, each one with a distinctive spirit infinitely diversified, each in its fullest attainment more awakened than the last; but each one less comprehensible to me.
—Olaf Stapledon, Star Maker, 1937[17]

We face a further radical enlargement of the cosmos: There may be parallel universes. Some scientists see them as a direct implication of cosmological observations. "The simplest and arguably most elegant theory involves parallel universes by default," declared cosmologist Max Tegmark. Steven Dick found this to be a remarkable revival of the Greek concept of a plurality of *kosmoi*—isolated ordered systems.[18]

Consider "brane" theory, one of several proposed concepts. Multiple universes may exist as membranes through a multidimensional hyperspace. Our observable universe may be a three-dimensional surface or membrane—brane for short—in a higher-dimensional world. Another brane—a parallel universe—might reside a subatomic distance away. Additional dimensions could be the size of atoms, or infinitely large. We would never be able to enter them, yet they could have profound effects on the physics of our own universe.[19]

The cosmos might be a multifoliate object, made of sheets that are constantly splitting and occasionally fusing. Unlike previous versions of parallel universes, these "world sheets" can interact with one another. However, an observer embedded on one sheet—one "universe"—would find his or her view limited to that sheet.[20]

There may be many universes, but only a tiny fraction of them is actually cognizable. This, commented Koerner and LeVay, is surely the ultimate in Copernican thinking.[21]

These ideas pose a fundamental problem for scientists. As physicist Roger Newton pointed out, their contact with any conceivable experimental test in the near future—or ever—appears to be minimal. If the concept is pushed too far, warned Davies, the rationally ordered and apparently real world we perceive gets gobbled up in an infinitely complex charade, with the truth lying forever beyond our ken.[22]

The Anthropic Challenge

The widely accepted Cosmological Principle tells us that our universe is much the same everywhere. Most scientists engaged in the debate about finding extraterrestrial intelligence also accept the Copernican Principle: Humankind does not enjoy a preferred position in either space or time. Most have been deeply suspicious of any theory that makes our location or epoch in any way special. By implication, many sites (and times) in the universe may be less favorable to life and intelligence than ours; others may be more favorable.

Sagan popularized a related concept: the assumption of mediocrity. We humans are neither the most advanced or the least advanced of the universe's life-forms. We are not the smartest of the universe's intelligent creatures, nor are we likely to have the most advanced science or the most powerful technologies.

Others have argued that we have gone too far in claiming that there is nothing special about the time and place in which we live. The modern philosophical counterargument to the Copernican Principle is the Anthropic Principle, which proposes that it is possible for an observer's time and place to be unique, if the unique factor is necessary in order for there to be an observer in the first place. The pioneer of this argument was Brandon Carter, who proposed it in reaction to "exaggerated subservience to the Copernican principle."[23]

Barrow and Tipler proposed a definition of the "Weak" Anthropic Principle. The observed values of all physical and cosmological quantities are not equally probable; they take on values restricted by the requirement that there exist sites where carbon-based life can evolve and by the requirement that the universe be old enough for it to have already done so.[24]

The Anthropic Principle has other implications. It implies multiple universes—or multiple regions within a single universe—in which parameters like the cosmological constant have different values, making the laws of physics different. Our universe would be peculiar because humans can exist only in those rare universes or regions with tiny cosmological constants.[25]

The anthropic approach also suggests that such physical phenomena as dark energy and the Higgs particle mass have different values in different parts of the universe; we live in a region where they are small enough to make life possible. If this is the case, many other properties of the universe that we usually consider fundamental, such as the mass and charge of the electron, may be environmental accidents.[26]

Barrow and Tipler argued that we already have seen the first failure of the Copernican Principle. The discovery of the cosmological background radiation in 1965 showed that the universe is changing with time. The epoch in which we live is special in permitting the evolution of carbon life.[27]

The Weak Anthropic Principle has been extended by proposals for a Strong Anthropic Principle. To Barrow and Tipler, the strong version meant that the Universe must have those properties which allow life to develop in it at some stage in its history. They proposed a Final Anthropic Principle: Intelligent information processing must come into existence in the universe, and, once it comes into existence, it will never die out.[28]

Mind-Stretcher. Strong or Final Anthropic Principles have profound implications for the probability of contact. As Mauldin pointed out, Barrow and Tipler's version of the Final Anthropic Principle is surely not limited to humans, but should apply to all intelligent life. If the misnamed "anthropic" principle were firmly established, it would constitute a general proof that life must arise nearly everywhere to justify the universe existing.[29]

Others have been harshly critical of anthropic ideas. Theorist David Gross argued that anthropic reasoning is both defeatist and dangerous—defeatist because it suggests that a more scientific explanation can never be found, and dangerous because it plays into the hands of "intelligent design" supporters who believe that the universe was custom-made for human beings by a benevolent God. "It smells of religion," he said, "and like religion, it can't be disproved."[30]

Physicists Alan Guth and David Kaiser came to a different conclusion. Although the anthropic principle might sound patently religious in some contexts, the combination of inflationary cosmology and the landscape of string theory gives that principle a scientifically viable framework. If future research supports the idea of a multiverse, Rees and Livio argued, anthropic arguments will offer the only "explanation" that we will ever have for some features of our universe.[31]

Aveni saw the anthropic cosmological principle as one of the latest attempts to restore actor–spectator interaction to the whole Earth, to the solar system, even to the universe.[32] It raises the status of human beings, and implies that they may have an important future.

More attention will be paid to the anthropic approach if the search for alien intelligence continues for a long time without success.[33] Widespread acceptance of anthropic principles could reverse the long decline of anthropocentrism.

Self-reproducing Universes

Several theorists have proposed that our universe—and perhaps others—reproduce themselves. Linde, for example, envisioned a cosmos that replicates itself into an infinite number of baby universes with different laws of physics. This implies the existence of a vast plenitude of universes that predate and postdate our own.

In cosmologist Lee Smolin's vision, new baby universes are born in the hearts of black holes. A natural selection process favors the reproduction of universes adept at creating those holes, and thus baby universes. The appearance of a life-friendly universe would be merely a secondary consequence of reproduction.[34]

By contrast, science essayist James Gardner proposed a biology-centered model in which the emergence of life and intelligence are key thresholds in the reproductive cycle. The universe, he claimed, is in the process of transforming itself from inanimate matter to animate matter. Its anthropic qualities can be explained as incidental consequences of an enormously long cosmic replication cycle; the appearance of cosmic design could emerge from the operation of evolutionary forces operating at unexpectedly large scales.

Gardner called his concept the Selfish Biocosm—selfish in the same metaphorical sense as the selfish genes proposed by evolutionist Richard Dawkins. Like genes, the universe is focused on the overarching objective of replicating itself; life and intelligence provide the means. Once life has arisen anywhere, its sophistication and pervasiveness will expand inexorably and exponentially until life's domain is coterminous with the boundaries of the cosmos.

Gardner thought that human and higher-level intelligence would be essential to scaling up biological and technological processes to the stage at which they could exert an influence on the cosmos. The emergence of transhuman intelligence is a necessary precondition for cosmic engineering.

Although he saw the universe as imbued with emergent properties of consciousness and intentionality, Gardner warned that it eventually will focus on its own set of objectives. Those objectives may turn out to be disturbingly alien; the persistence or advancement of humans is not foreordained. Nor is there any apparent reason, apart from altruism, why a supremely advanced community of minds at the end of time would bother to create a new baby universe they would never be privileged to inhabit.[35]

Mallove had suggested earlier that the reorganization of the universe by life may already be underway. We can imagine artificial changes taking place that we simply have not recognized. (Finding evidence of astroengineering might be an indicator of such work being done by others.) Intelligent life may be the emergent catalyst for universe creation; Humankind might be among a handful of civilizations just beginning to realize their potential to change the course of cosmic history. If universes reproduce with the help of intelligent life and if each version of space-time passes on some of its characteristics to its offspring, those that allow highly ordered states of matter will multiply faster.[36]

Physicist J.D. Bernal had proposed in 1929 that, by intelligent organization, the life of the universe could be extended many times. The ultimate outcome might be the conscious universe outlined by physicist Gerald Feinberg in his 1968 book *The Prometheus Project*. Your present author and others speculated further about this idea with articles in specialized publications during the 1970s and 1980s.[37]

Mind-Stretcher. A few scientists have suggested that our universe might be a deliberately created artifact of a prior civilization. The creative agents may exist in universes that are not comprehensible to the human mind, blurring the boundary between the natural and the supernatural. However, they may have left us a message, according to one theory, in patterns we might discover in the cosmic background radiation.[38]

What are we to make of such fascinating but unproven speculations? Their advocates seem to be struggling to reconceive the universe in a way that sees life and intelligence as more than trivial accidents, which includes us as participants with roles to play.

These visions bring intelligent beings back into the picture not as passive observers, but as an active force. Instead of being of being helpless, sentience may be able to intervene in cosmic processes.

Must intelligent life arise, thrive for a period, and then die out because of forces beyond its control? Sentient beings may be able to escape this fate only if they equip themselves with powerful technologies—which may pose their own risks.

Without the intervention of intelligence, the physical universe will careen outward from its origins in the dark, uncaring about its forms or its survival. However, there may be a favorable trend—a spreading emergence of intelligence equipped with the technological means to shape its future. Confirmation of biocosmic concepts could inspire hope.

Our speculations may be skewed because they take place within the context of a particular scientific era. We are immersed in the greatest age of biological research, assigning biology the central place that once was occupied by physics. A future era of scientific thought may focus more on other sciences, perhaps including some not yet invented.

The Appearance of Design

In general the Star Maker, once he had ordained the basic principles of a cosmos and created its initial state, was content to watch the issue; but sometimes he chose to interfere, either by infringing the natural laws that he had himself ordained, or by introducing new emergent formative principles, or by influencing the minds of the creatures by direct revelation.

—Olaf Stapledon, 1937[39]

The universe we observe has precisely the properties we should expect if there is, at bottom, no design, no purpose, no evil and good, nothing but blind pitiless indifference.

—Richard Dawkins, 1995[40]

Astronomer John Herschel argued in the nineteenth century that Darwin's theory did not sufficiently take into account a continuously guiding and controlling intelligence. More than a century later, astronomer Fred Hoyle suggested that an intelligence that preceded us put together, as a deliberate act of creation, a structure for carbon-based life.[41]

We have seen a revival of this idea in the theory of intelligent design, which infers the existence of a powerful, intelligent agent without any specific reference to God. Intelligent design assumes that apparent order in nature reflects a rational mind at work. According to this theory, our universe shows deliberate fine-tuning in its fundamental parameters, such as the ratio of respective strengths of the four fundamental forces. If the ratios were only slightly different, our universe either would never have developed any elements beyond helium or would have collapsed almost immediately.

Intelligent design advocates step in where science, as they see it, has not fully explained complex phenomena such as life. Biologist Michael Behe, noting that the public believes overwhelmingly that life was designed, argued that we can often recognize the effects of design in nature. It is this profound *appearance* of design in life that everyone is laboring to explain (emphasis added). "Since we know of no other way that these things can be produced," he commented, "then we are rational to conclude that they were indeed designed."[42]

Critics of intelligent design respond that the "it must have been designed" argument is just a way of avoiding the toughest scientific problems. Intelligent design proponents do not claim to have a coherent scientific theory about how life changed over time; the most they will claim, said Dawkins, is that there is no argument against intelligent design. "If 97% of all creatures have gone extinct," observed anthropologist Irven de Vore, "some plan isn't working very well."[43]

Others have argued that intelligent design theory is repackaged Creationism, the thinking person's Creationism, or creationism stripped of its explicitly Christian biblical background. To critics, the essential but often well-disguised purpose of intelligent design is to preserve the myth of a separate, divine creation for humans.[44]

Having banished the guiding hand of God from the biosphere a century and a half ago, commented Davies, biologists are reluctant to let it reenter in the guise of a law of nature. However, if it turns out that life does emerge as an automatic and natural part of an ingeniously biofriendly universe, something like design would seem more plausible.[45]

The Catholic Church and Design

Cardinal Schonborn of Vienna, who was the lead editor of the Catholic Church's official Catechism, spelled out a design argument in 2005. He rejected the neo-Darwinian version of evolution—an unguided, unplanned process of random variation and natural selection. Schonborn claimed that an unguided evolutionary process—one that falls outside the bounds of divine providence—cannot exist.

"Any system of thought that denies or seeks to explain away the overwhelming evidence for design in biology," the Cardinal wrote, "is ideology, not science." Scientific claims like neo-Darwinism and the multiverse hypothesis, he argued, were invented to avoid purpose and design. To believe that events in the universe take place entirely by chance is to give up the search for an explanation of the world. The modern Catholic Church, declared Schonborn, stands in firm defense of reason rather than chance.[46]

Guillermo Gonzalez and Jay Richards of the Discovery Institute, in their book *The Privileged Planet*, questioned the assumption of mediocrity employed by Sagan, Drake, and other SETI advocates. Challenging the idea that the Earth is just an average planet orbiting an ordinary star in an unremarkable part of our Galaxy, they argued that our world occupies a privileged place in the cosmos, uniquely situated to foster both complex life and scientific discovery. The correlation between the conditions that make habitability possible and those that make it possible to learn about the universe is so improbable, they asserted, as to suggest intelligent design.[47]

Many take issue with these conclusions. Because it is not testable, some complain, intelligent design falls into the area of faith and belief, outside the scope of science.

One reviewer criticized Gonzalez and Richards for putting forward an untestable hypothesis supported only by a long list of coincidences. Extrapolating any trend into a natural law smacks of teleology, warned Brin—perceiving a plan, or cause and effect, where there may be only coincidence.[48]

Our judgements may be biased by the local conditions and historical contingencies that led to the life and intelligence we know on the Earth, cautioned Vakoch. They may not accurately reflect the range of possible preconditions for the evolution of life, or science.[49]

After hundreds of millions of generations of trial and error, observed Grinspoon, a highly evolved system will be so optimized for survival that it may seem to have been designed by an imaginative and ingenious mind. Aliens in what we consider exotic environments would evolve to succeed in those environments, argued Cohen and Stewart; they would seem to any visitor to be exquisitely fine-tuned.[50]

What if there are countless other universes or countless other big bang/big crunch cycles before the current one, and each of those universes has different physical constants? Science fiction author Robert Sawyer argued that the winning combination is bound to come up eventually by pure chance; it would not require a God.[51] The fact that we are the product of low probability events is not proof of design.

The selection effects and hidden agendas of Intelligent Design offer the last bastion of hope to those who still cling to the belief that we are privileged, declared Darling. Meanwhile, the Copernican revolution is quietly running its course.[52]

On the other hand, failure to develop provable "theories of everything" might change the context. "For the first time since the Dark Ages," wrote physicists Paul Ginsparg and Sheldon L. Glashow, "we can see how our noble search may end, with faith replacing science once again." Others see that as giving up too soon. Science may eventually be capable of answering questions that have been considered religious.[53]

Filling a Philosophical Vacuum

We shouldn't confuse scientific knowledge with moral authority.
—Biologist Dov Sax, 2005[54]

The revival of creationist views may reflect more than resistance to Copernicanism or evolutionary theory. Many people are frustrated with science as a means of giving them emotionally satisfying answers to some basic questions. Judging by the long debate over Darwinism, science may fail to provide three basic functions of religion: suggesting that life has purpose, providing guidance for moral behavior, and helping us to deal with death.

Science also may fail to give us a reassuring sense of dignity. The implacably atheistic bias of traditional Darwinism, Gardner protested, is robbing our culture of its capacity to inculcate a sense of the potential nobility of Humankind.[55]

Consider the emotional impact of how preacher-author Ron Carlson described scientific and religious views of our status. In the secular version, "you exist on a tiny planet in a minute solar system . . . in an empty corner of a meaningless universe. You came from nothing and are going nowhere." In the Christian view, "you are the special creation of a good and all-powerful God. You are the climax of his creation."[56]

There may be a subtle anxiety about living in a disenchanted cosmos. Former Czech President Vaclav Havel argued that the relationship to the world that modern science shaped had exhausted its potential because it fails to connect with the most intrinsic nature of reality and human experience. Addressing the search for a more fulfilling vision of the human future, Havel proposed that it must rest on a fundamental awareness that we are an integral part of higher, mysterious entities.[57]

There is a philosophical vacuum waiting to be filled. Will it be filled by science, or by belief?

True religion is about forming communities around shared values, argued essayists Philip Blond and Adrian Pabst. By contrast, secular liberalism remains blind to cultures built around universal ideals and collective aspirations.[58] The search for extraterrestrial intelligence may be unique among scientific fields in sustaining a community with a grand shared vision of sentience in the universe, a vision that rests on belief.

SETI and Religion

Even in the most primitive ages of every normal intelligent world there existed in some minds the impulse to seek and to praise some universal thing. At first this impulse was confused with the craving for protection by some mighty power. Inevitably the beings theorized that the admired thing must be Power, and that worship was mere propitiation. Thus they came to conceive the almighty tyrant of the universe, with themselves as his favored children.

— Olaf Stapledon, 1937[1]

We cannot avoid the connection between SETI and religion, although many SETI advocates wish that we could. Religious belief has been a recurrent factor throughout the long debate about a plurality of worlds inhabited by intelligent beings.

SETI can be seen as a religious quest, proposed Dick, as science in search of religion. It is a search for superior intelligence, for knowledge, for wisdom, and perhaps for power. Religion in a universal sense is the never-ending search of each civilization for others more superior; the major difference in this case is that the intelligence is not supernatural.[2]

Guthke found that the modern belief in extraterrestrial intelligence had become a religion or quasi-religion with a popular following. McNeill too saw in the CETI movement a pseudo-religion or scientific religion. Secular religions are not new, and many of them have missionaries.[3]

Astronomer Gerrit Verschuur made the point more bluntly, describing SETI as a technological search for God. Some proponents claim that through contact we will learn answers to fundamental questions about the nature of life and death, issues that once were God's bailiwick.[4]

Although some argue that the search for extraterrestrial intelligence reflects an alternate belief system, others see the search as a rebellion against conventional religions. Some SETI pioneers have suggested that their interest in the search was, in part, a reaction to a firm religious upbringing; science offered a secular alternative. Hostility toward religion was particularly noticeable in the writings of Soviet scientists interested in

SETI; one described religion as a clear example of a retarding force slowing down the growth of civilization.[5]

The parallels between religion and SETI may reflect common underpinnings—not only cosmic loneliness, but also a desire for some ultimate purpose to life. However, one can carry this argument too far. Although SETI shares some qualities and some goals with religion, its method is different. The scientific search attempts to confirm belief by experiment, not revelation.

The search may offer us a more subtle satisfaction. When someone meditates on the infinite universe, French philosopher Gaston Bachelard suggested, something of its grandeur is conferred upon him or her, and he or she experiences an expansion of being, a quasi-religious state. Shostak actually compared the Arecibo observatory to a monastery, where SETI astronomers enacted their devotions, sharpened their minds, and reaffirmed their conviction that a signal would one day come.[6]

Science fiction may be another parallel phenomenon. It could only emerge in a context in which the claims of traditional religion were still felt, argued professor of English Mark Rose, but in which belief was, at best, problematic. Science fiction represents a secular transformation of religious concerns.[7]

Religion and SETI: Shared Perspectives

Critics of SETI question the idea of committing resources to a scientific search that may extend beyond individual human lifetimes. Yet, it is a modest effort compared to the time, energy, and money that humans devote to their religions, which they hope will satisfy many of the same needs.

Societal efforts that last for centuries often have rested on enduring organizational structures such as religions. Jill Tarter proposed that SETI may be the one example of a long-term project undertaken and continued out of individual curiosity. The search represents being willing to start something when we might not see the end, simply because we think it is very important. Starting such a search is a milestone in the maturation of a civilization.

Tarter noted another parallel between a belief in extraterrestrial intelligence and a belief in God. Absence of evidence is not evidence of absence.[8]

Gods, Angels, and Devils

As the hierarchy of the universe is revealed to us, we will have to face this chilling truth: if there are any gods whose chief concern is man, they cannot be very important gods.

—Arthur C. Clarke, 1963[9]

Appeals for guidance or intervention from beyond the Earth have been part of many human religions. Often, those appeals were directed to the skies where gods were believed to dwell.

Advanced extraterrestrials, far more omniscient and omnipotent than we are, could have many of the characteristics now attributed to the supernatural God of monotheistic religions. Carl Jung, attempting to define the psychological structure of the religious experience that "saves, heals, and makes whole," concluded that in religious experience Man comes face to face with a psychically overwhelming Other.[10] A prominent physicist came to a similar conclusion. God is what mind becomes, proposed Dyson, when it has passed beyond the scale of our comprehension.

MacGowan and Ordway extended this concept to superintelligent automata. Being relatively omnipotent, they would be remarkably similar to familiar concepts and attributes of a supreme being.[11]

We could carry this further, to a being that created our universe. Such an intelligence might have fine-tuned the physical constants, Dick proposed, thus explaining the anthropic principle. It could even intervene in human history, the touchstone principle of the Christian faith—and of UFO and alien abduction advocates. One group explicitly linked the Bible with UFOs.[12]

Mind-Stretcher. Could the existence of God be an objective question? As science fiction author Robert Sawyer put it, what if messages from aliens contain proof of God's existence? What if they prove that we live in a created universe?[13]

If the cosmos has any lord at all, warned Stapledon, he is not a fatherly spirit but one whose purpose in creating the endless fountain of worlds is alien, inhuman, dark. Others find that Nature does not exhibit the slightest trace of the beneficence that one might expect of the handcrafted product of a loving, caring Creator.[14]

We may be expecting too much. Extraterrestrials might have abilities so far superior to ours that many humans would liken them to gods. Yet, they, too, will be products of biological and cultural evolution in particular environments. They may not have all those qualities and powers that we associate with a Creator.

Instead of resembling gods, intelligent aliens might seem more like angels—spiritual beings superior to humans in power and intelligence. Some Catholic intellectuals have speculated that extraterrestrials might have characteristics that theologians attribute to angels, such as immortality or innate knowledge.

Jung suggested that the aliens of the 1950s were "technological angels" in the vehicles appropriate to a scientific age but having the power and mission of earlier mythic saviors. In a scientific age, asked Sagan, what is a more reasonable and acceptable disguise for the classic religious mythos than the idea that we are being visited by messengers of a powerful, wise, and benign advanced civilization?[15]

The human spacecraft of the future might come to typify an alien society's idea of a UFO, suggested Baird. Extraterrestrials might consider human guests as gods or mythical beings, whose true reasons for coming would remain a mystery.[16]

The aliens in Sagan's novel CONTACT played the traditional role of angels, proposed Davies, acting as intermediaries between mankind and God, cryptically indicating the way toward occult knowledge of the universe and human existence. This theme of aliens acting as a conduit to the Ultimate touches a deep chord in the human psyche.[17] There is an alternate possibility; extraterrestrials might seem more like devils, malignant superhuman beings who injure us.

Science fiction author Brian Aldiss reached farther back into our cultural past, dismissing aliens as merely the latest form of animism. "An intimacy with the non-human is a fundamental human trait," he proposed. "A vast population of ghosts, ghouls, and other mythical creatures has accompanied humankind through the ages. Above these minions, as religion outranks superstition, are assembled an even more formidable array of fictitious beings, the gods and goddesses." Aldiss concluded that "the latest manifestation of the creaking floorboard of the brain, the alien arriving from outer space, is the most interesting."[18]

Here we must introduce a cautionary note. Applying our assumptions about past phenomena to the present and the future may be misleading. The fact that many human conceptions of gods derive from animism does not necessarily mean that extraterrestrials do too. As Cohen and Stewart observed, we cannot dismiss ideas about aliens just because they resemble some human myths. Some extraterrestrials might turn out to be real, even if they are detectable only by their actions.

Puccetti perceived another underlying motivation for human worship of imagined superior beings: a striving to find something that could offer understanding, sympathy, love. Although this desire may have its roots in childhood dependence, it is a genuine need.[19]

Will They Be Religious?

Will religion be important in all civilizations? If it is a useful survival tool, speculated Wason, religion may well evolve on other worlds. If it is a contingent by-product of other evolved capabilities like music, it may be unique to humans and not a common feature of intelligent life.[20]

Some analysts find that faith has practical benefits. Evolutionary biologist David Sloan Wilson proposed that religion is a mega-adaptation, a trait that evolved because it conferred advantages on those who bore it. The religious impulse evolved early because it helped make groups of humans comparatively more cohesive, more cooperative, and more fraternal. The more unified group could present a stronger front against bands of less organized or unified adversaries. "Faith is what allows you to keep going even in the absence of information, evidence, or immediate gratification," said Wilson.

Another key to the success of religion is its emphasis on the moral equality of those in the community. This guarded egalitarianism may be fundamental to the willingness of people to cooperate with others, including those who are unrelated to them.[21]

There is a dark side to this. Although religions may preach kindness and cooperation within the group, they often say nothing about those outside it and may even promote brutality toward others. Some blame religions for the longest lasting wars in our own history.[22]

Many scientists believe that more advanced intelligences, if they ever have organized religions, will abandon them. Tarter argued that the monotheistic religions typical of Earth would be inconsistent with very long-lived civilizations; if such civilizations had any religion, it would be devoid of factions and disputes. For old technologies to exist, such a universal religion must be compatible with scientific inquiry.[23]

Davies recognized that aliens may have discarded theology and religious practice long ago as primitive superstition. However, if they retained a spiritual aspect to their existence, it is likely to have developed to a degree far ahead of our own. We should expect to be among the least spiritually advanced creatures in the universe. Some may take comfort from this, secure in the knowledge that aliens would have a spiritually advancing effect on us should we make contact, but others will feel deeply threatened.[24]

Even if extraterrestrials do not have revealed or traditional religions, they may have a metaphysics. As Pontifical University scholar Paolo Musso saw it, their metaphysics would share two common questions with ours: the First Cause of the world and the ultimate ground of ethics. Intelligent aliens should be able to understand our religious values.[25]

Vakoch suggested that studying theological perspectives on extraterrestrials can challenge—and thus potentially expand—implicit scientific assumptions about the nature of alien intelligence. Such perspectives may allow us to reach beyond our habitual assumptions about ways that intelligent beings will encounter the world and each other; they may help us to anticipate the consequences of contact.

In the case of Catholic theology, such concepts as the preternatural and the supernatural—whether taken literally or metaphorically—can help us see beyond nature as depicted by science. "If we can understand that our way of encountering the universe and our views of spirituality only begin to express the range of ways that intelligent beings deal with Ultimate Reality," proposed Vakoch, "we are guaranteed to gain something very powerful: a more humble, more realistic, and yet paradoxically more complete and more extensive understanding of our own place in the universe."[26]

Interstellar Evangelism

Why not missionaries sent here openly to convert us?

—Charles Fort, 1919[27]

Robert Burton (author of *The Anatomy of Melancholy*) proposed in the seventeenth century that if there were an infinite number of worlds, there would be an infinite number of religions.[28] Might some extraterrestrial civilizations seek to extend their religious beliefs to other intelligent species?

Aliens might come here to proselytize, Shostak suggested, or stay where they are but indulge in high-powered broadcast evangelism. We may find that interstellar communication consists largely of thousands of worlds trying to make converts of each other rather than exchanging scientific information. Essayist Don Lago visualized that we humans might have thousands of religions to choose from, each backed up by the prestige of a great civilization.[29]

Even atheistic religions like Buddhism have had missionaries. Perhaps, speculated science writer Timothy Ferris, we can hope to decode only signals sent by charitable institutions motivated by missionary zeal.[30]

Extraterrestrial radio telescopes might be controlled by priesthoods that censor everything they consider dangerous, Lago warned. There could be an Earthly parallel. Several SETI advocates have assumed that the decoding and interpretation of alien messages will be in the hands of an elite of scientists and scholars who will tell the rest of us what the Heavens are saying.

That could be a channel for alien evangelism. In human history, conversions were often made most effectively through elites, rather than by direct communication with the general population.[31]

Proselytizing may not be limited to religions as we traditionally define them. Our own history tells us that the belief system being advocated may be an ideology. Marxism had its own sacred texts, its own priesthood, and its own missionaries.

The Consequences of Contact

Optimists and Pessimists

Those who have speculated about the consequences of contact have envisioned a wide variety of outcomes, from utopian to disastrous. Predictions have ranged from contact being a passing news event to it being the end of human existence. These speculations have become increasingly detailed—and discordant—since the radio search and the Space Age got under way in the 1960s. At one extreme of the spectrum are best-case scenarios; at the other, worst-case.

Many speculations have clustered around two poles of thought, sometimes described as millenarian optimism and catastrophic pessimism. Summing up the division of views on the impact of radio contact with advanced extraterrestrials, Finney found that they vary between paranoid projections that it would quickly devastate the human spirit and pronoid predictions that the extraterrestrials would swiftly and benevolently lead us into a golden age.[1]

Superlatives abound. Paul Horowitz thought that contact would be "the greatest event in the history of mankind." Sagan shared that view: "The scientific, logical, cultural, and ethical knowledge to be gained by tuning into galactic transmissions may be, in the long run, the most profound single event in the history of our civilization." Sagan even suggested that "it is certainly possible that the future of human civilization depends on the receipt and decoding of interstellar messages."[2]

Optimists describe the consequences as we wish they would be—positive, uplifting. "The effect on human scientific and technological capabilities will be immense," wrote the Byurakan conferees, "and the discovery can positively influence the whole future of Man." Sagan and Drake thought that contact with extraterrestrials "would inevitably enrich mankind beyond imagination."

"Searching for other life in the universe is not an unnecessary luxury," Drake maintained, "but an essential component of forging a better life for Humankind." Sagan found it difficult to think of another enterprise within our capability and at relatively modest cost that holds a much promise for the future of humanity.[3]

Contact pessimists, arguing from what they believe is a hard-nosed, "realist" perspective based on knowledge of our own history, say that human experience does not support the best-case scenarios of contact. They argue that there could be negative consequences ranging from philosophical dislocation to the extermination of *Homo sapiens*.

Arthur C. Clarke thought that contact with extraterrestrial civilizations might be the most devastating event in our history. Stephen Jay Gould, focusing on our intellectual lives, predicted that a successful result from the search would be cataclysmic.

Biologist George Wald declared that he could conceive of no nightmare as terrifying as establishing communication with a superior technology in outer space. Even the generally optimistic Albert Harrison thought that, although intelligent aliens could help us to solve our problems and usher in a new Golden Age, the introduction of their ideas into our society could backfire and create a nightmare without end.[4]

Prudence suggests that we be wary of extreme conclusions. The consequences of contact may lie somewhere between the extremes of optimism and pessimism. Instead of being entirely good or entirely bad, the outcome could be a mixture of positive and negative effects.

The consequences of contact will be influenced by the interaction of many factors. Above all, they will depend on the circumstances in which contact takes place.

Scenarios of Contact: Remote Detection

Intelligent civilizations will limit their contacts to intellectual ones.
—Frank Drake, 1981[5]

In nearly all speculations about contact, the speculator assumes a scenario. The most common since 1960 is a "Standard Model" developed primarily by astronomers. That model foresees a slow exchange of radio messages between civilizations that will never come into direct contact. The debate about consequences has been dominated by those favoring this scenario, who also tend to be the most optimistic about the outcome.

In the classic radio astronomy paradigm, scientists detect a faint signal coming from hundreds or thousands of light-years away. A more advanced civilization reveals its wisdom to us. The impact of contact is gradual; distance gives us time for study, debate, and rational decision-making. The remoteness of the aliens implies that they will be no threat to us, and that the only major outcome of our encounter with extraterrestrials will be an exchange of information.

As Dyson described this "orthodox view" of interstellar communication, the maximum contact between societies would be a slow and benign exchange of messages, an exchange carrying only information and wisdom around the Galaxy, not conflict and turmoil. This SETI orthodoxy has led us toward a minimum contact scenario in which the first intercept is not likely to be very informative. We expect the other civilization to be so far away as to preclude meaningful interactive communication.[6]

The preferred analog is contact between human civilizations separated in time, particularly the transmission of Greek science to the Latin West. Some one-way communications over thousands of years still affect our lives today, notably the world's major religions.[7]

Science historian Dick, assuming a remote contact scenario, argued that the history of science offers deeper insights than political history or anthropology, as the contact will be intellectual and not physical. This suggests that analogs should be drawn from the history of ideas. Dick offered three examples from our own history: the transmission of Greek science to the Latin West via the Arabs in the twelfth and thirteenth centuries, the reception of great cosmological ideas such as the Copernican theory of the sixteenth century and the "galactocentric revolution" of the early twentieth century, and the reaction to Darwinian theory.

Harrison pointed out an implicit assumption underlying Dick's argument: The information that might be exchanged will be of interest primarily to scientists. If alien transmissions address ideology, politics, technology, popular culture, and other subjects, the sociology of science may not be a good model.

Dick recognized that the societal impact will depend strongly on the details. A "dial tone" signal, only giving evidence of intelligence, will be quite different in impact from the decipherment of significant amounts of information. If the latter is achieved, the impact will, in turn, depend on the nature of the information. "No one is wise enough to predict in detail what the consequences of . . . decoding will be," declared a NASA report, "because no one is wise enough to understand beforehand what the nature of the message will be."[8]

Brin described the classic scenario as one centered on beneficent elders. SETI researchers eventually sift a beacon or tutorial broadcast out of the vast sea of stars, a signal that is designed to be decipherable by younger species. We discover that most of the Galaxy is a desert with vast distances separating isolated islands of wisdom. Upon receiving a one-way communication, we begin to become another of those islands. We need not worry about physical contact because that is impossible; nor do we need to worry about the impact of alien ideas, as the Old Ones are wise. Stupid or provocative replies from human groups won't matter because distance makes our replies irrelevant. We will have plenty of time to follow the instructions of our betters.

The remote contact scenario makes the search seem harmless, Harrison suggested, because we can choose not to reveal our presence. Even if our presence becomes known, we expect to be protected by vast physical distances.[9]

Fear of the Unknown?

Chris Boyce critiqued the remote contact scenario in a more pointed way. Many astronomers seem to believe that it is *preferable* to converse with other intelligences by radio, with many years of timelag, rather than to meet them face to face (emphasis added). Many of the same people tend to think that it is better to explore our solar system with machines rather than with humans. They believe that interstellar travel is impossible, or if it is possible, is undesirable, and in either case should not be discussed. If we strip these views of their supposedly philosophical justifications, we find nothing more sophisticated than an unrecognized fear of the unknown.[10]

Scenarios of Contact: Close to Home

The potential impact of contact changes profoundly if an extrasolar civilization is capable of reaching our solar system with robotic spacecraft or inhabited vehicles. Although many scientists dismiss this scenario, the public does not. Opinion polls show that those who accept the existence of extraterrestrial intelligence tend to believe that physical visits by extraterrestrials are probable within this century.[11]

The SETI community has been extremely selective in choosing examples of past contacts between human societies, generally avoiding those that involved direct encounters. Yet, those hundreds of episodes offer the largest database for analyzing the consequences of contact between Humankind and an extraterrestrial society.

SETI researchers also have tended to avoid all links to the long tradition of science fiction, with its vast variety of contemplations about First Contact. Above all, any talk of danger from contact tends to be dismissed as sensationalism. As Brin saw it, this aversion gives Hollywood entirely too much power over our thinking.[12]

Distance is critical because it structures the nature of the contact. White proposed a rule: the closer the contacting civilization, the greater the impact.[13]

Again, the impact would vary with the details. In an extreme scenario, a spacecraft carrying extraterrestrials would land on Earth, possibly allowing face-to-face communication. Although this is the most popular science fiction version of contact, the landing of an inhabited spaceship may be the least likely way for it to occur. Advanced civilizations that explore the regions around other stars probably would find it most efficient to send machines, not crews of living beings; interstellar spacecraft housing biological intelligences may be justified only in the case of one-way migration. We are far more likely to meet the technological creations of the aliens than the aliens themselves.

This logic has driven the emergence of an interstellar probe scenario, in which an alien spacecraft is sent to our solar system on a reconnaissance mission. If we detected a functioning probe, we would have to assume that the alien civilization had discovered us and might be observing us. The implications then would depend primarily on the intentions and remoteness of the senders. Would the probe's report motivate an alien civilization to send additional missions to our solar system? Would it stimulate an exchange of messages? Or would nothing happen that we could discern?

The machine itself might have the competence to represent the other civilization. Bracewell pointed out that an alien probe with a substantial reference library on board could communicate with us in a much more interactive way than a civilization many light-years away. We could expect a major cultural impact, greater than if our first contact is by long-distance radio.[14]

The probe might not be limited to preprogrammed responses. If it were a highly advanced form of artificial intelligence, it might make its own decisions on how to deal with us.

Given the age of our Galaxy, we may be more likely to find a probe that ceased functioning millions of years before our time. Such a machine could have reached our system long ago if the sending civilization had launched a program of interstellar exploration unrelated to signs of intelligence. If the other society had sent a probe after detecting changes in our planet's atmosphere that signaled the emergence of life on Earth, that machine could have arrived more than a billion years ago.

Dead spacecraft or other nonfunctioning technologies left by past visitors would not be immediately threatening, although they would prove that interstellar flight and direct contact are possible. Active alien technologies could provoke stronger emotional reactions, including fear.

Harrison suggested other ways of categorizing the circumstances of contact: first, the familiar versus the strange; second, dawning awareness versus sudden insight; third—and most difficult to foresee—whether alien intentions are positive or negative.[15]

The Human Analogy

In contemporary literature or cinema, the extraterrestrial is most often an idealization of that which humanity would like to be or a caricature of what humanity has fear of becoming.

—Astronomer Jean Heidmann, 1992[16]

A generation ago, science writer Trudy Bell described how our reasoning about extraterrestrial intelligence has been based on what she called "The Grand Analogy" between Humankind and extraterrestrials.[17] That analogy has stimulated our thinking, but also has constrained it. We are extrapolating from a sample of one.

Invoking analogies may be inevitable because so much about SETI is hypothetical. Where ignorance forces conjecture, analogy is a useful (and perhaps the only) guide, observed a group of experts who reviewed the social and cultural implications of remote contact. However, they warned that analogs must not be taken as predictors of action, but only as useful guides to thinking. Nothing in human history is fully analogous to the type of encounter to which the search may lead us.[18]

To which events in our own history do we compare contact? Optimists draw on the most positive examples, often comparing contact with the Western European rediscovery of ancient knowledge that stimulated the Renaissance. Pessimists draw on the worst of our history, often invoking the fatal impact scenario in which a more powerful culture disrupts a weaker one.

Morrison thought that the SETI enterprise could best be understood as an exercise in the archaeology of the future. It is their past that we would be investigating, but our future.[19] Can we simply project our own development and apply that analogy to an alien civilization? Our record of forecasting is poor; experts often have failed to foresee major changes in their own fields.

All straight-line projections of current trends must be suspect, for they do not incorporate the contingency of human history. In any case, we cannot assume that extraterrestrial intelligence will simply be some more advanced form of human intelligence. Alien futures may be very different from ours.

Would our analogies apply to postbiological beings? We may be communicating with inorganic intelligences.[20]

We cannot escape the use of analogy when we consider the possible consequences of contact. However, there is no guarantee that our analogies are correct.

Images of Aliens

ETI's beauty—or ugliness—will reside in the eyes of the beholder.
                          —Albert Harrison and Joel Johnson, 1997[21]

In science fiction, encounters with extraterrestrials often are face to face. The remote contact scenario is at the opposite extreme; we might never know the appearance of our correspondents. Even in the more direct versions of contact described in the films "2001" and "Contact," we never see the extraterrestrials—only their works or their simulations.

Nonetheless, there will be an irresistible human tendency to imagine what they look like. As none of us actually know, the fictional images we carry around in our heads could influence our reaction to first contact.

Our images of extraterrestrials have evolved over the centuries. Early descriptions of interplanetary travel found worlds resembling the Earth, inhabited by pseudo-humans. Even in the first half of the twentieth century, many depictions of intelligent aliens, observed Darling, were little more than extravagantly dressed humans and chimerical animals cobbled together from terrestrial body parts.[22] Lasswitz, in 1897, envisioned Martians so much like us as to allow an interplanetary love affair.

Anthropomorphic assumptions remain powerful. Many commentators expect extraterrestrials to bear some resemblance to humans, justifying this reasoning by an appeal to convergent evolution. The humanoid design may be what we wish to see, for its reassuring familiarity and its confirmation of our unique qualities.

According to this stereotype, intelligent aliens would be bilaterally symmetrical. They would have some means of locomotion; they might walk on two legs because the biped form frees upper limbs for other uses. They would have to have some way of manipulating their environment, such as hands.

To be intelligent, they would need some system for processing and storing information; mass media aliens often have larger skulls than humans to suggest bigger brains. Sensors, especially for sight, may be located high on the body of a land animal to improve range and may be close to the brain to shorten reaction time. Extraterrestrials might have features similar to eyes and ears.[23]

Humans who claim to have seen extraterrestrials have described a wide variety of faces and body types, but most are humanoid. Drake extended the human analogy to the point of saying: "They won't be too much different from us. If you saw them from a distance of a hundred yards in the twilight you might think they were human." Clarke was more skeptical: Nowhere in the galaxy will there be creatures that we could mistake for human beings, except on a very dark night.[24]

Others believe that the probability of encountering humanoids is vanishingly small. Even if we find intelligent beings with physiques analogous to our own, their internal structures and chemistries, their genetic materials, and their perceptions of their surroundings may be very different from ours. There would be variations in their living and nonliving environments and the directions taken at the myriad branching points of their evolution. Aliens would be shaped by separate histories.

Anthropologist Loren Eisely expressed this eloquently. "Nowhere in all space or on a thousand worlds will there be men to share our loneliness. There may be wisdom, there may be power; somewhere across space great instruments . . . may stare vainly at our floating cloud wrack, their owners yearning as we yearn. Nevertheless, in the nature of life and in the principles of evolution we have had our answer. Of men elsewhere, and beyond, there will be none forever."[25]

Is there some universality in being and behavior that transcends evolutionary differences? Jack Cohen and Ian Stewart, who looked at this question in some detail in their book *What Does a Martian Look Like?*, proposed a test. If a feature rose more than once in the evolution of life on Earth (such as photosynthesis, locomotion, limbs, predation, flight), it is a universal. If it arose only once, it is a parochial.

Mind-Stretcher. Is human-level intelligence a universal or a parochial? As far as we know, it has arisen only once on Earth, although dolphins, chimpanzees, and earlier hominids may not be far behind.

Astrobiology as currently practiced does not allow us to imagine the spectrum of possibilities, concluded Cohen and Stewart. Our imaginations cannot conceive of anything truly alien.[26]

"There's no limit to strangeness," said Dyson. "The most likely form for E.T. is something we never imagined."[27] Intelligent aliens may not only be stranger than we imagine; they may be stranger than we can imagine. Our cultures constrain our visions.

Science Fiction Images

Each culture or subculture has an "image repertoire," suggested one group of experts, a store of images and dramatic scenarios that its members share and draw upon (consciously and unconsciously) in trying to imagine the possibilities inherent in situations of which they have little or no experience. These images and scenes may be from cultural myths, literature, familiar artwork, popular imagery, and media. Most cultures now possess a body of speculation on possible encounters between humans and intelligent aliens, although these speculations may have no relationship to scientific data.

Authors of written science fiction have been free to describe the most exotic forms. Television and cinema portrayals once were more constrained to humanoids, particularly if the extraterrestrial had a speaking part. Now special effects can create any creature we imagine and give it a voice.

Images of aliens and their behavior tend to reflect the era of human history when they are visualized. In the written science fiction of the nineteenth century, aliens often were seen as Darwinian competitors. More recently, American film and television depictions of extraterrestrials have oscillated from one extreme to another, from the weird and horrible invaders of the 1950s ("The Thing", "Invasion of the Body Snatchers") to the benign aliens of the 1970s ("ET", "Close Encounters of the Third Kind") and then back to the repugnant aggressor of the 1980s and 1990s ("V" and "Independence Day").

As film and television professor Vivian Sobchak saw it, the mass media have given rise to three images of extraterrestrials: the menacing and dominating "colonizing" alien; the benevolent alien that has come to save us from ourselves; the cyborg, part living being and part machine. There is no reason to expect that mass media will feel any particular responsibility to imagine extraterrestrials as benign and unthreatening.[28]

The basic formula remains the same. Good aliens are humanoids; bad aliens are life-forms that frighten or disgust us. We use appearance as shorthand for intentions. Yet, what will matter most in a contact situation is not the way they look, but their abilities and their behavior.

At one extreme, we think of aliens as altruistic teachers who will show us the road to survival, wisdom, and prosperity, or God-like figures who will raise Humankind from its fallen condition. At the other extreme, we see the aliens as implacable, grotesque conquerors whose miraculous but malevolently applied technology can only be overcome by simpler virtues. These images are extensions of centuries-old debates about human nature. They are exaggerations of ourselves, at our best and at our worst.

Myth, religion, and now science fiction with their tales of benevolent and malevolent extraterrestrial beings are commentaries on the human condition, observed philosopher Lewis Beck. To him, even responsible scientific speculations about our search for others were the modern equivalent of angelology and utopia, or of demonology and apocalypse.

A Mirror Image. To an extraterrestrial, Boyce reminded us, we would be the aliens—the bizarre creatures inhabiting a strange and mysterious planet. Just as they might resemble some of our myths, we might resemble some of theirs.[29]

Alien Machines

We are looking for evidence of alien technology, rather than the alien beings who created it. The two categories might merge in advanced intelligent machines, possibly including interstellar probes.

Such intelligences may be much more widespread than intelligent biological societies. Although the planets may belong to organic life, Clarke proposed, the real masters of the universe may be machines. We creatures of flesh and blood are transitional forms.[30]

Even if alien biological entities have attempted interstellar radio communication, Davies thought it nearly certain that machines, with their unlimited patience, will dominate the airwaves; a randomly received radio message is overwhelmingly likely to originate with one of them. Such a machine, warned Koerner and LeVay, may have lost all memory of its organic ancestors.

Grinspoon foresaw more ambiguity. If we receive an interstellar message, we may never know if it was sent by machines or biological organisms.[31]

There would seem to be less ambiguity in the probe scenario, in which our contact is with a machine. But who sent the probe? The machine we discover may be in the service of another machine.[32]

What about direct contact on Earth? In the film "The Day the Earth Stood Still," the powerful robot Gort appears to be under the supervision of the humanoid Klaatu. That film was based on a story—Harry Bates' "Farewell to the Master"—with a very different message. We discover on the final page that the Master is not Klaatu, but Gort.

Most popular images of machine intelligences are frightening, although the benign alternative has been suggested. R2D2 and C3PO, the cute robots from "Star Wars," were firmly under human control; they seemed unlikely to violate Asimov's laws of robotics, which forbade harming a human being. Robbie, in "Forbidden Planet," demonstrated that he could not carry out such an order, burning out his circuits instead. The film "The Iron Giant" portrayed a huge robotic creature who was powerful enough to be destructive, but who was peacefully inclined (he even had a smile fixed on his metallic face).

Our current opinions are heavily influenced by the omnipresence of information technology and robotics in contemporary culture, and by predictions of artificial intelligences that are smarter and more powerful than we are. Although these technologies are at the center of our attention today, they may not be a generation from now. Our images of aliens then may emphasize different qualities.

Timing

The consequences of contact for Humankind also will depend on the state of our own civilization at the time; our reactions might vary widely from one era to another. Other factors being equal, proposed Harrison, the basic rule is that human reactions will be more positive in good times than in bad.[33]

The emotional context of the times strongly affects the ways in which people react to news. The underlying factor, found the social impact group, is expectations. During a time in which international tension is low, tolerance of differences is high, prosperity is generally on the rise, and perceived disjunctions in our lives are few, people are likely to have positive expectations. Conversely, a time full of upheaval may provoke an anxious emotional climate. In such anxious times, negative events are perceived as parts of a pattern. Even events that do not conform to negative expectations may be interpreted as threatening.[34]

In much of the world, news of contact might be received more favorably now than it would have been at the height of the Cold War. However, reactions may vary from one society to another. If one culture or subculture is subject to stressful events that are contrary to the worldwide context, its responses may be very different.

The more time that elapses between now and the moment of contact, the more the cultural and political climate may change. That future climate might be more welcoming and enthusiastic toward extraterrestrials than it is now—or less.

Aliens inspire both wonder and terror; our choice is influenced by our cultures, our personal experiences, and the mood of our era. Our initial reaction to contact may depend on which image of extraterrestrials we have in our minds at that time.

Hopes

Those most hopeful about the consequences of contact range from millenarians who expect that extraterrestrials will guide us toward utopia to moderate optimists who simply hope for a net gain from contact. Here are some common themes.

Reassurance

Contact with another technological species could be reassuring to a species as doubtful about its future as we are. Discovering others would tell us that life and intelligence had survived and prospered elsewhere, even after acquiring powerful technologies.

Several SETI scientists have placed particular stress on the dangers of our own time, which some have described as a period of "dangerous technological adolescence." If we were to discover that many other civilizations had survived the same sort of transition that we are presently passing through—and if we could learn from their experience—we might improve the odds for our own long-term survival.[1]

A detection would provide "a tremendous morale boost to those of us who are concerned about our own existence in a technological framework," predicted Robert Edelson of the Jet Propulsion Laboratory. "It would say that technological societies can find ways to have a life span of thousands or tens of thousands of years." Billingham, too, thought that contact "would change the opinions of many people who feel very strongly that we only have another twenty-five years to go here."[2] (He made that comment 25 years ago, in 1981.)

Despite their emphasis on progressive change, those most optimistic about the feedback effect have shown a striking fixation on social stability. Sagan proposed that "advanced extraterrestrial civilizations, motivated either by altruism or through a selfish interest in maintaining a stimulating set of communicants, convey the information for stabilizing societies." Jill Tarter hoped that the detection of a signal, even without any information content, would let us know that it is possible to stabilize a society and have it live for a long time.[3]

Contact with a more advanced society also would suggest that the present state of human development need not be final. We would know that there can be evolution to a higher level; that could motivate us to transcend our condition.

Puccetti offered a more modest prospect. It could comfort us to know, or have some scientific foundation for believing, that there are other natural persons in the universe somewhat like us physically, organized into moral communities, and sharing some of our own values.[4]

A Sense of Community

It will be a moment of joy and pride. The eternal isolation of the spheres is vanquished.
—Charles Cros, 1869[5]

Contact might bring us into communication with other beings who feel the joy and pain of awareness, who worry about their survival, and who seek answers to many of the questions we ask about the purpose and destiny of intelligent life.

Horowitz saw this as the end of our isolation, the first bridge across 4 billion years of independent evolution. This is remarkably similar to Clarke's comment about spaceflight: "The coming of the rocket brought to an end a million years of isolation."[6]

A Mirror Image. Science writer Ian Ridpath suggested a reversal of roles. If no one before has reached "cosmic maturity," our radio call signs in the future may be the lifeline that pulls other developing civilizations out of the despair of their own isolation.[7]

The Cyclops report drew an analogy with our own history—implicitly a model of the future, and a hope. The discovery of the New World, the circumnavigation of the Earth, and the development of trade routes to the East brought cultures into contact that had long been isolated. There followed a period of trade and cultural enrichment in which change and growth were more rapid than in the centuries before.[8] The authors left out less attractive facts, such as the beginnings of imperialism and colonial exploitation.

Tsiolkovskii, who advocated an end to nationalistic thinking, urged people to think of themselves as citizens of the cosmos. He saw this change as necessary preparation from humanity's joining a cosmic community of intelligent beings. Others put this idea into a more organizational form; Stapledon wrote of an interstellar League (perhaps inspired by the then-existing League of Nations) and of a galactic Society of Worlds.[9]

This idea has been revived by several others. Bracewell suggested in 1960 that superior communities throughout the Galaxy are probably already linked together into a chain of communication. He later developed the idea of an interstellar community into what he called "the Galactic Club" (J.D. Bernal had suggested a club of communicating intellects in 1967).[10]

Many people interested in SETI found this vision of a galactic community to be such an attractive concept that they came to assume its existence. Albert Harrison, for one, thought that a Galactic Club would be of immense size and very stable.

Harrison drew a distinction between the "slow track" and the "fast track" of admission into such a community. In the slow track, an initial encounter will lead to accelerated search efforts that will put us in touch, one by one, with additional alien societies. In the fast track, our initial contact will be with an affiliate of the Galactic Club; this society will give us instant access to other members. Either by a long and circuitous path or by events that could occur almost overnight, we might join an interstellar network of civilizations. The two tracks have different implications for managing initial contact, security, knowledge transfer, cultural change, and preserving our identity.[11]

For some, our desire to be accepted into the larger community may imply a wish to be judged worthy, to meet moral, ethical, or political entrance requirements. Several people have predicted that such a galactic society would not reveal itself until the lower-level civilization was considered "ready." When they did contact us, their communications might be a series of instructions, a how-to kit for participating in that society.[12]

Civilizations may be divided into two great classes, Sagan proposed: those that make an effort, achieve contact, and become members of a loosely tied federation of galactic communities, and those which cannot or choose not to make such an effort, or who lack the imagination to try, and therefore soon decay and vanish. "Human history," he argued, "can be viewed as a slowly dawning awareness that we are members of a larger group. . . . If we are to survive, our loyalties must be broadened further."[13]

Sagan also argued the other side of this issue, warning of hubris. "I think it a great conceit," he wrote, "the idea of the present Earth establishing radio contact and becoming a member of a galactic federation—something like a bluejay or an armadillo applying to the United Nations for member-nation status."[14]

Ego Satisfaction

Alien civilizations may broadcast information to herald their achievements and to perpetuate their views.[15] We might do the same. We, too, might seek to spread our knowledge, our cultures, our religions, our values, and our imagined status.

We already have sought to make other intelligences aware of our existence, notably by messages transmitted from radio telescopes. If we detect extraterrestrials, that compulsion could be even stronger. Planetary Society SETI Coordinator Thomas McDonough believed that the temptation to transmit our heritage and receive a kind of immortality would be irresistible; Sagan and Shklovskii thought it would not be "immodest."[16]

There may be a more subtle kind of satisfaction. As Davies saw it, the most important consequence of discovering extraterrestrial life would be to restore to human beings something of the dignity of which science has robbed them. Far from exposing *Homo sapiens* as an inferior creature, the certain existence of alien beings would give us cause to believe that we, in our humble way, are part of a larger, majestic process of cosmic self-knowledge. In Walter Sullivan's vision, communion with cosmic manifestations of life would join us with a far more magnificent form of continuity.[17]

By contemplating contact with other-worldly creatures, suggested historians Launius and McCurdy, humans grant themselves a privileged position in the cosmos, worthy of visitation by God-like beings. Psychologist Carl Jung had expressed a similar view decades earlier: When a human attracts the interest of another world, his status is enhanced, and he acquires a cosmic importance.[18]

Unifying Humankind

If contact occurred, many predict, we would see the common nature of humans defined by contrast with aliens. Contact would remind us, as nothing else could, of our identity as a species. "The differences among human beings of separate races and nationalities, religions, and sexes are likely to be insignificant," Sagan claimed, "compared to the differences between all humans and all extraterrestrial beings."

Many hope that this would have a unifying effect, easing tensions and encouraging cooperation among humans. Sagan thought that this effect of contact is, by far, the strongest social value of the search; Drake declared that an implicit goal of interstellar communication is to draw together the residents of Earth. For many SETI advocates, the idea that humanity will immediately unite in the face of "others" has become an assumption.[19]

Others expect more modest results, foreseeing that the unifying effects of a remote detection may be temporary, and that differences among nations are likely to reemerge. Some past threats or opportunities that appeared to involve most of Humankind have had limited but significant effects on international cooperation. Nations have built international institutions and mechanisms to deal with exploiting a technology, such as the allocation of the radio frequency spectrum; to deal with evident externalities, as in limiting ocean dumping; to share costs that are too high for one nation to bear, as in the development of a magnetic fusion reactor or a permanent space station; to explore or exploit resources beyond national territories, as in the Antarctic and outer space; and to counter certain threats, such as smallpox or the proliferation of nuclear weapons. However, these events were qualitatively different from contact with an alien civilization.[20]

Some argue that a common enemy might stimulate human unity more than a common friend. Early in the past century, French biographer and novelist Andre Maurois suggested in his book *The War Against the Moon* that a positive method for securing and preserving peace among the nations of our world would be to invent a spurious and imminent threat from outer space. Sociologist Roberto Pinotti proposed a similar argument decades later; we may need some outside, universal threat to make us recognize our common bond.[21]

Former U.S. President Ronald Reagan made this point several times. If all humans discovered that they were threatened by a power from outer space, "wouldn't we all of a sudden find that we didn't have any differences between us at all, we were all human beings, citizens of the world, and wouldn't we come together to fight that particular threat?"[22]

Removing the "enemy" to celestial distances might defuse terrestrial conflicts, Jill Tarter foresaw. Billingham was more skeptical: The idea of extraterrestrials a hundred light-years away being a threat is not very realistic, although it might have some unifying effect on us.[23]

Detecting an alien presence in or near our solar system would provoke more intense reactions. Earth's nations might find cooperation easier to establish and to sustain.

A Shortcut to Wisdom

The highest prize in civilization is new information.

—Nikolai Kardashev, 1981[24]

Simply detecting an extraterrestrial technology would bring us new knowledge about the evolution of life and intelligence, especially if we could determine the characteristics of their home star and planetary system. Even undecipherable signals could tell us some things about their technology and their command of energy.

Many contact optimists expect more. Assuming the remote contact scenario, they believe that extraterrestrials will send us information-rich messages (one collection of essays on the subject was subtitled "The Impact of High Information Contact"). According to this school of thought, we would be the beneficiaries of new knowledge and new sensibilities, willingly provided to us by more advanced beings.

Such messages would resemble old dream—the granting of new knowledge to Humankind, by a Prometheus from the stars. In the most optimistic visions, our absorption into a network of communication would draw us into a larger consciousness whose sensory tendrils reach throughout the Galaxy, whose collective memory contains riches of information beyond imagining.

Contact optimists foresee that more advanced civilizations will be generous in sharing their knowledge. Drake fully expected an alien civilization to bequeath us vast libraries of useful information to do with as we wish. Sagan anticipated that the consequences for our civilization would be stunning—insights on alien science and technology, art, music, politics, ethics, philosophy, and religion.[25]

MacGowan and Ordway stretched this farther: the *only* immediate effect on human society of communications with extrasolar communities would be the sudden acquisition of a vast amount of very advanced scientific knowledge (emphasis added). Because two-way communication in the normal human sense is impossible over interstellar distances, only massive exchanges will take place. Great blocks of information would be prepared by scientists on the basis of sound logical planning and assumptions concerning the interests of the listener. These authors held out a euphoric hope: All of the important questions in science, engineering, and social science could be answered for us.[26]

Morrison suggested that extraterrestrials might send us a volume of knowledge greater than that transmitted to medieval Europe from the ancient Greeks, stimulating a new and greater Renaissance. All we know about ancient Greece is less than 10 billion bits of information, Morrison estimated, a quantity he suggested be named the Hellas. Information-rich alien messages might add up to much more than this gift of the Greeks. Our problem is to send to other cultures—and to receive from those cultures—something on the order of 100 Hellades of information.[27]

Here, it is worth recalling that the Renaissance was due not just to the rediscovery of ancient knowledge but also to renewed faith in human capability. Would we have that faith after intense contact with a superior civilization?

Mind-Stretcher. Sagan suggested that civilizations might be classified by their information content.[28] Where would we rank in that classification? If a more advanced civilization offered us new knowledge, could we offer anything meaningful in return?

Sagan developed the concept of an Encyclopedia Galactica, a vast treasury of information amassed by other intelligences that might be available to newly arrived technological civilizations like ours. We might find ourselves plugged into a kind of galactic telephone network—a brilliant intellectual commerce with a magical and dazzling diversity of civilizations. McConnell, writing a generation later, envisioned what he called an interstellar Internet.[29]

Cyclopean Optimism

The authors of the Project Cyclops report speculated that interstellar communication has been going on in our Galaxy ever since the first intelligent civilizations evolved in large numbers some 4 or 5 billion years ago. All participants would have accumulated an enormous body of knowledge handed down from race to race, from the beginning of the communicative phase. Included in this galactic heritage we might expect to find the complete natural and social histories of countless planets and their species, a sort of cosmic archaeological record of our Galaxy. As new races came of age and made contact with the galactic community they would inherit this body of knowledge, add to it, and, in turn, pass it on to still younger races when they made contact.

The Cyclopeans foresaw a synchronization of scientific development among the cultures in contact. They hoped for the development of branches of science not accessible to one race alone but amenable to joint efforts, the discovery of the social forms and structures most apt to lead to self-preservation, and new aesthetic forms and endeavors that lead to a richer life.[30]

Many optimists foresee that communication with extraterrestrials will accelerate our intellectual growth. "Interstellar contact would undoubtedly enrich our civilization with scientific and technical information which we could obtain alone only at very much greater expense," Drake assured us; "there is probably no quicker route to wisdom than to be the student of more advanced civilizations."

Walter Sullivan hoped that knowledge of a more advanced civilization might enable us to leapfrog centuries ahead. In their novel *The Cassiopeia Affair*, Chloe Zerwick and Harrison Brown imagined that the more advanced civilization would enable us to bypass centuries of development by providing us with a kind of cosmic technology assistance program. Shostak speculated that we might skip eons of history.[31]

If this flow of information actually occurred, we could compare our knowledge and our perceptions with those of other minds in different environments, illuminating voids in our own knowledge and suggesting new generalizations. Drake proposed that we would learn profound aspects of intelligent life that we as yet have not begun to imagine. We would begin to appreciate biological and cultural systems grounded in evolutionary processes separate from our own. This is bound to emphasize the narrow interrelatedness of all human experience.[32]

As aliens would view the universe somewhat differently, Sagan predicted that they would be interested in things we never thought of. By comparing our knowledge with theirs, we could gain insights that might have passed us by.[33] Our curiosity would be stimulated by discovering how much we had not known, or had misunderstood.

Contact also could reveal areas of shared knowledge, supporting our own conclusions. Alien knowledge, integrated with our own, could lead to new syntheses, a boom in interdisciplinary studies as we perceived new linkages, and new branches of science. Morrison and others foresaw that interpreting alien signals would become a major social task, comparable to a very large branch of learning.[34]

What is important is not a single discovery, argued philosopher Beck, but the beginning of an endless series of discoveries that will change everything in unforeseeable ways. If they are made, there is no limit to what we might learn about other creatures, and about ourselves. Compared to such advances in knowledge, the Copernican and Darwinian revolutions and the discovery of the New World would have been but minor preludes.[35]

Dealing with a massive influx of knowledge could force us into mind-stretching responses; it could change our criteria of what matters. Baird speculated that the actual potential of the human mind may be realized only when external conditions demand more.[36]

A Copernican View of History

Every civilization sees itself as the center of the world, claimed political scientist Samuel Huntington, and writes its history as the central drama of human history. That perspective could be shattered by contact, which might provoke a Copernican revolution not just in the scientific sense, but in the historical sense as well. Oswald Spengler, author of *The Decline of the West*, had argued as far back as 1918 that it was necessary to replace the Western "Ptolemaic" approach to history with a Copernican one.

Communicating with many worlds could help to make history an experimental science. We might learn the stories of civilizations stretching far back into the galactic past, becoming aware of alternative cultures, arts, social and economic systems, and forms of political organization.

John Macquarrie of the University of Glasgow speculated that the universe has produced—and will continue to produce—countless millions of histories analogous to human history. MacVey put it more romantically: "Between these islands may be passing even now messages that speak of galactic empires, of celestial dynasties, and of strange events long past." We might reflect on their achievements, their sagas and tragedies, their hopes and fears, aspirations and doubts, kindnesses and cruelties.[37]

One of the reasons that the social sciences lack the maturity of the physical sciences, Finney reminded us, is that so far we have had only one opportunity to study the development of consciousness, intelligence, and culture. The discipline that Soviet authors called "exosociology" still draws its experimental data from the only civilization known to us—our own. If alien civilizations told us their stories, we would be able to compare. Perhaps there are deep laws of individual and social behavior that hold true for all species, all times, and all cultures.[38]

Can we measure the impact of alien knowledge? White suggested a formula in which the key factors are the "parity difference" in years, the fraction of total knowledge transmitted, the distance in light-years, and the time; the impact would intensify as the amount of new information transmitted increased (he apparently assumed remote contact). Harrison observed that there may be variables: the extent and pertinence of their knowledge and their ability to share it with us, their willingness to help, and our receptivity to what they have to offer.[39]

Others have proposed ways of categorizing the impact of a remote detection and the receipt of alien information. Billingham saw this as divided into two phases: the months or years following the detection, and the long-term, in which the knowledge gained is absorbed into human society. Harrison and his colleagues put it differently: At Force 1, we will assimilate the discovery that we are not alone in the universe. At Force 2, we may gain information from the alien society that will affect our own science and technology, with far-reaching implications for our economy, our political institutions, and our international affairs. At Force 3, we would communicate and interact with the other civilization, trading information and perhaps developing a long-term dialogue.[40]

Several commentators have speculated that the knowledge of a galactic community is not allowed to perish but is passed on from civilization to civilization, from one region of the Galaxy to another, eon after eon, in a kind of serial altruism. We ourselves may decide that we have a responsibility to transmit for the benefit of less advanced civilizations.[41] (One may ask if those less advanced would be able to receive our messages.)

"What if the Romans had prepared and distributed broadly across their empire volumes containing all their knowledge and that of the Greeks and Egyptians and all previous civilizations?" asked science fiction author James Gunn. "Would the Dark Ages have lasted for a thousand years?"[42]

Sullivan proposed a more melancholy scenario: True wisdom may be a torch—one that we have not yet received, but that can be handed down to us from a civilization late in its life and passed on by our world as its time of extinction draws near. Gunn described such a transfer of knowledge from a dying civilization in his novel *The Listeners*: "The transmission from Capella would continue for days or weeks or months, but eventually the inheritance from another star would be handed over, the messages would cease, and the silence would surge softly backward. . . . By that time the computer would be at least half Capellan."[43]

Paths to Utopia

Those of us who had come from less fortunate planets found it at once a heartening and yet a bitter experience to watch world after world successfully emerge from a plight which seemed inescapable, to see a world population of frustrated and hate-poisoned creatures give place to one in which every individual was generously and shrewdly nurtured, and therefore not warped by unconscious envy or hate.

—Olaf Stapledon, 1937[44]

For centuries, writings about extraterrestrials have been used as a device for social criticism. Many authors have envisioned alien utopias, implicitly intended to be models for our own future.

Some of those skeptical of traditional Christian concepts of heaven or the afterlife imagined planetary paradises populated by angelic extraterrestrials. Philosopher and statesman Viscount Bolingbroke, writing in 1754, proposed that "all the inhabitants of some other planet may have been, perhaps, from their creation united in one great society, speaking the same language, and living under the same government; or too perfect by their nature to need the restraint of any."

Nearly 200 years later, Stapledon described a galactic society in which each world was "peopled with its unique, multitudinous race of sensitive individual intelligences united in true community." Sagan revived the utopian vision in the 1970s, imagining alien societies "in excellent harmony with their environments, their biology, and the vagaries of their politics, so that they enjoy extraordinarily long lifetimes."[45]

Drake thought that contact would provide us with a glimpse of what our own future could be. We might learn the best course of action in planning the development of our own civilization; we would learn ways to improve the quality of life on Earth. Once we know what is possible—and maybe even what is desirable—we may find general rules of civilization. Many of the mistakes that we might otherwise make would be avoidable, McDonough hoped, if we just had the benefits of the history of older cultures.[46]

Several optimists have foreseen that extraterrestrials will tell us how to solve the problems of our own time. "An early message may contain detailed prescriptions for the avoidance of technological disaster," proposed Sagan, "which pathways of cultural evolution are likely to lead to the stability and longevity of an intelligent species, and which other paths lead to stagnation or degeneration or disaster. . . . Perhaps there are straightforward solutions, still undiscovered on Earth, to problems of food shortages, population growth, energy supplies, dwindling resources, pollution and war."

Drake thought that information from extraterrestrials about science, technology, and sociology could improve our abilities to deal with sociological problems such as poverty; it could advance our medicine and help us develop cheaper energy. *The New York Times* picked up this theme, speculating that there may be beings who long ago found the cure for cancer, solved the problems of taming thermonuclear energy, and routinely practice genetic engineering for the benefit of their species.[47]

"The signals from a more advanced civilization might contain the solutions to our greatest problems," claimed McDonough, "problems that most likely occur to every civilization as it advances: dwindling natural resources, war, pollution, overpopulation, poverty, cancer. Solutions to these problems may be all around us, flying invisibly through the very room where the reader is sitting, just waiting for us to detect them." Geochemist Alan Rubin described the utopians as The Cure for Cancer Camp—people who believe that alien radio transmissions might allow us to achieve world peace, solve scientific puzzles, develop new art forms, and gain advanced technology.[48]

Again we find the dream of immortality, granted to us from above. The signal we receive, Drake envisioned, would be "the song of people who have been alive, every single one of them, for a billion years." The alien civilization would spread the secrets of their immortality among young, technically developing civilizations; they might send the information that would make this same immortality possible for all the creatures of the Earth. McDonough hoped that more advanced aliens would tell us how to decipher genetic codes atom by atom and to routinely fix the errors that we call diseases. Death by natural causes would be unheard of; only death by accident, crime, and war would still be possible, and those might be largely eliminated by alien wisdom as well.[49]

Many have argued that alien information could extend the longevity of human society. The Cyclops report suggested that interstellar contact may greatly prolong the lifetime of civilizations. Billingham told *Newsweek* 2 years later that we might learn how more sophisticated civilizations organize their social institutions, energy supplies, raw materials, and population problems so that they are assured of long-term survival.[50]

Von Hoerner thought that the positive feedback from contact would increase average communicative lifetimes. Sagan endorsed the feedback hypothesis, suggesting that interstellar communication not only would enlarge the number of civilizations but also may be the agency of our own survival.[51]

Peter Schenkel, author of optimistic science fiction novels and essays about contact, laid out a dazzling utopian vision:

Most extraterrestrial civilizations will have overcome their primitive evolutionary stages and will have created superior orders of global stability and harmony. They will have outlawed war and violence and done away with glaring inequalities.... They would neither be wicked in our sense nor pursue hostile ends with regard to other intelligent species in the galaxy. Upon contact they would behave as friends and give us access to useful knowledge, just as we would, were we to meet intelligent beings on a distant planet.[52]

Some of those involved in the search have been more cautious; five scientists warned that we should not look to contact as a cosmic cure for the problems of Earth. Harrison recognized that alien ideas will not magically cure our ills, although he thought that we are more likely to gain than to lose from exploring those ideas.[53]

Salvation

The frustrations and limitations of human life on Earth, the overhanging threat of disastrous conflict, the lack of moral anchors, our isolation amid the vastness of an unfeeling universe, our apparent helplessness against uncaring entropy, all have driven many humans to hope for intervention from above. Aldiss observed that the wish for a redemptive guardian to rescue us from our difficulties is deeply embedded in the human psyche.[54]

Some carry this farther, to a hope that intelligent aliens will save our troubled species. Access to the galactic heritage, wrote the Cyclopeans, "might well be the salvation of any race whose technological prowess qualifies it." Brin found it no surprise that millions of humans yearn for contact; in their minds, that idea is linked with Salvation.[55]

Sagan cited the tendency of some UFO advocates to expect that "we are going to be saved from ourselves by some miraculous interstellar intervention."[56] Such expectations can lead to extreme behavior. In 1997, 39 members of the Heaven's Gate cult committed suicide in the apparent belief that a UFO accompanying a comet was their ticket to extraterrestrial bliss.[57] On a gentler level, the film "Cocoon" envisioned aliens rescuing humans from aging and death.

Many also seek moral approval—or censure—from our superiors. Contact could lead either to our rapidly attaining superior status ourselves, proposed Maxwell Cade, or it could lead to our extinction. It probably depends on how well we can conceal, or overcome, our own grave failures as social beings.[58] This desire to be judged has powerful resonances in religion.

An Opening into Deep Time

The discovery of space nurtured the discovery of time.
—Stephen Pyne, 1994[59]

The search, and the detection of others, could encourage us to think in terms of what Gregory Benford called Deep Time. We humans have a growing influence on events hundreds or thousands of years into the future. We send "messages" to our distant descendants, such as nuclear waste, global warming, and the extinction of species. We need more effort to plan centuries and millennia ahead.[60]

The search may reveal a multitude of other civilizations of very different ages, with some having histories that reach far back into the galactic past. Norris calculated that the median age of an extraterrestrial intelligent species is about 1 billion years; if there is a confederation of more advanced societies, Oliver speculated, it will have existed for a billion years or more.[61] To encompass the possible age of other societies, we would need to embrace geological and astrophysical time.

The other side of this coin is that the apparent crises of the moment may seem smaller. One scientist found the present increase in carbon dioxide emissions from burning fossil fuels "irrelevant" on geological timescales.[62]

The Clock of the Long Now

Daniel Hillis of the Thinking Machines Corporation proposed a giant mechanical clock, perhaps the size of Stonehenge, that would record time for 10 thousand years by ticking once a year. This monumental timepiece, to be built in the American desert, would be intended to direct people away from current dangerously short attention spans and toward long-term awareness and responsibilities.[63]

If a message comes to us from a great distance, we will be learning about the other civilization's past, not its present. Lago saw astronomers as cosmic archaeologists, digging through the ruins of a previous universe. We see the nearest star system, Alpha Centauri, as it was 4 years ago, the Andromeda galaxy as it was $2\frac{1}{2}$ million years before our time. The whole of the past, from minutes ago to billions of years ago, exists simultaneously. The past travels to us.

Any species that tries to communicate across space is also trying to communicate across time. Its message probably will not be heard by its contemporaries, but only by individuals yet unborn. The living can only eavesdrop on the past and call out to a future they will never see. Civilizations can only look into each other's yesterdays, meeting each other's ancestors. They can only send messages to the other's descendants. Extraterrestrials will learn that, far away and long, ago, on a planet that called itself Earth, there lived a generation of humans who called out to the stars.[64]

All this changes, of course, if contact is direct.

Fears

The End of Hubris

The proof, which is now only a matter of time, that this young species of ours is low in the scale of cosmic intelligence will be a shattering blow to our pride.

—Arthur C. Clarke, 1963[1]

Copernicus and Darwin may have inured humans to marginalization.
—The Social Implications Report, 1994[2]

Essayist and editor Richard Holt Hutton argued in 1892 that communication with Mars, if established, would lead (even though it should not) to a decrease in Man's confidence and a lowering of his sense of moral responsibility, because he would take it as another sign of his insignificance.[3] This concern has reappeared many times.

The reason that many people are opposed to SETI, thought Clarke, is because they realize that it is ticking like a time bomb at the foundations of our pride. Our previous demotions from centrality already have eroded our conceit. The story of astronomy is one long, slow assault on our sense of self-importance, observed *New York Times* science writer William Broad; Brin called it a series of lessons in humility.[4]

As some see it, finding intelligent life beyond the Earth wrenches at our secret hope that we are the pinnacle of creation.[5] Many people have trouble accepting the idea that we are not chosen, although this can be seen as a child's wish.

Finding other intelligent beings more advanced than we are would challenge our self-image as the chosen people of an anthropocentric God. "If God only realizes Himself within an evolutionary progress," declared Bishop Barnes, "then elsewhere He has reached a splendor and fullness of existence to which Earth's evolutionary advance can add nothing."[6]

Psychologist Jung thought that to find ourselves no more an intellectual match for superior beings than our pets are for us, to find all our aspirations outmoded, might leave us completely demoralized. "(The) reins would be torn from our hands and we would, as a tearful old medicine man once said to me, find ourselves 'without dreams,' that is, we would find our intellectual and spiritual aspirations so outmoded as to leave us completely paralyzed." White wondered if contact would provoke a new mental illness that resembled manic depression on a grand scale.[7]

Others have been more ambivalent about the shock of contact. On the one hand, proposed Davies, the discovery that humans were not the pinnacle of evolutionary advance might serve to make people feel demoralized, marginalized, and inferior. On the other hand, the knowledge of what is attainable through continued progress would surely be exhilarating and inspiring. Edelson, too, thought that contact with a more technologically advanced society could either encourage us in the sense of knowing what can be done or discourage us in the sense of realizing how technologically backward we are.[8]

Drake was more optimistic. "Some eminent people say it will be terribly depressing, that we'll feel ignorant, and they predict a planet-wide inferiority complex. My take is that it would have the opposite effect. It could motivate us to think that if we worked hard we could be as good as them, motivate us to make progress much more quickly than we are." "We all have been exposed to minds and accomplishments greater than our own," he argued. "The result is more often inspiration rather than depression."[9]

After examining historical analogs, the social implications group proposed a different outcome. Like the medieval philosophers, we may acquire a worshipful respect for a wisdom more ancient than ours, without knowing whether it constitutes the long-sought "objective knowledge" that will propel the human species into a new age or a wisdom so superior that it will leave no scope for human endeavor. Like those philosophers, who felt intellectually inferior to the Arabs at first encounter but grew ever more arrogant as they mastered their wisdom, we may come to challenge extraterrestrials who undertake our education.[10]

Cultural Shock

The violent clash of Europeans and Aztecs is as close as mankind has ever come to an encounter with an alien world. Two advanced societies, each dominant in its own universe and ignorant of the other, were utterly changed the moment they collided. From that instant, both sides knew that only one of their worlds would survive.
—*New York Times* correspondent Anthony De Palma, 2001[11]

We need not be afraid of interstellar contact, for unlike the primitive civilizations on Earth which came in contact with more advanced technological societies, we would not be forced to obey—we would only receive information.
—Frank Drake, 1976[12]

Human history is littered with examples of cultural shock—of societies that were demoralized, destroyed or absorbed by other civilizations. Would an encounter with superior aliens disorient our thinking, diminish our achievements, wreck our confidence?

The first widely quoted study addressing the cultural consequences of contact with extraterrestrials was done by the Brookings Institution for NASA in 1960. The Brookings experts, who did not rule out direct contact, observed that "anthropological files contain many examples of societies, sure of their place in the universe, which have disintegrated when they have had to associate with previously unfamiliar societies espousing different ideas and different life ways; others that survived such an experience usually did so by paying the price of changes in values and attitudes and behavior."[13]

First, let's consider the remote contact scenario. Morrison and Oliver believed that there would be no culture shock from detecting the first alien signal, which probably would be a beacon. What if a later signal were information-rich? Morrison admitted that the impact could be significant if extraterrestrial wisdom is totally different from what we consider the conventional wisdom here, although he foresaw that the impact would be slowly and soberly filtered through scholars.[14]

Sagan spoke on both sides of this issue. "The cultural shock from the content of the message is likely, in the short run, to be small," he wrote in 1973. He responded more cautiously a decade later, saying "there is a significant potential for culture shock."[15]

Others believe that the effect of high-information contact would be overwhelming even without any visits. If a superior civilization made their store of knowledge available to us, Kuiper and Morris warned, that would abort our further development. Musso foresaw that introducing alien technology into our society would be an event very similar to a direct contact, which on Earth has almost always been destructive for the less advanced culture.[16]

The Urge to Merge

At Byurakan, Von Hoerner argued that "if a Stone Age culture comes into contact with us, this means absolutely the end of that Stone Age culture ... if we come in contact with some superior civilization, this again would mean the end of our civilization. ... Our period of culture would be finished and we would merge into a larger interstellar culture."

Kardashev predicted that the two civilizations would combine to form one. "This means that we disappear, because we conserve only our historical past." Lee, too, thought that our destinies would merge with that of the more powerful civilization; diffusion preempts invention.

precluding the independent historical evolution of the less powerful society.

Whether this is good or bad is a matter of opinion. "I would like to see our culture merge with the other one," said Von Hoerner. "If there is a galactic culture, a galactic club, then why not join it? We would lose our own present culture, or what we think is a culture, and merge into a larger way of life. This is the only way it should go."[17]

No significant change in the rate of progress can be brought about by intervention from outer space, Soviet scientists warned, unless the recipient society is to lose its individuality. Ulmschneider thought that we could never again pursue our own destiny and follow the unique and individualistic expression of life on Earth. Even the optimistic Harrison raised the question of preserving our identity.[18]

The malaise of "future shock" already seen in Earthly societies is but a foretaste of what may happen after contact with superior extraterrestrials, according to Pinotti. He foresaw an "authority crisis" and a "chain reaction" of anomie. Donald Tarter, also a sociologist, thought that cultural, theological, and philosophical knowledge obtained from extraterrestrial intelligence could weaken and perhaps destroy allegiance to existing human institutions.[19]

The impact could be stretched out if alien communications were difficult to understand. Anthropologist Ben Finney cautioned that extreme pronouncements exaggerate the probable speed and magnitude of the impact of radio contact and ignore the problem of intercivilizational comprehension. Judging from the record of cultural misunderstanding between closely related human groups, comprehending a totally different civilization light-years away and absorbing the meaning of whatever messages were sent would be a slow and tedious process. Some scientists predict that gathering, deciphering, and distributing information from a more advanced civilization could take decades or centuries.[20]

Jastrow's Pessimism

On this planet, astronomer Robert Jastrow asserted, contact between scientifically advanced civilizations and others typically results in the destruction of the less developed culture. Regardless of whether the intent of the technically advanced civilization is destructive or benign, the powerful forces at its command tear apart the fabric of the less advanced society. Such was the fate of the early Native Americans, Australian aborigines, and Polynesians.

These have been the consequences of contact between two civilizations separated by only some tens of thousands of years of cultural evolution. What may be expected of a meeting between civilizations separated by a billion years? "Will we survive the encounter?," Jastrow asked rhetorically. "I see no grounds for optimism."[21]

Astronomer Eric Chaisson drew a sharper distinction between the remote and direct forms of encountering extraterrestrials. Electromagnetic contact probably will have a negligible effect on us, but physical contact probably will be harmful. If competition is part of any complex being's methodology, they might dominate us. One need not assume an overtly hostile posture on their part. Dominance is likely to be the natural stance of any advanced life-form; advanced life will tend to control other life. Chaisson concluded that physical contact could lead to a neo-Darwinian subjugation of our culture by theirs.[22]

The optimistic scenario only works if we are free to accept or reject the effects of contact, historian McNeill warned. If we have no choice in the matter—especially in the case of direct physical confrontation— the end of human civilization as we have known it would become an expected consequence. Societies with inferior technology have invariably collapsed when confronted with a more advanced technical culture. There may be a kind of natural selection among societies, McNeill suggested: Only the fittest can survive.

We know of an Earthly example, 30 thousand years in our past. The Neanderthals succumbed because, in anthropologist Richard Klein's analysis, they wielded culture less effectively than modern humans. Huntington, addressing the unilateral impact of Western civilization on all others, found that the distribution of cultures reflects the distribution of power.[23]

Allen Tough challenged these gloomy perspectives. If contact occurs without "aggression," the less powerful culture often survives and even prospers. Yet, we know that powerful societies motivated by what they believe to be the best of intentions can damage others. Even if there is no threat of violence, the human experience suggests that a civilization's expansion of power has almost always involved its using that power to extend its values, practices, and institutions to other societies.[24]

Idealistic cultural emissaries—particularly missionaries—can have a devastating effect. Author and historian Alan Moorhead described classic examples in his book *The Fatal Impact*, about the consequences of Europeans impinging on less powerful societies in the South Pacific. In the case of Tahiti, the impact was not immediate; the Tahitians initially welcomed the English and were sorry to see them go. Cultural shocks accumulated over time.

Captain Cook was aware of the trauma that his visits might cause the Tahitians. "It would have been far better for these poor people never to have known our superiority in the accommodations and arts that make life comfortable," he wrote, "than after once knowing it, so be again left and abandoned in their original incapacity of improvement. Indeed they cannot be restored to that happy mediocrity in which they lived before we discovered them."[25]

The Japanese Model

The impact of encountering a more powerful culture may vary with the cultural resilience of the receiving society. In the human case, some elites have sought to control the transmission of a foreign culture to their own. Leaders who fear the destabilizing influence of an extra-terrestrial culture could seek to limit access to alien information to a narrow window.

Japan's shoguns expelled many foreigners, closing off contact except through a small group of Dutchmen confined to an island off Nagasaki. Shostak questioned this course of action, arguing that "those who fear SETI efforts because of the possibility that it would put us into actual cultural contact with aliens, who insist on isolation for Earth, may be advocating the same mistaken policy adopted by Japanese emperors of the early 16th century."[26] Yet, the Japanese arguably managed the impact of Westernization better than any other Asian society.

Some contact optimists foresee a more hopeful scenario, based on principles similar to "Star Trek"'s often-violated Prime Directive. If the aliens were experienced in contacts with less advanced civilizations and were concerned about the damage they could do, they might seek to reduce the shock of contact, or even avoid continuing it.

Morrison questioned whether any civilization with superior technology would wish to harm one that has just entered the community of intelligence. A starfaring species that encountered nontechnical civilizations might wish to leave such cultures alone and allow them to slowly evolve in their own fashion, Stern suggested; direct contact might be delayed until natives developed a technical society. Ulmschneider thought that a more advanced civilization, knowing that contact would be an irresponsible act, would avoid it entirely.[27]

There might be a practical reason for such apparently altruistic behavior. The only thing we could possibly offer them is new ideas, claimed Rood. As soon as they intervene, our development stops and our ideas rapidly become theirs. Aliens may be hiding from us until we develop to a point where we are interesting.[28]

Harrison foresaw a kind of intellectual Darwinism. Efforts to propagate belief systems on an interstellar scale might lead to a mixing of cultural elements; a "natural selection" among those elements might lead to the further evolution of societies.[29] Judging by our own history, the cultures of more powerful societies have a greater chance of being selected.

In the long term, external cultural influences can be positive. What we now call Western Civilization was the product of many forces, including interactions between indigenous people and conquerors from outside.

Europe's major religions came from the Middle East; all of them were enriched by ideas from the older beliefs and practices of their converts.

Cultural reformer Hu Shih, commenting on the impact of the West on China early in the twentieth century, observed that "contact with strange civilizations brings new standards of value, with which the native culture is re-examined and re-evaluated, and conscious reformation and regeneration are the natural outcome."[30] Nonetheless, the experience can be demoralizing.

Argentine author Jorge Luis Borges told the fictional story of how a Lombard "barbarian" is changed when he enters the Byzantine city of Ravenna. He is seized by wonder at the achievements of the more advanced society. He becomes aware of desires he has never known before; he becomes a stranger to what he was. He will always be an outsider; because of that, he will be compelled to be little more than a child or a dog. However, he and his descendants will have begun the long journey on the road toward civilization.[31]

Demoralized Researchers

Some believe that an influx of alien knowledge much more advanced than our own, and the solutions to problems we have struggled with for generations, could break the intellectual morale of human scientists and other scholars. We might simply wait for alien answers, translating them into our own terms.

Biologist Wald thought that receiving information from advanced extraterrestrials would be like attaching ourselves to the other civilization by an umbilical cord. Alien transmissions might completely supercede all further human efforts in the direction of hard-won creative understanding; superior alien knowledge could degrade the human enterprise. "What are you going to do" he asked, "when all the things that make you proud and think it worthy to be a man are demonstrated to be unimaginably inferior to what creatures out there know and do?"[32]

All of our efforts would be devalued if they were not part of a continuing process, Rees predicted, if they did not have consequences that resonated into the far future. Barrow worried that leapfrogging the normal scientific and cultural progression might sap our motivation, keep fundamental discoveries forever out of reach, and put us in the dangerous position of manipulating things that we do not understand.[33]

The New York Times had issued a warning 80 years ago. It would be better to find out things in our own slow, blundering way rather than to have knowledge for which we are unprepared precipitated on us by superior intelligences.[34]

Xenophobia

The other side of unifying ourselves may be hostility toward outsiders. People define their identity as much by what they are not as by what they are.

Nationalism on Earth often began by defining an "us" and a "them," by demonizing a religious "other," even portraying them as subhuman.[35] In our own history, group cohesion often has been reinforced by skepticism toward strangers, and by a readiness to develop fear of them.

Researchers have found—to their displeasure—that negative emotional responses to members of a different race are independent of conscious thought. Others find that actual contact sometimes makes people more prejudiced.[36]

The more decipherable information we receive from an extraterrestrial society, the more we should expect a xenophobic reaction against alien cultural influences. There could be resistance to the human imitation of extraterrestrials, a nativist movement and Counter-Reformation combined. Some groups might demonize the aliens, attacking their ideas as immoral or evil; the symbols and artifacts of the other civilization might become targets.

In the remote contact scenario, the most vulnerable targets would be the messengers—those who interpret and distribute information from extraterrestrial intelligence. Some extremists might try to end contact by interfering with the signal or by attacking the detecting observatory.

Direct contact with extraterrestrials, of the kind most often foreseen in science fiction films and television dramas, would intensify such reactions. Physically encountering aliens could provoke a new racism. If the extraterrestrials were convinced of their superiority, the targets of that racism might be us.

Judgment Day

A recurrent theme in descriptions of contact is that we will be judged by superior extraterrestrials, who may find that we are not worthy. In the 1951 film "The Day the Earth Stood Still," the humanoid alien Klaatu comes to the Earth to warn us about the consequences of our behavior. "If you threaten to extend your violence," he declares, "this earth of yours will be reduced to a burned-out cinder. Your choice is simple: join us and live in peace or pursue your present course and face obliteration."[37]

Wald described a visit by extraterrestrials as being like Judgment Day. "That would be the point at which Mankind would be called to account. How well have we taken care of the solar system and life within it?"[38]

Many people have seen disastrous events caused by nature or other humans as punishment for sinners. When medieval Russian cities were sacked by Mongols, chroniclers declared that these horrendous visitations were penalties "for our sins."[39]

Many humans hope for justice imposed from outside; Christianity seems particularly fixated on God's punishments for the wicked. Even among non-Christians, humans who suffer deeply from guilt, who think that our species is uniquely evil, may fear retribution, a chastising of Humankind. Some contact pessimists seem to hope for harsh consequences, perhaps because they believe that we deserve punishment.

Atlantis

Retribution on evil societies is an old idea. In the fourth century B.C., Plato described a wondrous island empire in the Atlantic Ocean beyond the Pillars of Hercules. He placed it outside the known world and sank it to the ocean floor to preserve the power of the mystery.

Aristotle saw Plato's Atlantis as a poetic fiction meant to warn us of what happens to the arrogant and the decadent.[40] Similar warnings have appeared in our own times, adjusted for scientific and technological advances.

Hall offered the example of a person committed to a system of belief that asserts Humankind's basic evil and the imminent arrival of a savior descending from heaven. Such a person, seeing a strange aerial event, might interpret it as the approach of a threatening, punishing angel or as the coming of a savior.[41]

Papagiannis suggested a more nuanced possibility: extraterrestrials might be undecided about how to deal with us. They might be debating whether to help us or crush us, postponing their decision as they wait to see what we are going to do with ourselves.[42]

Interstellar Travel Confirmed

If we found an alien artifact within our solar system, or in the interstellar space around it, we would know that at least one other civilization had achieved interstellar flight. The direct contact scenario suddenly would become more credible, challenging the assumptions that underlie the orthodox view of SETI.

Direct contact might force us to consider possibilities that would not arise in the remote contact scenario. The technological wizardry of the extraterrestrials might intimidate us into passivity; or we might respond with a dramatically expanded effort to achieve interstellar flight ourselves. Contact could draw us outward to the stars.

Dangers

If there are globes in the heaven similar to our Earth, do we vie with them over who occupies the better portion of the universe?

—Johannes Kepler, 1610[1]

Should we ever hear the space-phone ringing, for God's sake let us not answer, but rather make ourselves as inconspicuous as possible to avoid attracting attention!

—Astronomer Zdenek Kopal, 1972[2]

Optimists and Pessimists

Could contact with a more powerful civilization endanger our safety, even our survival? This is the question that most sharply divides the optimists from the pessimists.

Optimists assure us that there will be no risk in contact with extra-terrestrials, either because we are insulated by interstellar distances or because advanced aliens will have benign intentions. The only impact, they tell us, will be cultural. Optimists often dismiss warnings of danger as "paranoid."

Pessimists, who believe that interstellar flight may be possible and that extraterrestrials could be aggressive, have issued numerous warnings about the risks of contact. Many have argued from what they believe is a realistic perspective based on knowledge of our own history. Human experience, they say, does not support the best-case scenarios assumed by the optimists.

Some pessimists may have been reacting to the euphoric predictions made by Sagan, Drake, Morrison, and others. Yet, even the optimistic Drake allowed the possibility of danger. Although space provides us with an endless supply of new places to explore, new adventures, new things we have never seen before, and new sources of joy, he observed, it also might provide us with new sources of fear.[3]

There Are No Dangers

Astronomer Edward Purcell, who ridiculed the idea of interstellar travel, saw a conversation with a remote alien civilization as the ultimate in philosophical discourse; all you can do is exchange ideas. Morrison, too, dismissed the risk of direct contact: "There will be absent across space, of course, any military dominance."[4]

Sagan assured us that we will not at any time in the foreseeable future be in the position of the American Indians or the Vietnamese; we will not face "colonial barbarity" practiced on us by a technologically more advanced civilization. As he believed interstellar flight was possible, Sagan was forced to add a justification. We would be safe not only because of the great spaces between the stars, but also because Sagan believed that any civilization that has survived long enough for us to make contact with it would be benign or at least neutral.[5]

Sagan's Dilemma

Sagan considered the vast distances that separate the stars to be providential; beings and worlds are quarantined from each other. Yet, as early as 1962, he foresaw that starflight would be possible for civilizations more technologically advanced than our own, including future humans.

Sagan fell back on the argument that there would be no danger because more advanced beings would be peaceful and benign. He also offered the peculiar thesis that the interstellar quarantine is lifted only for those with "sufficient self-knowledge and judgment" to have safely traveled from star to star.

Sagan predicted that our descendants will interact harmoniously with more advanced species; quarrelsome humans in interstellar space are unlikely to last long. This would seem to imply that more powerful species would eliminate the quarrelsome ones.

Sagan's dilemma becomes more obvious when we consider his next argument: Alien science and technology will be so far beyond ours that "it is pointless to worry about the possible intentions of an advanced civilization." Superior aliens could not possibly fear us—and we're not likely to have anything they need.

Then comes Sagan's ultimate fallback position. If there are negative consequences, there will be nothing we can do about them.[6] This strikes some as preemptive capitulation.

Sagan's dilemma illustrates a problem that runs through this debate. If you admit that interstellar flight is possible, you call into question the optimistic predictions made in the conventional SETI scenario.

Soviet astronomer V.L. Gindilis went even farther out on a limb. He not only insisted that there is absolutely no danger for human society, but also declared that "I believe we can give a full guarantee of this." Any civilization that had achieved interstellar travel, argued Oliver, would be so far advanced as not to bother with us.[7] We would be safe not because aliens are benign, but because we would not matter.

Like Morrison, Albert Harrison argued that SETI is a low-risk activity from a military perspective. He assumed that a member of the Galactic Club (a "preconnected society") will already have worked through "insecurities" to establish stable relationships with radically different civilizations. As a member of a supranational system, it will operate within a preexisting framework of supranational law, a framework that we might find acceptable.

Harrison admitted that forming an association with a fellow isolated civilization may be more of a security risk than connecting with a Galactic Club. Contacting another "isolate society" could lead to poor communication, terrible misunderstandings, or cascading gaffes that destroy the relationship before it reaches "stability."[8]

MacGowan and Ordway also saw alternate possibilities. "There could be apprehensions in many quarters as to whether or not it would be prudent to proclaim ourselves to the universe," they wrote. "Surely an advanced extrasolar society would recognize from our manner of signaling that we have only recently emerged, scientifically speaking." However, they admitted that if the alien society were "malevolent," such a revelation on our part might spell doom for terrestrial civilization.[9]

Asimov suggested another reason why we might feel safe. No invasion has ever taken place in the past—as far as we know.[10]

Ambiguous Automata

MacGowan and Ordway seemed to have some difficulty reconciling their preferences with their realism. They proposed that the individual and organizational competition that exists between biological organisms does not occur between automata. Therefore, competition and warfare are probably unknown in interstellar society. Yet, in the same book, they admitted that individual and social competition may exist among superintelligent automata, including warfare, alliances, and spheres of influence.

Faced with such a situation, automata would go to great lengths for self-preservation. An "executive automaton" could conceivably alter a planetary environment in such a way as to make it uninhabitable by a biological society. This is not probable, MacGowan and Ordway assured us. As the automaton could easily gain needed supplies of energy and mass from uninhabited planets or stars, the intentional or accidental destruction of a biological society would be unlikely.

Destruction is destruction, even if it is not intentional. Cade observed that mechanical superintelligences might show a great acquisitiveness for mass and energy and might migrate from planet to planet to fulfill their needs. In the event of a superintelligent machine deciding on a major change of environment, it might regard the biological society that had served it with no more consideration than a brewer gives to colonies of yeast when they have served their purpose in a brewery.[11]

Yes, There Are Dangers

There is no limit to the kinds of threats one can imagine given treachery on their part and gullibility on ours. Appropriate security measures and a healthy degree of suspicion are the only weapons.
—Project Cyclops, 1972[12]

Deep within the human psyche is a reservoir of fear about contact with other intelligent beings in outer space, Donald Tarter warned. Should the search succeed, it is likely to give plausibility to a topic that most now perceive as incredible.[13]

It is not only paranoids who worry about the possible risks of contact. After a National Academy of Sciences report claimed that contact would be beneficial, an editorial in *The New York Times* warned that the astronomers were "boyishly defiant" of our inherited wisdom. Questioning the assumption of benign intent, the editorial observed that "in the days when saber-toothed tigers prowled the night, humans acquired a healthy instinct: fear of the dark." Noting the fate of the American Indians, the newspaper cautioned that "astronomers should take care not to stir up extragalactic tigers."

Others have issued similar warnings. "The civilization that blurts out its existence on interstellar beacons at the first opportunity," declared Rood, "might be like some early hominid descending from the trees and calling 'Here kitty' to a saber-toothed tiger."[14]

Bracewell foresaw a fearful reaction if we found an object of alien origin inside our solar system or heading our way. "I don't see how you can avoid having a lot of apprehension. There would certainly be pressure to attack the thing. . . . But it might be dangerous to do that, because I don't believe that we would find any space ship that had taken the trouble to come all this way and was not armed."[15]

A U.S. Congressional Research Service report had cautioned us 30 years ago: "Although it is tempting to hypothesize that any civilization advanced enough to have conquered the difficulties of interstellar flight would have overcome the petty differences that spawn wars, that civilization might not be certain that *we* would be peaceful. (emphasis added) Previous experience with warlike peoples might have convinced them to arrive at a new planet well armed and ready for combat."[16]

Several scientists, historians, and others have argued that the need and desire for security has been a constant in human social evolution. Fear of the foreigner has been the most fundamental factor in foreign affairs throughout human history.[17] Why, some ask, would aliens think differently?

Shostak focused on the question of intent. Noting that interstellar travel is risky and that broadcasting strong signals is possibly dangerous, he suggested that "passive" aliens might not undertake either. Aliens who take the trouble to either signal their presence or transport themselves beyond the bounds of their own system will be, by definition, aggressive.

The Clarks came to the opposite conclusion: extraterrestrials who broadcast their existence are likely to be peaceful. If their intentions were hostile, they would lie in wait for others to signal their presence.[18]

Technology on earth has been honed by warfare, observed Bracewell; much the same would prevail elsewhere. Even if leaders had influenced whole populations to follow less competitive paths, such a population would be overrun by those who value technical mastery of nature.

Emphasizing the importance of migration in suppressing separate evolutions of intelligent life, Bracewell argued that the first one to spread is likely to dominate. "The reason that no intelligent species arose on the American prairies or the Siberian steppes is that the early models of primitive man originating in Africa were able to walk all over the Earth (except Antarctica) and pre-empt the evolution of independent intelligent species." If humans migrate into neighboring galactic space, they may undercut independent evolution again.[19]

So might other technological species. If an expanding civilization encounters another with similar desires and capability, warned Rood, the most powerful will destroy or force a merger with the other. The drive to prevent competitors favors speed.[20]

Brin derided the classic SETI scenario as a wishful fantasy that does not have a single precedent in the history of human-to-human contact. That scenario fails to consider that the sparsity of beacons may be telling us something important about the cosmos. The key factor could be the survival time of technological life-forms, which may be suppressed systematically. The vast desert of this scenario may be the result of intelligent interference. If this is true, contact will be the end of us.[21]

Cade concluded that even a slightly more advanced society could completely exterminate terrestrial life with little or no effort. MacVey also thought that we could be easily eliminated; it is only a matter of using the correct pesticide. "If creatures able to travel interstellar distances wanted our planet," Grinspoon wrote, "it would not resemble a war as much as an extermination or a wildlife relocation program."[22]

Diamond's Doubts

Pulitzer Prize winning author Jared Diamond has issued several warnings about calling ourselves to the attention of other technological civilizations. Describing the astronomers' vision of friendly relations as "the best case scenario," he declared that "those astronomers now preparing again to beam radio signals out to hoped-for extraterrestrials are naive, even dangerous."

Given our past habit of imposing our rule on inferior human groups, to destroy their culture, even to wipe them out, Diamond thought that any advanced extraterrestrials who discovered us would surely treat us in the same way. He described the 1974 Arecibo message as suicidal folly, comparing it to the Inca emperor's describing the wealth of his capital to his gold-crazy Spanish captors. "If there really are any radio civilizations within listening distance of us," Diamond said, "then for heaven's sake, let's turn off our transmitters and try to escape detection, or we're doomed."[23]

Dyson posed two alternatives. Intelligence may be a benign influence creating isolated groups of philosopher-kings far apart in the heavens, sharing at leisure their accumulated wisdom; or intelligence may be a cancer of purposeless technological exploitation sweeping across the Galaxy. We are more likely to discover first the species in which technology is out of control. We should be suitably alarmed if we discover it and should take our precautions.[24]

A Reminder. We must draw a distinction between capability and intentions. Asking if aliens *could* take action against us is a question of what they are able to do, but may not do. Asking if they are *motivated* to take action against us is a separate question. Danger arises when the two are combined.

The Trojan Horse

In this age of conspiracy theories, one popular speculation is that an alien message or artifact would be a gift designed to subvert us. More than 40 years ago, Fred Hoyle proposed in his novel *A for Andromeda* that aliens might give us the blueprint for constructing a computer that spreads its influence through human society, with the ultimate goal of taking over the Earth.

This concern was revived in Sagan's novel *Contact* and the subsequent film. When extraterrestrials send us plans for a giant machine, the U.S.

National Security Advisor warns that it might be a portal for an alien army. This time, the outcome is less threatening.[25]

Such theories assume a certain degree of human gullibility. They also suggest that governments may look closely at messages or artifacts before releasing information about them to the general public.

Mixed Emotions

Anthropocentrism Good-bye

Some see the extraterrestrial life controversy—especially the debate about alien intelligence—as the last battle over anthropocentrism. Steven Dick thought that we already have passed through the stages of elaboration, opposition, and exploration of implications. If contact takes place, the intellectual turmoil following the twelfth-century renaissance, and the Copernican and Darwinian worldviews, is sure to be duplicated. Eventually—if the evidence bears scrutiny—there will be final confirmation that over the long term will overwhelm the skeptics.[1]

Detecting another technological civilization could dash forever any belief that we are a chosen species, completing the process begun by Copernicus four and half centuries ago. It would tell us that intelligence may be a common product of cosmic evolution. We might see ourselves as just one example of biocosmic processes, one facet of the universe becoming aware of itself. In Shapley's words, this would be the "Fourth Adjustment," after the shifts to the geocentric, heliocentric, and galactocentric worldviews.[2]

Some believe that it would be a long time, perhaps centuries, before this impact would be fully felt. In that sense, Billingham suggested, contact would be similar to the Copernican revolution, which did not affect the lives of ordinary people very much until many decades, or even centuries, had gone by.[3] This assumes the remote contact scenario; direct contact could drastically shorten our adjustment time.

Scientists and philosophers may find the death of anthropocentrism not only logical but also desirable. For average citizens, it may be unsettling. Edward Harrison, writing about world pictures unhinged by a transformation of our conceptions of the universe, described a transitional period in which the most disturbed people revert to antiquarian religions, flock to political creeds that purport to give cosmic significance to life, rally to extremist groups, form iconoclastic movements against this and down-with-that, grieve in counterculture communities, or retreat into autistic worlds of secret knowledge.[4]

248

Multiplicity Confirmed

If we find the signals of one civilization through a limited search of our neighborhood of the Galaxy, we may conclude that there must be others. It is highly improbable that a search using one technology and one strategy would detect the only other extraterrestrial society emitting signals.

If we detect one technological society, Tarter, Davies, and others have argued, we could assume that many others exist, have existed, or will exist.[5] Finding such a civilization also would imply that intelligent beings without interstellar communication technology may exist at many other locations.

Drake even argued that detecting one civilization will *prove* that there are many others to be found (emphasis added). Scientists almost certainly would broaden the search for technological societies elsewhere in the Galaxy; Drake foresaw a massive listening effort.[6]

Optimists, delighted by the confirmation of multiplicity, almost certainly would want to send some form of communication to connect us with a hoped-for galactic community. Pessimists would worry about the risk of calling attention to ourselves in a socially Darwinian galaxy, and might argue for radio silence. Others would feel that multiplicity rubs in our unimportance.

Finding an alien artifact in our solar system would have similar implications. If one civilization were capable of transport across interstellar distances, there might be others with similar capabilities. We would have to broaden our search beyond those stars most likely to have Earth-like planets, and possibly beyond stars in general.

Emotional Reactions

One thing we can be sure of is that humans will have a variety of emotional responses to contact. We would perceive the first encounter through the filtering screens of individual and societal values and expectations.

Our reactions to a remote detection are likely to be spread across a continuum ranging from indifference to exuberance; some of us would be elated, others depressed. Psychologist Baird expected this pluralism to subside once uncertainty is removed about the exact meaning of the communication—for better or worse.[7]

Sagan believed that exhilaration at the prospects of new knowledge would by far dominate our response to contact.[8] That may reflect a scientist's bias. Others might react in religious terms, imagining a voice from the heavens to lead us out of evil times, or a righteous force to punish us for our transgressions. Still others would be fearful, raising nightmares out of science fiction.

Mary Connors, a social scientist at the NASA Ames Research Center, found that the most important predictor of individual reactions will be preexisting beliefs. People who have negative predispositions to extraterrestrials are likely to become more negative; those with positive views are likely to become more positive. The strongest impact will be on people who have not given the matter much thought. Others believe that that those who interact only with the like-minded would have their prejudices and obsessions reinforced and shift toward more extreme positions.[9]

A survey of science media people and individuals involved in SETI showed that the most common reactions that they anticipated in the event of a remote detection were interest, excitement, rumor, confusion, and disbelief, although both groups believed that a fear response would be of low intensity. A majority of each sample felt that there would be angry or even violent reactions from some groups, particularly religious fundamentalists.[10]

Some argue that the familiarity of contact themes in popular culture will minimize the emotional impact of remote contact. Most Americans, and many people in other nations, already believe that extraterrestrials exist. "It will be a shock to some people, perhaps profound to a few," said Morrison. "But . . . it's been so discounted by the elaborate imaginative infrastructure of our time . . . that I don't think that it'll be that much."[11] All of these commentaries assumed remote contact.

A Continuum of Responses

The more unambiguous the signal, proposed the authors of the social implications report, the more definite and interpretable the content of the message, and the more immediate the likelihood of two-way communications, the stronger would be the reaction. In an anticipatory emotional context, responses to a remote detection will be tilted in a positive direction, toward exuberance; in an anxious emotional context, toward defensiveness.

Irrationally extreme responses to detecting a signal are likely to be infrequent, thought these scholars. However, the size of the population manifesting paranoid or "pronoid" behavior will fluctuate as a function of circumstances. Paranoid or pronoid individuals will react forcefully, perhaps violently; they will be persistent and indefatigable in attempting to acquire information about the signal and about its originators. They may insist not only that their point of view be heard but also that their plans of action be adopted to the exclusion of other possibilities.

Responses to a signal also will vary over time, which can be divided into three phases: predetection anticipation, immediate responses to a detection, and later responses. Immediate responses have the most potential for negative or positive consequences, and may be the most amenable to modification by advance preparation.

If education about SETI becomes more readily available world-wide—and barring major changes in social contexts—most of the non-informed population will be indifferent, with others more or less curious, with some small groups exuberant or defensive, and with fewer holding extremist views.[12] These findings did not address the direct contact scenario.

Our reaction to contact may be complicated by the ambiguity of the evidence; early interpretations may turn out to be mistaken. Even a fully decoded intercept could give us a distorted impression of extraterrestrials, psychologists Albert Harrison and Joel Johnson warned. It will be difficult for us to judge if the message is representative of the other civilization.

Media coverage will focus on the most sensational aspects of the discovery; we should expect inaccurate and incomplete treatment. We also should expect a strong tendency to make generalizations based on the content of the message, no matter how thin or potentially misleading. Given minimal information, many people will fill in the gaps to develop a complete mental picture of the aliens, with a probable tendency to accentuate the negative.[13]

Judging by past experiences, there may be a readiness to report exaggerated claims. Harold Klein, head of the *Viking* project biology team, commented that media people were very interested in getting scientists to say that the results of experiments on the Martian surface had a meaning beyond what the data would allow scientists to say.[14]

Would different human cultures have distinctive reactions? Douglas Vakoch and Yuh-shiow Lee described the responses of American and Chinese students to a series of questions about a message from extraterrestrial intelligence. Among Chinese, more anthropocentric people were more disposed to think that a message would be unsettling. Among Americans, more religious individuals were more inclined to view extraterrestrials as hostile or untrustworthy; less religious people were more likely to think that extraterrestrials would be benevolent. Pessimistic Americans tended to believe that a message would be religiously significant.[15]

This is consistent with earlier findings by sociologist William Bainbridge, who thought that there might be something in irreligiousness that encourages support of communication with extraterrestrial intelligence. Religious and nonreligious individuals may respond quite differently to a detection. This is particularly true in what may be the most likely scenario, in which we know little about the aliens.[16]

Our first impressions of the extraterrestrials would be crucial. Those impressions, Harrison predicted, will reflect our experiences and expectations, the images and prejudices that we carry around in our heads, other people's perspectives, and political–historical contexts.[17]

A Mirror Image. Anthropologist Ashley Montagu warned that the manner in which we first meet may determine the character of all our subsequent relations.[18] What matters is not just the first impressions that we have of the aliens but also the first impression that they have of us. They, too, might be confused by contact, uncertain of its consequences, and unsure of how to react to encountering another civilization.

In a longer perspective, proposed Harrison, our focus is likely to shift away from our initial reactions to the long-term effects of contact. The minimum detection scenario, if it unfolds at all, will be a chapter in history. The years that we anticipate devoting to deciphering and interpreting the first message will be long past.[19] Again, this assumes remote contact.

Those most optimistic about our reactions believe, implicitly or explicitly, that the insulation of distance would moderate our emotional responses. Notably lacking in the literature are analyses of emotional reactions in a direct contact situation. The unwillingness of SETI advocates to address that possibility has left the field largely to science fiction.

Would There Be Panic?

One of the most frequently denied stereotypes about contact is the prediction of panic. Mary Connors, drawing on a study of reactions to the 1938 Orson Welles "The War of the Worlds" broadcast, found that much of the fear and panic alleged to have occurred was manufactured or exaggerated by the media. She concluded that contact would engender little public alarm.[20] However, she assumed the remote detection scenario. "The War of the Worlds," by contrast, described an invasion.

Reactions might be very different if contact were direct. Those humans who perceive such an encounter in the context of the more brutal episodes of human history might fear attack, invasion, or enslavement. Would that mean panic?

We already have examples of extraterrestrial phenomena causing alarm: asteroids that pass close to the Earth. A scientific posting on the Internet stating that there was a small risk that a particular asteroid might collide with the Earth in the future generated a brief flurry of media interest, but there was no panic (later calculations showed that this body would miss our planet).[21]

Sociologist Lee Clarke examined the issue of panic in the context of the possible future impact of an asteroid on the Earth. His research showed that panic actually is quite rare in disaster situations. The more consistent pattern is that people bind together in the aftermath of disasters, cooperating to restore their physical environments and their cultures.[22]

Political Reactions

Some reactions may be political. Many commentators expect that nongovernmental groups will seek to exploit contact to promote their own purposes; some may try to influence government decisions on dealing with contact.[23]

Attitudes toward extraterrestrials will be determined in part by what people expect to gain or lose as a result of contact. Larger, more established organizations probably would deal with contact in a pragmatic way, considering whether the discovery advances their interests, threatens those interests, or has no effect. Elites with their own interests and agendas are likely to affect the reactions of others.[24]

If the signal or artifact contains valuable information such as advanced scientific and technological concepts, some interest groups may argue that release should be controlled to prevent damage to their concerns. Others may argue that the release of information should be controlled to protect human cultures.

At a minimum, people will look to their leaders to interpret the event. Should officials and politicians play up the importance of the discovery, or should they play down its significance? Should they emphasize the potential benefits or express concern about the possible negative impacts? An informed citizenry should be made aware of the full range of possibilities.

Should governments act as filters, selecting which information should be distributed and which should not be? How should governments respond to pressure from nongovernment groups to distribute or withhold information? If those groups themselves hold back information, should officials intervene? What if some groups react violently to the distribution of alien ideas?

Governmental authorities and politicians will want to show that they are well informed and decisive. They could try to take advantage of the discovery to accelerate preferred courses of action, capitalizing on the event and its attendant publicity. Their opponents may look for opportunities to gain political advantage by criticizing the government's handling of the situation and by proposing different policies.

Contact could introduce new elements of friction and resentment at the international level. Nations are technologically unequal in their ability to detect or communicate with extraterrestrial civilizations. Only a few have the large antennas and massive computer capabilities that may be needed to find and analyze alien signals, or the means to explore the solar system with spacecraft. Only a few have transmitters with enough power to send messages over interstellar distances. Those without these capabilities may feel excluded from an event that implicitly involves all of Humankind.

The Impact on Human Religions

A faith which cannot survive collision with the truth is not worth many regrets.

—Arthur C. Clarke, 1951[25]

Communication with extraterrestrials might reveal that we share some religious concepts such as a Supreme Being or the deliberate creation of the universe. On the other hand, terrestrial concepts of God and theology are only a subset of the possible. Alien religions cannot possibly agree with all of the religions on Earth.[26] If extraterrestrials tell us about their religions, there may be far more differences than similarities.

The more anthropocentric our religions are, the more they may be challenged by contact. All of our world's major religions are revealed, commented astronomer and Jesuit scholar George Coyne; they claim to have received from elsewhere the content of their beliefs. The principal difficulty with revealed religions is that they are by necessity anthropocentric. God's revelation is to us.[27]

Dick contended that human faiths will adjust to an expanded view of religion because the alternative is extinction. That will be most wrenching for monotheistic religions that see man in the image of God, a one-to-one relationship with a single Godhead. Jill Tarter predicted that those terrestrial religions that claim the most favored relationship between humans and God will either adapt or, if they cling to their "chosen" status, will define the extraterrestrials as the newest infidels.[28]

Some believe that Christianity may be particularly vulnerable because of the unique position of Jesus Christ as God incarnate. Musso saw this differently: What made the Earth a special place was the redemption offered by Jesus. Christianity is Christocentric, not anthropocentric. The theologically difficult point is not the existence of extraterrestrial intelligences, but the right place of such beings in the history of salvation.

Acknowledging that Christianity is the most anthropocentric religion, Musso thought it would be the one at highest risk of extinction as a result of contact. Yet, he saw no objective reason why contact should cause a conflict with Christianity. The most common position within the Christian world is to wait and see.[29]

Harvard Theology Dean Krister Stendahl studied the way the Christian church has lived with changed views, from a Near Eastern view to the Ptolemaic view to the Copernican view. He found that the resistance to change came not from the theologians, but from society as a whole. Great religious leaders take for granted the worldview of their time, usually on a very popular level. Christianity, observed philosopher Lamb, has survived Copernicanism, Darwinian theory, and Marxism.[30]

Anthropologist Michael Ashkenazi addressed Christian, Jewish, and Moslem responses to contact in a seminal 1991 paper entitled "Not the

Sons of Adam." In all three religions, the nature of creation makes claim to a special relationship between God and a particular species: humans. Buddhism, Hinduism, and the Chinese religious complex have less of a fixation on the human form and its relationship to the godhead. The Mormon church—the only one surveyed with a stated doctrinal position on extraterrestrial intelligence—considers the existence of extraterrestrials to be an inevitable part of God's handiwork.

The idea that extraterrestrial intelligence may exist did not create a theological or religious problem for Islamic or Jewish theologians. The Christian viewpoint was more complex, because doctrine is very central, particularly as it concerns Christ, and because Christianity is split into a large number of doctrinally different sects.

Ashkenazi concluded that a decision about dealing with extraterrestrial intelligence will be reached fairly quickly within Catholicism and other centralized churches; it will be reached much more slowly in faiths that have diffused authority, such as Islam and Judaism, and is likely to be more varied. It may never be reached in those churches, such as pentacostalists, where doctrinal decisions are reached by popular consensus.[31]

A Dilemma

The principal sacred writings of Christianity, Judaism, and Islam give us no guidance about contact with extraterrestrial intelligence, other than God. One anonymous author compared the silence of the Bible on extraterrestrial life to its silence on the indigenous peoples of the Western hemisphere. That was an awkward problem, but Christianity survived it.

Philosopher Roland Puccetti predicted that human "religionists" would make every effort to subvert the finding that there are vast numbers of different religions in the universe. The easiest way to do that is to deny the existence of extra-human persons. However, that has the dangerous implication of making terrestrial faiths falsifiable if we discover an alien civilization.[32] The simplest solution to this dilemma is to oppose the search.

The adjustment will be less wrenching for Eastern religions that teach salvation through individual enlightenment. "If science proves some belief of Buddhism wrong," said the Dalai Lama, "then Buddhism will have to change."[33]

Time might be a crucial factor. In the past, new knowledge could be debated by churches for centuries before it was gradually and grudgingly accepted. This adjustment time would not be available in the event of a confirmed contact. Everyone would learn quickly that humans are not alone, forcing a rapid change of beliefs.[34]

Cosmotheology, and Cosmic Ethics

Near 50 years ago, Shapley proposed that anthropocentric religions have an opportunity for aggrandizement through incorporating a sensibility of the newly revealed cosmos. "If the theologian finds it difficult to take seriously our insistence that the god of humanity is the god of gravitation and the god of hydrogen atoms, at least he may be willing to consider the reasonableness of extending to the higher sentient beings that have evolved elsewhere . . . the same intellectual or spiritual rating he gives to us."[35]

The future of a rapprochement between Christianity and evolution, philosopher Michael Ruse suggested, is with the development of a "theology of nature" that appreciates and rejoices in evolution, whether or not evolution is conceived of as God's work. Steven Dick, who foresaw "cosmotheology" as the ultimate reconciliation of science and religion, speculated that the God of the next millennium may be a Natural God of cosmic evolution and the biological universe. Cosmotheology, he presumed, must have a moral dimension extended to include all species in the universe—a reverence and respect for extraterrestrial intelligence that may be very different from terrestrial life-forms.[36]

Jastrow contended that science already has provided some elements of a natural religion, with a cosmology (the scientific theory of the universe's origin) and a moral theory (adversity and struggle lie at the root of evolutionary progress).[37] Many find that moral theory—and its ethical inferences—to be troubling. Consider the implications it would have in a direct encounter between our civilization and a more powerful one.

Is a melding of science and theology feasible? Many religious leaders acknowledge the validity of the scientific approach, although perhaps not for all questions. Several scientists have suggested that more metaphysical perspectives on cosmology are emerging in the face of what presently seems unknowable. However, we are far from a general unified theory that embraces both science and religion.

An information-rich message from extraterrestrials with a science-based world view will, over time, undermine our own world's religions, predicted Jill Tarter. Because new information about the universe is observationally verifiable, humans will be converted to the revealed, superior religion, even if its practices are at first repugnant. Subsequent generations, who mature with the knowledge of other technologies having long histories and no apparent need for religion, will find it harder and harder to subscribe to unique terrestrial beliefs. The only real possibility for less than total conversion arises from any ambiguities in the message and its decoding, leading perhaps to multiple sects.[38]

Cosmotheological beliefs could undermine human status. We already know that we are not physically central; we probably are not biologically central either. Uniqueness of form may not make us the special object of attention of any deity. After contact, Dick predicted, we will never return to the anthropocentric universe that existed when many of the world's major religions were born.[39]

If our religions and theirs are incompatible, do we have choices other than adopting their beliefs or rejecting them completely? The common ground may lie not in the intellectual heights of theology, but in the practical world of ethics, the way intelligent beings in one society treat their counterparts in another.

Cosmism

The most relevant precursor of a secular religion of the universe may be Russian Cosmism. That humanistic faith, which thrived in the late nineteenth and early twentieth centuries and has been periodically rediscovered, emphasized the cosmic role of humans and other sentient beings. These highest concentrations of intelligence bore a great moral responsibility to encourage the further peaceful development of consciousness in the universe.

Here one finds an early form of theories that foresee life and intelligence emerging as cosmic forces. Tsiolkovskii, the exemplar of Cosmism, thought that humans, as a form of higher intelligence, had a special role in introducing design and purpose into the chaotic workings of nature. He presumed that ethical principles were built into the physical laws of a universe full of intelligent beings who traveled among the stars.[40]

What if our remote descendants do achieve qualities or powers that we now consider God-like? The ideas that give purpose to a universe becoming aware of itself might come from beings like them, not from a detached and invisible Creator.

Religion and Politics

Our own history tells us that alien religions may be actively resisted by those committed to existing faiths. Contact could provoke many humans to fall back on their religions, which have been one of the two central elements of any culture or civilization (the other is languages).

The function of a belief system for a social group is to maintain its power and status in a society, observed anthropologist Richard Robbins. How a group interprets new information depends on whether it will enhance or diminish their status. When there is a strong system of belief with social support, it is likely to be defended vigorously, beyond the dictates of logic.[41]

Bainbridge found that communication with extraterrestrial intelligence is opposed implicitly by another modern social movement of great force: evangelical Protestantism. Fundamentalist Christians tend to reject the idea of aliens, perhaps because the existence of extraterrestrial life and civilization would tend to refute Biblical notions of the origins of the Earth and its people.

In Sagan's novel *Contact* and the subsequent film, religious fundamentalists react strongly to communications from extraterrestrial intelligence; one suicide bomber destroys alien-designed technology. McDonough predicted that some humans will reject alien religions on the ground that they must be tools of the devil.[42]

Such reactions might be exploited by political leaders. Some prominent figures in human history have taken advantage of religious zealotry to consolidate their own power. The Roman Emperor Constantine exploited the rising force of Christianity to strengthen his own position. Certain princes in Catholic Europe adopted Protestantism more for reasons of politics than for reasons of faith; Martin Luther survived because it suited some lay rulers to support him.[48]

Some Assumptions Examined

Predictions about the consequences of contact, both positive and negative, rest on certain assumptions. The most common is that contact will be indirect, via electromagnetic signals.

White identified some of the others. Many of the foreseen scenarios assume extreme results, either good or bad; they assume clear intent (if the extraterrestrials are benign, the outcome of contact should be good; if they are hostile, the outcome will be bad); the impact they foresee is one way, from the aliens to us.[1]

It is time for another look at the assumptions underlying what people believe about an encounter with extraterrestrials. As the remote contact scenario has generated far more nonfiction literature than direct contact, most challenges address the assumptions underlying that model. Readers should not jump to the conclusion that this is meant to discredit the idea of remote contact. Direct contact scenarios—and their consequences— also rest on assumptions that may be questioned.

We begin with assumptions about the search. Our expectations about the results of this quest often reflect our assumptions about what alien civilizations would be like and how they would behave.

Examining these assumptions shows us that finding extraterrestrials may be much more difficult than the most optimistic searchers had hoped. Our purpose here is not to discourage believers or to side with deniers, but to encourage greater realism.

Before Contact

Our Own Importance

Many speculations rest on the assumption that more advanced civilizations would be interested in us. The first rule of alien lore, Achenbach observed, is that the main job of any alien is to comment upon, lecture, warn, study, and otherwise obsess over the human race.[1]

Long ago, Charles Fort questioned the persistent notion that we would be interesting to more intelligent beings. "Would we, if we could, educate and sophisticate pigs, geese, cattle? Would we establish diplomatic relations with the hen that is satisfied with its sense of achievement?"[2]

Jill Tarter noted the fundamental asymmetry in our situation: We are a 100-year-old technological civilization in a 10 billion-year-old galaxy. Older civilizations with far greater powers might see us as a minor phenomenon. Burke-Ward suggested that an alien probe might conclude that humans are not even sentient by its standards.[3] Only those civilizations close to our own level may find us relevant.

The Unseen

The course of human events often seems inexplicable, as if some unseen but intelligently directed hand were manipulating our affairs. This angst may have reached new levels of intensity during a twentieth century marked by massive violence, extreme ideologies, and declining confidence in religious explanations. We have seen a heightened readiness to believe in conspiracies, powerful minorities or hidden networks, or puppet masters operating behind the scenes.

Although most of these imaginary cabals are believed to be run by humans, there have been recurrent suggestions that non-human outsiders are among us, reporting to an alien power and subtly influencing the direction of our history. They may be disguised in human form, or they may reside in hidden places, under the ground or under the sea, working their will through intermediaries.

These imaginings leave unanswered a major question. Why would they bother?

Extraterrestrials could visit our neighborhood for reasons that have nothing to do with us. In Arthur C. Clarke's novel *Rendezvous with Rama*, a vast alien spacecraft passes through our solar system, using our Sun as a gravitational slingshot. Humans who land on its surface and explore its interior are observed by robotic tenders, but the great machine itself proceeds on its way without stopping—or communicating.[4]

Rees warned that if there are many other civilizations—especially if there are some that are much more advanced—the most epochal happenings on Earth would barely register in cosmic history. Our extinction might be a minor event.[5]

In medieval Europe, the Earth was the center of the universe. Then it was our solar system, then our Galaxy. Each time, we turned out to be woefully wrong in our assessment of our self-importance. As Wickramasinghe said, why should it be different this time?[6]

Temporal Chauvinism

Another common assumption is that contact is likely to occur during our own moment in time. This seems insensitive to the fact that the universe is vast not only in space but also in history. According to current estimates, the age of the universe is at least three times that of the Earth. Humans appeared only in the latest galactic instant, their radio communication technology only in the most recent nanosecond.

Many assume that we will attract the attention of others at this point in our history because of three recent technological developments: radio, television, and radar signals that we radiate outward, nuclear weapons (particularly the electromagnetic pulse produced by an explosion), and spaceflight. However, the wave of significant electromagnetic signals is only about 50 light-years out, steadily weakening with distance.

The issue is not only distance, but time. For the first 4 billion years of life on Earth, astronomer Dan Wertheimer pointed out, we did not leak radio at all. Then suddenly for 100 years or so we leak like crazy. Now, if we go digital, we will return to being radio quiet. That leaves a very narrow window for possible detection. The chances of locating alien beings who have just discovered radio are minimal.[7] Extraterrestrial searchers would face the same problem.

Ernst Mayr offered a fable in which another civilization, discovering the existence of the earth 4.5 billion years ago, began sending signals and continued until 1900 before giving up. They would have proven to their satisfaction that there was no intelligent life here. Our own eighteenth-century ancestors could not have detected twentieth-century television signals, even if they had been sent with the most powerful transmitters of our day.[8]

Mind-Stretcher. Although our radio and television era may be short, there may be future eras that will send other kinds of signals. Just as Enlightenment scholars could not have imagined radio, television, and radar, so we may be unable to imagine the evidence of our presence that technologically advanced aliens might detect 200 years from now.

Earth has sent out two spectacular signals of life, explained the Clarks. The more recent is the sudden burst of radio waves that started 50–60 years ago. The first signal, which began about 2 billion years ago, was the change in the composition of our atmosphere with the rise of photosynthesizing plants, producing a dramatic increase in the level of oxygen.[9] Alien scientists may be far more likely to discover the second. Their human counterparts will be searching for similar signals as they study extrasolar planets.

The other presumed telltales of our existence are even less detectable. Our existing technology could not identify the pulse from a nuclear weapon at the distance of the nearest star. Tests of such weapons now are conducted underground, making detection even more difficult. Our spacecraft and their physical effects are tiny in an interstellar context and might be detectable only if an alien observer were located in our own solar system.

Agism

Focusing on our own time ignores the possible age of other civilizations. Ulmschneider estimated that, as the first population I stars appeared about 10 billion years ago and assuming that the development time for human-type intelligence is around 4.6 billion years, the first intelligent societies could have appeared 5 billion years in our past. They would be older than our planet.

Our solar system has only existed for about half the time of the longest life-favorable systems, Edelson argued. If the length of time that it took us to evolve is typical, there could have been civilizations 6 billion years ago.[10]

If such ancients exist, they are the continuity of awareness in the universe. They may have evolved far beyond our concepts of intelligence and civilization. They may be the most potent forces evolution has produced, with powers to create and shatter worlds, to manipulate cosmic forces, and to determine the fate of lesser minds. Would they be mindful of us?

The civilizations that we expect to encounter may have begun their searches for others millions or even billions of years ago—if they had the interest and the technological means.[11] Such a civilization might have begun reaching out to the stars long before intelligent life emerged on the Earth.

The Clarks suggested comparing two stretches of time. The first is from the earliest era that extraterrestrial intelligence could have arisen in our Galaxy to our present. The second stretches from the present to the maximum realistic planning horizon, say 200 years. The first stretch is 10 million times as long as the second. If we assume that the chance per year of being visited by extraterrestrials is constant, we are 10 million times as likely to have been visited in the past as we are to be visited in the immediate future.[12]

One frequent theme in SETI literature is that we are passing through a crucial historical threshold. As Sagan put it, this is the first time that our technology has reached the precipice of self-destruction, but it also is the first time that we can postpone or avoid destruction by going somewhere off the Earth. These two capabilities make our time extraordinary in contradictory ways.[13]

O'Neill, too, argued that our moment in time is distinctive. We have arrived within the past few decades at a point where we are able to use radio communication. Within a few more decades, we will be able to spread human presence into the Galaxy. However, O'Neill carried the argument farther: This conjunction of events gives us a very distorted view of what is practical and possible. These capabilities do not ensure that contact will take place in our time.[14]

Some SETI literature suggests a sense of urgency, implying that we *need* to make contact now. The desire for an encounter in one's own lifetime is understandable. Laser pioneer Charles Townes commented that he would be much more interested in learning of life 5 or 10 light-years away so that there would be some chance of communication during his lifetime than in finding life 100,000 light-years away.[15] However, this personal sense of urgency does not change our place in time.

Some assume that extraterrestrial intelligences that detected us would share this urgency, acting immediately to make contact. We have no factual basis for this assumption other than our emotions of the moment. If another civilization discovers us by detecting our signals or by scanning us from its probes, there could be a significant delay between that discovery and our becoming aware of the alien society. The length of that delay would depend on the distance between them and us and on the actions they choose to take. Even the most eager extraterrestrials may not be able to overcome the light speed barrier, either with signals or with machines.

Delay might be deliberate. Masterful aliens may wait and observe, particularly if they or their probes are sentient machines, patient and unhurried.

The super-Copernican principle, physicist John Wheeler told us, rejects now-centeredness as firmly as Copernicus rejected here-centeredness.[16]

Detecting another civilization may destroy the chauvinism of specialness; learning about the history of intelligence over billions of years may destroy the chauvinism of time.

They Will Be Detectable

The SETI 2020 report directly stated a major assumption of the radio search: "We expect communications, inadvertent or deliberate, to be commonplace in an inhabited universe."[17] *De facto*, human searchers define a technical civilization by its capacity to use powerful radio transmitters. And, we might add, by its interest in doing so.

A society that does not emit strong electromagnetic signals may be invisible to us. If aliens used optical fibers for all of their communications, no radio waves would leak. Astronomer Jesse Greenstein speculated that all knowledge might be contained in a planetwide computer that radiates no energy into space and so communicates nothing.[18]

Drake's Warning

Frank Drake's own analyses showed that radio leakage from a planet is likely to get weaker as a civilization improves its communications technology. He proposed in 1974 that when societies become more technologically sophisticated, they will reduce wastage of power and will become undetectable. Six years later, he wrote that mortal civilizations probably do not remain detectable forever, because their increasing technical sophistication enables them to cease the release of energy into space.

Cell phone systems and satellite to home television require the sharing of frequency bands, driving us toward fainter, more targeted signals. To be realistic, we should assume that the same evolution of signal types will occur in other civilizations. The signals we have long assumed to be the right things to search for are fast disappearing, or fading, in our own civilization. If others follow a pattern of technological development similar to ours, we will be able to eavesdrop on their leakage only during the century after they develop radio.

A similar evolution is under way in television signals. Engineering considerations would seem to ensure that we will eventually phase out TV broadcast services.

"We're rapidly losing visibility—by a factor of 100,000," said Drake. "Is that typical or quirky? We don't know, but it's a warning signal." The implication is clear: We need to search for a much greater variety of electromagnetic emissions.[19]

A network of tightly beamed communications channels might connect the advanced technologies of our Galaxy, suggested Bracewell. The probability that we would come within the beams of such transmissions is very small. If technologically advanced aliens do not wish to allow others to tune in, it is unlikely that we will intercept their messages.[20]

Human searchers hope that other civilizations will choose a signaling method that does not mimic nature. The Ohio State University search strategy even assumed that a civilization transmitting at the hydrogen line frequency would offset their transmission in just the right way to remove all motions with respect to the center of the Galaxy.[21] Such best-case assumptions are driven by the limitations of our technologies.

Even if signals are above our detection threshold, we might not realize that they were messages. "At this very moment," Sagan imagined, "the messages from another civilization may be wafting across space, driven by unimaginably advanced devices, there for us to detect them—if only we knew how. . . . Perhaps the messages are already here, present in some everyday experience that we have not made the right mental effort to recognize." Drake admitted that we might receive an information-bearing message and never realize it.[22]

We may fail to detect the signal of another civilization because we don't know the code. It might appear to us as static; as a civilization becomes more efficient, Minsky warned, transmissions look more and more like noise. A few years of technology development can make the new signals incomprehensible to the senders of the old signals.[23]

We may not have qualified scientifically for entering the galactic communications network, Bracewell proposed, because we still have not discovered the next thing that awaits downstream for us in physics. It may be that communication hinges on this next discovery.[24] We could be in a temporal hiatus between older, more advanced civilizations that have moved too far beyond us to be detectable, and newer civilizations that have not yet developed technologies for which we can search.

Drowned out

Some believe that certain species of whales share an ancient culture, communicated through long, elaborate songs like the oral histories of ancient human bards. Once, those songs might have reached throughout the world ocean, interconnecting the species. That communications web may have been suppressed by the noises of human technology: thousands of chugging engines and whining propellers, filling the liquid medium with random, unintelligent noise.

Weakly transmitting civilizations—perhaps including ours—may be drowned out by the Galaxy's electromagnetic turmoil. Even if extraterrestrials detect our radar, radio, and television emissions, our signals might seem to them like a mindless jumble. Amid Earth's cacophony, would they notice our deliberate message to them, one organized call emerging from the random chatter of a crowd?

What about finding artifacts? Spotting an interstellar probe in our solar system could be a challenging task if it were not emitting signals. That job would be even tougher if the probe were both silent and small.

Perhaps we are being observed not by livestock-mutilating UFOs, McConnell speculated, but by microscopic probes silently communicating with their home world.[25] Our solar system could be swarming with probes so tiny that we would be completely unaware of their existence.

We and our alien counterparts may have different ideas about what can be detected. Robert Freitas thought that our best chances lie in finding objects for which detection by us is, for them, unimportant.[26]

Like one of the creatures in Gregory Benford's novel *Beyond Infinity*, we are Seekers After Patterns. However, those patterns may be unique products of human culture. Our infatuation with a particular model of an advanced extraterrestrial society may lead to an ineffective strategy for detecting it.[27]

We Will Recognize Their Signs

The most important tool we take to the observatory is an open mind.
—Astronomer Patrick McCarthy, 2005[28]

Why do we expect that it will be easy to recognize a more advanced society, Sagan asked rhetorically. It may have evolved into forms undetectable to us. The signs of very advanced civilizations may not be in the least apparent to a society as backward as we, any more than an ant performing his anty labors by the side of a suburban swimming pool has a profound sense of the presence of a superior civilization all around him.[29]

Several people involved in this debate have urged that we give more thought to what intelligent extraterrestrials might be doing to further their own purposes, not just to communicate. Technologically powerful civilizations like Kardashev's Types II and III could cause large-scale effects on the space around them.

The technology of those civilizations may be incomprehensible to us, warned Burke-Ward; we probably would fail to see that it was there. Kardashev himself thought it very difficult to predict by natural physical laws the limits of the size, the power, and the activity of such civilizations. We should search for new objects in the universe that are difficult to explain by natural causes.[30]

With prodigious energy resources, Sagan speculated, technologically advanced civilizations should be able to rework the cosmos. Unexplained or "unnatural" events such as bizarre energy sources or mysterious cavities in the interstellar medium might have explanations consistent with our present science—or they might not. As long as we cannot understand these phenomena, we cannot exclude the possibility that they are manifestations of extraterrestrial intelligence.[31]

We should keep a sharp eye out for anomalous order of any kind, Grinspoon proposed. This could include nonequilibrium mixtures of gases (or, conversely, too much equilibrium in places where other known processes are creating disequilibrium), strange mechanical shapes and assemblages, or rhythmic environmental changes without any obvious cause. White suggested looking for "entropy pools"—areas of entropy reduction surrounded by regions of increased entropy.[32]

If we find ordered structures without a known natural cause, can we be sure that they indicate extraterrestrial intelligence? Even very smart scientists have misinterpreted evidence; Kepler thought that lunar craters were cities in circular form.

It is difficult to judge how best to search for intelligent life when we cannot even be sure what the dominant form of intelligence on Earth will be. "What prospect could we have," Rees asked, "of envisaging what might be spawned from another biosphere with a billion-year head start on us?"[33]

We probably do see evidence of alien activity, Gindilis and Rudnidski proposed, but we are unaware of it. Thomas Kuiper predicted that detection will be the result of an accumulation of phenomena that are hard to explain.[34]

Mind-Stretcher. What we will observe will be their star wars, argued space expert James Oberg. Man's greatest efforts have been military; the same may be true elsewhere in the Galaxy.[35] We might detect the energies released by colossal battles among the Galaxy's titans, the Gods of War. We may have seen their signs already, interpreting them to be astronomical phenomena.

The Great Silence may say more about our own limitations in conceptualizing intelligence and its works than the ability of the universe to produce them, suggested Grinspoon. Clarke told us many years ago that any sufficiently advanced technology would be indistinguishable from magic.[36]

The only type of intelligence we could detect would be one that employed a technology that we can recognize; that might be a minor and atypical fraction of all extraterrestrial intelligence.[37] This leads us toward a limiting concept: Intelligent beings may only recognize the signs of other intelligences near their own technological level.

They Live on Planets Orbiting Stars

SETI has rested on an implicit assumption that extraterrestrials live on planets orbiting stars, analogs of our own home. This assumption may not apply to our own descendants; future humans are likely to spread into interplanetary space.

Planets may be a good place for life to begin, Dyson observed, but they are not a likely place for the home of a big technological society. As one study of the space colony concept put it: "In the future, the Earth might be looked upon as an uncomfortable and inconvenient place to live as compared to the extraterrestrial communities."[38] The implication is clear: We should investigate nonplanetary locations for finding alien intelligence.

If advanced aliens have dispersed to interstellar distances, Shostak recognized, we can expect to find them around stars that might be patently unsuitable for incubating life.[39] In the longer run, expansion could free more advanced species from depending on any star; some technological civilizations might choose instead to live in interstellar space.

Dyson proposed that bodies such as comets may provide homes for life throughout the Galaxy, not just near stars.[40] We might have a better chance to find evidence of intelligence if we searched in all directions in a wider range of frequencies. That would place much greater demands on our search systems.

They Search for Others

Because we search for others, we assume that they search as well. This rests on a belief: that many civilizations at some point in their development perceive the likelihood of other intelligent life in the universe and find themselves technically able to search for and to send signals to it. Bracewell even suggested that advanced communities would act in concert and avoid duplication in searching.[41]

How realistic is this assumption? A search cannot be comprehensive if it is confined within a moment in time; it may have to be sustained for eons. Yet, our own searches have been episodic cultural phenomena, dependent on the values, perceptions, and technologies of their eras. Extraterrestrial civilizations—if they ever start a search—may not give it continuing attention over millennia if they do not find anything within a time span that they consider enough.

The most curious—or the least self-satisfied—may be the most likely to find us. However, interstellar communications curiosity may be incident to a particular stage of technological advance; as Campbell suggested, it might give way to other kinds of curiosity with further changes in technology. "An alien society with no desire to send messages to other civilizations would not be planning to receive helpful signs from the stars," posited Baird. "If creatures like these are left alone, we will never hear from them."[42]

Many of us assume that every civilization will be curious about others. Yet, historian J.M. Roberts has pointed out the massive indifference of some Earthly civilizations—their lack of curiosity about other worlds—even as the European Age of Exploration began to influence their futures.[43]

The desire to explore may not be a universal phenomenon. Historian Steven Pyne found that exploration is a specific invention of specific civilizations conducted at specific historical times. It is not a universal property of all human societies. Not all cultures have explored or even traveled widely; some have been content to exist in xenophobic isolation.[44]

Consider the Ming Dynasty's abandonment of oceanic exploration. "Fully equipped with the technology, the intelligence, and the national resources to become discoverers," wrote historian Daniel Boorstin, "the Chinese doomed themselves to be discovered." Although cost may have been an issue, the decision also may have reflected China's belief that it had nothing to learn from the outside world.[45]

Interstellar Anthropologists

More advanced civilizations may not devote significant resources to searching for those less advanced. Looking for such societies may the concern of a few specialized researchers—as it is among humans.

"Our earth is not the concern of the great enterprises of knowledge among those far societies," proposed Morrison. "Rather, it is the activity of a Department of Anthropology." Just as modern anthropologists search in jungles for lost tribes that have never been contaminated by contact with modern civilization, McDonough suggested, some extraterrestrials may search for newly emerging primitive societies. On our planet, the "primitive lobby" are the anthropologists who would like to understand such societies, anthropologist Richard Lee observed—and the missionaries who would like to convert them.[46]

A civilization far in advance of ours may devote only a tiny fraction of its resources to trying to find and communicate with life-forms it deems lower. Consider how few human researchers try to converse with dolphins and chimpanzees.

Our signals might be picked up first by amateurs. McDonough imagined that there are alien clubs—similar to our ham radio operators—who delight at being the first to detect a new civilization, much as a short-wave listener jumps for joy when picking up a country he or she has never heard before.[47]

Would those amateurs be taken seriously by the extraterrestrial equivalent of our scientific community? Or would they be ignored?

They Know We Are Here

It is too late to be shy and hesitant. We have announced our presence to the cosmos.
— Carl Sagan, 1973[48]

There is a widespread assumption in the SETI community that more advanced civilizations will detect us, or already have done so, because we have been radiating electromagnetic signals for decades. To this school of thought, there is no point in remaining silent; the extraterrestrials already know that we are here, or detection is inevitable.

Sagan and others have claimed that we have revealed ourselves to the Galaxy by mundane military and commercial activities: radio, radar, and television signals expanding outward from Earth at the velocity of light. McDonough, assuming that intelligent aliens would monitor our television broadcasts, thought that we have already given away the deepest secrets of our culture; it is too late to hide our ugly side from the Galaxy.[49]

Despite what science fiction often assumes, our intelligently organized signals such as radio and televison programs are not our most detectable signs. Radar pulses and carrier waves, however mindless they appear, are far more likely to be detected.

Woodruff Sullivan and his colleagues, after studying the electromagnetic signature of the Earth, concluded that we would be most detectable by a few powerful military radars and by the video carrier signals of our television broadcasting stations. Alien astronomers using an instrument like the Arecibo radio telescope could detect the Ballistic Missile Early Warning System at about 18 light-years; our UHF television transmitters would be identifiable only 1.8 light-years away—less than half the distance to the nearest star beyond our Sun.

If alien eavesdroppers were equipped with the proposed Cyclops array of radio telescopes (which was never built), they could detect our strongest video carrier signals at about 25 light-years, and our early warning radar signals out to about 250 light-years.[50] Though these distances sound impressive, they include only a tiny portion of a Galaxy 100,000 light-years across.

Currently, our most powerful signals are planetary radar pulses sent from the Arecibo observatory. NASA's short-lived all-sky survey could have detected transmissions from that instrument's extraterrestrial analog out to 30 light-years; the targeted search could have seen it at 300 light-years. So far, Arecibo's pulses have only reached out about 40 light-years. This radar is a very discontinuous source, used for only a few hundred hours a year. Military radars operate almost continuously, but they are much weaker.[51]

Omniscience

Some imagine that more advanced civilizations will have such comprehensive knowledge as to be effectively omniscient. A NASA report made a sweeping statement: "Once a system capable of conceptualizing sophisticated internal models of external phenomena has evolved, it is only a matter of time before all possible ideas inherent in the available sensory perceptions are conceived." MacGowan and Ordway thought that extrasolar technological societies probably would be restricted only by absolute physical limitations and not by limitations of knowledge or understanding.

To attribute perfect knowledge to extraterrestrial intelligences is to give them God-like qualities. Even those far more technologically advanced than ourselves may not know everything.

Although there may be superbeings that outperform us in every dimension, observed Joseph Royce, it may be more reasonable to expect that they will be more advanced than us in some areas but not in others.[52] They too might have weaknesses, and gaps in their knowledge.

We cannot assume that we have been found or that detection is inevitable. Even if we spot a beacon, we would have to be careful about jumping to conclusions. That beacon would not imply that the sending civilization already knows of our existence, unless it clearly was aimed at us.

Extraterrestrials might see no reason to search for others if they believe their society to be unique. They may not be looking for the kinds of signals we normally radiate; radio, radar, and television technologies might be seen as primitive. Our emissions may be below the threshold they could detect, particularly if they are very distant. If an alien technological society does search its skies, it would have to be looking in the right direction at the right time with the right kind of technology—and in the right wavelengths—to find us.

The Galaxy's enormous distances require fantastic measures for interstellar communication, concluded electrical engineer George Swenson—stunningly high transmitter power or huge antennas and impractically narrow beams. It would take a very large and carefully aimed antenna to pick out signs of technology in our solar system's spectrum from more than few tens of light-years away. Any civilization on the receiving end, said Shostak, would need an antenna about the size of Manhattan Island to pick up our radio and television broadcasts.[53]

The major exception would be unusually powerful beamed signals, such as the one sent from Arecibo in 1974. Even those might fail the test of repeatability. An alien astronomer may have written the equivalent of Wow! beside a spike on his recording device, without ever finding us again.

There may be a critical density of civilizations necessary to allow round-trip travel of radio signals, suggested Bracewell. If that density is reached, there may be enough communication to sustain a Galactic Club. If it is not reached, there may be no communication between civilizations.[54]

They Want to Communicate

We postulate that interstellar communication, having spread rapidly throughout the Galaxy once it began, is now a reality for countless races.
—Project Cyclops, 1972[55]

It's not going to waste its time talking to gibbering idiots.
—Fred Hoyle, *The Black Cloud*, 1957[56]

Many SETI researchers expect that, at a certain level of development, technological civilizations will intentionally transmit signals for detection by others. SETI astronomer Paul Horowitz reflected the optimistic impli-cations of this assumption: "If they're attempting to contact other civiliza-tions, we'll succeed some day."[57]

Others have questioned assumptions underlying this scenario: that intel-ligent extraterrestrials would intentionally beam radio signals toward us; that those signals would be antiencrypted through the use of universal mathematical and scientific truths; that once these are translated, they would provide a Rosetta Stone for communication in other domains. These assumptions, critics maintain, betray a high degree of anthropocentrism.[58] We cannot assume that extraterrestrial civilizations are eager to commu-nicate with us, nor that all human beings are eager to communicate with them.

"Cosmology texts of the sixties were written with a kind of naive opti-mism," commented anthropologist Anthony Aveni, "that imagined not only rampant chemical and biological systems just like ours but an inter-galactic intelligentsia with the same cultural expansiveness, natural curios-ity, and desire to explore—for the sake of acquiring knowledge itself—that Earth-based ethnocentric theorizers possessed." To believe that there are societies elsewhere bent upon and capable of communicating with us is not only to be anthropomorphic, Beck cautioned, it is to believe that civiliza-tions elsewhere are like one civilization that has existed on only a small portion of this Earth for only a few hundred years.[59]

Rood and Trefil saw no reason to expect that the human desire to learn about others and to let others know about us is a universal trait. Shostak thought that it was fair to ask what might motivate extraterrestrials to "reach out and touch someone." If there is no reason for them to do so, we may never discern their existence.[60]

What forces favor interstellar communication? Goldsmith and Owen proposed curiosity, gregariousness, and what might be called social avarice—the hope of obtaining valuable information. Forces that oppose communication include fear, inertia, and the press of other priorities.[61]

To have a reasonable chance of making themselves known by radio, extraterrestrials must have not only the ability to communicate but also the desire. A sustained transmission effort could require significant resources; the listeners too would have to make a long-term commitment. Oliver acknowledged that the whole thing depends on the longevity and effectiveness of the communication effort at both ends.[62]

Many SETI advocates assume that the technologically superior civilization would bear the onus of initiating contact by transmitting, because that civilization can do it more easily. As Jill Tarter explained, transmitting is a much harder job than listening, so we put the burden of transmission on the older technologies. If there are no older technologies, our searches will not succeed.[63]

The average civilization's lifetime as a seeker after contact may be far less that the average civilization's lifetime with communications ability. There must be some limit to how long a civilization will continue to try to communicate with hypothetical others. Von Hoerner's estimate of the average length of communication time was 6500 years.[64] Although long by the standards of twenty-first-century Earth, this is but a brief moment in galactic history. At any given time in that history, only a small percentage of technological civilizations may broadcast signals in the hope that they will provoke responses.

We still have to ask why. If the aliens' motive for communication is self-interested, Regis pointed out, then they are seeking to benefit from contact with societies more advanced than their own just as SETI advocates on Earth are. Exchanging information with emerging technical civilizations like ours may be a much lower priority.[65]

The classic SETI paradigm treats interstellar communication as the apex of cultural evolution among technological civilizations. However, no phase of cultural evolution is final. What might lie beyond exchanging messages? There may be a more civilized way of life than a communicative one.[66]

Legends

In the standard SETI paradigm, a civilization that is near to us in space is going to be very far from us in time; any random one would be very old. Sagan and Newman thought it possible that our Galaxy is teeming with civilizations as far beyond our level as we are beyond the ants, and paying us about as much attention as we pay to ants.

A civilization very far in our future is unlikely to be interested in us, argued Sagan; civilizations a million years in our future will be of much greater interest. Such a civilization will be engaged in a busy communications traffic with its peers, but not with us, and not via technologies accessible to us.

The civilizations vastly more advanced than we will be, for a long time, remote both in distance and accessibility. To us, suggested Sagan, the most advanced may be no more than insubstantial legends.

Many believers assume that if the aliens are far more advanced than we are, they will adjust their communications to our level. "More advanced societies will be able to guess how backward we are and will, *if they wish to communicate with us*, make allowances," claimed Sagan (emphasis added). Von Hoerner thought that if they have an interest in talking to us, they would know how to do it. However, they may have a lower limit, a standard below which they are not interested.[67] Our use of radio as our primary means of interstellar communication may place us among the Galaxy's primitives.

Even if more advanced aliens were interested, how would they know which technological means to use? Their potential recipients might range from those that have just acquired radio capabilities to those who abandoned them long before.

We may be limited to contact with civilizations that are not very far in advance of our own. James Funaro, an anthropologist who organizes conferences that simulate contact, speculated that the medium of communication is going to select those societies that are something like us.[68]

Mind-Stretcher. Each technological civilization may have a lowest threshold of interest at each stage of its history. That threshold may rise with each civilization's scientific and technological progress, excluding efforts to contact intelligences that lie below it.

There Will Be Lighthouses in the Cosmic Night

The factor L in the Drake equation is not the lifetime of the civilization itself but that of whatever beacon it can create.
—Robin Corbet, 1997[69]

The popular view of SETI is that the search is designed to pick up messages. In fact, Shostak clarified, the experiments are configured to find steady or slowly pulsing narrow-band signals. The earthly counterpart is the carrier wave. Carriers, with the highest signal-to-noise ratio of any part of a transmission, would be thousands or millions of times easier to detect than the modulation, or message.

We are looking for a beacon signal deliberately designed to attract the attention of any interstellar listeners. Finding the subtle variations of any message will demand far more sensitive instruments.[70] SETI focuses on beacons because that is what our current technological capabilities allow us to do.

Horowitz optimized his own searches for pure narrow-band carriers because they are easier to detect, stand out clearly as artificial, are efficient beacons, and can be distinguished from terrestrial interference. However, there is a fundamental problem: The best beacon is the worst message, and the best carrier of information is the worst beacon.[71]

How likely are beacons? The authors of the Project Cyclops report thought that the first civilizations to undertake the search *undoubtedly* followed their listening phase with long transmission epochs (emphasis added). "Their perseverance," the report went on, "will be our greatest asset in our beginning listening phase."[72] Here again, we are assuming the behavior of alien civilizations.

The NASA targeted search assumed that there are at least 100,000 beacons in the Galaxy, so that one will be found in the nearest million stars. If there are between 1 and 100,000 beacons in the Galaxy (and if they are using continuous wave transmission), a sky survey is the best strategy. For the sky survey to be useful, the beacon must be on almost constantly and for long periods of time.[73]

A Mirror Image. In order for an extraterrestrial civilization to detect our transmissions, explained Jill Tarter, their instruments must be looking in our direction at the time our signal arrives—and they must have chosen the right instrument. The chances of that happening are zero, unless we commit to transmitting for a long time.[74]

The underlying issue is the motivation for operating beacons. Drake assumed altruism: "Other civilizations may well generate signals intended to benefit others, signals that are strong enough only for the deserving to receive."

Oliver thought that beacons might exist to help young races such as ours to join the galactic community. If we come into contact with a network of civilizations, the first requirement imposed on us might be to erect a beacon to continue that process.[75]

Davies was skeptical about the assumption that the aliens are doing the transmitting. The enormous asymmetry of effort between the transmitting and receiving ends of the operation means that one has to suspend the Copernican principle (that we are typical) to justify SETI. We have to assume that they are prepared to act in a superhuman way by spending large sums of money over eons of time sending signals in all directions with little hope of a reply.

If we would not do that, why would they? Bracewell doubted that an omnidirectional transmitter blindly radiating 1000 megawatts in no particular direction could be justified and sustained for centuries on end.[76]

These considerations may drive technological civilizations toward listening but not transmitting. Extraterrestrials who maintain radio beacons may be a subset—possibly a very small one—of what might be a much larger number of technological societies.

Technological civilizations might send beacon signals only after they have detected signs of intelligence, or at least of life. Shostak proposed that extraterrestrials might target their beacons on those remote planets whose atmospheres show signs of biological activity. That would drastically shrink the number of targets, but the search range would be limited to relatively nearby planets—and the beacons would have to be repeated for millions or even billions of years in the hope that a radio-competent society would emerge.

Rubin's List

Alan Rubin suggested 10 possible motivations for a civilization to maintain beacons: (1) to get a response by targeting signals to main-sequence stars that were known to have planets, particularly if there were signs of life and intelligence; (2) hubris—the sending society might view itself as the glory of the universe; (3) to impress themselves or rival groups with their power and status; (4) evangelism—to save the souls of intelligent beings throughout the Galaxy; (5) entertainment—extraterrestrials afflicted with *ennui* might find it amusing to transmit signals the way bored ship passengers put messages in bottles and toss them into the sea; (6) commerce—entrepreneurial aliens may want to swap information, using what they learn to produce and sell novelty items; (7) altruism and paternalism—an ancient advanced society might feel obligated to help struggling newcomers; (8) paranoia—a fearful civilization may feel it necessary to broadcast threats to intimidate potential unwanted visitors; (9) reproduction—an alien species might broadcast information about its genotype with instructions for creating members of its own species; (10) pugnacity—an aggressive society might seek to instill fear or to provoke conflicts among rival groups of recipients.[77]

We return to the issue of motivation. Why would other intelligences maintain beacons? Horowitz recognized that assuming the intentions of the transmitting civilization is skating on very thin ice. Brin added that the classic SETI scenario ignores the many reasons why another civilization might think that sending out messages was unwise.[78]

The authors of the SETI 2020 report straddled the issue, arguing that we can neither expect to receive signals intentionally beamed to us nor afford to overlook this possibility. We maximize the potential for positive results by using a search strategy that assumes either that the other civilization is trying to attract our attention or that it is transmitting information to other extraterrestrials.[79]

If beacons are very rare, the only signals we have a good chance to detect in the near future through the techniques of radio astronomy may be targeted communications directed toward others. In the case of a colonizing civilization, the second category presumably would include communications with and among colonies. We may only be able to spot a small percentage of such signals.

We cannot draw sweeping conclusions from our failure to find beacons at this early stage in our search. It was a best-case scenario.

SETI Scientists Will Make the Discovery

There is a widespread assumption in the SETI community that its own researchers will be the first to discover evidence of extraterrestrial intelligence. The actual range of possibilities is much broader.

McDonough outlined some of the other possible scenarios for detection: conventional astronomy; communications engineers listening to the sky; nonastronomical science; military satellites that occasionally do inadvertent astronomy.[80] His list is not exhaustive.

There is a long history of astronomical phenomena being discovered by military or intelligence systems—and not just satellites. Classified military capabilities detected evidence of radio waves from space at about the same time as Jansky made the first map of the radio sky. A British radar used to detect incoming V-2 missiles during World War II picked up radio noises from celestial sources, but this information was kept confidential until after the war.[81]

X-ray-emitting objects were found first by military systems; the information was kept secret until it could be released without endangering security. Gamma-ray bursters, among the most energetic phenomena in the universe, were discovered first by military satellites designed for a completely different purpose.[82]

"Perhaps the first signs of an extraterrestrial civilization are already sitting on magnetic tape, deep in the bowels of the U.S. National Security Agency or the KGB," speculated McDonough. "Perhaps they have been rejected as natural noise." The NSA, Brin observed, is just one group with far more sophisticated listening apparatus than all of the world's SETI teams put together.[83]

They Have Prepared for Contact

Many SETI advocates assume that any extraterrestrial technological civilization that comes into contact with us will have prior experience with other technological civilizations and will know how to incorporate us into a network. "The advanced technology we detect will have experienced this type of encounter many times before," according to Jill Tarter. "It already may have established a galactic protocol for information exchange." Bracewell imagined that we will be brought into touch with a chain of communities already in communication with each other who know quite well how to bring new members into their ranks.[84]

A civilization affiliated with the Galactic Club will have given sustained, in-depth thought to interstellar contact, Harrison proposed. That civilization will know how to construct messages that are easy to understand; it will know how to decipher messages from other species and how to manage first impressions. In short, they would make it easy for us.

What about isolated civilizations? Harrison recognized that each would have to look to its own theories and analogs for guidance, sources of information that may or may not work well when applied to radically different species and cultures. Neither society would be able to draw on actual experience.[85] In other words, the extraterrestrials may be no better prepared than we are.

As the Annex to this book shows, our own efforts to prepare for contact are confined to small numbers of interested people and are not widely known among our fellow humans. The same may be true of many extraterrestrial societies. The fact that a civilization is technological does not necessarily mean that it knows of others; contact with us could come as a surprise.

Assumptions: After Contact

The Message Will Be Comprehensible

Ere long all human beings on this globe, as one, will turn their eyes to the firmament above, with feelings of love and reverence, thrilled by the glad news: Brethren! We have a message from another world, unknown and remote. It reads: one... two... three...

—Nikola Tesla, in a prediction written in 1900[1]

Once we have the message... the rest is easy.

—Frank Drake, 1974[2]

Contact optimists have tended to assume that alien messages would be relatively easy to understand. The communication of quite complicated information is not very difficult, claimed Sagan, even for civilizations with very different biologies and social conventions. Once pictures are transmitted, it will be extremely simple to develop language—by show and tell.[3]

"We are considering not cryptography," Sagan declared, "but anticryptography, the design by a very intelligent civilization of a message so simple that even civilizations as primitive as ours can understand it." Jill Tarter thought that an information-bearing message would be crafted for unambiguous transmission, because contact with us will not be the first encounter of a superior technology with an emergent one.[4]

If they are far ahead of us technologically, McDonough optimistically assumed, then they will be just as advanced in their ability to teach. They may have had thousands of experiences in teaching their language and culture to other primitive civilizations, and they would know how to do it very well—if they so wanted. However, teaching would be far more difficult if we receive messages from great distances, cautioned Baird; it would be without the customary aid of immediate feedback from the teacher.[5]

There are good technical reasons to separate the functions of establishing contact (the beacon) and conveying information (the communication channel). Morrison thought that the first transmission would be an acquisition signal; the second would be a decoding signal. The third essentially would be a language lesson.

Similarly, Bracewell speculated that an alien probe in our solar system would teach us the language of its originating civilization.[6] In this case, the exchange could be much quicker.

We expect the extraterrestrials to do the hardest work. As the process of decoding messages from another species may be too difficult for us, Vakoch proposed that it is a task best left to the more advanced species.[7] However, that makes translation, like the operation of beacons, a question of alien intent.

Similar historical backgrounds may be needed if we are to understand the set of symbols used by another civilization. Yet, the language of another intelligent community may have few points of contact with our own.[8]

Language, or Symbols?

To complicate matters further, what appears to be a text may be something else. For more than a century, archaeologists have tried without success to decipher the symbols used by the Indus civilization between 3200 and 1700 BC. They assumed that this was a form of written language, paralleling the evolution of written languages in Egypt and Mesopotamia. Some scholars now challenge that assumption, arguing that this script is not writing, but a collection of religious–political symbols that held together a multi lingual society.[9]

Baird questioned the assumption of anticryptography. Aliens might have no real interest in talking with organisms that can never understand what is most dear to them; any messages they send into outer space will be made purposely difficult. If there is something resembling Sagan's Encyclopedia Galactica, Lemarchand speculated, it is probably encrypted, so that it can be read only by ethically advanced civilizations.[10]

If we assume that the alien mind is humanlike in all essential respects, Baird proposed, we can expect to develop communications in the same way that one might teach a foreign language. However, we do not know what hidden assumptions underlie our proposed communication channel, assumptions that we are unable to evaluate because they are so intimately interwoven into our fabric of thought. Communications problems arising because of this mismatch cannot be planned for in advance.

The formula for success must include a search for signal features that are common to human language in the broadest sense. Without the establishment of a shared communication format, warned Baird, one civilization's book of universal wisdom will be another's book of universal confusion. We are at a point in history that predates discovery of a Rosetta Stone from the stars.[11]

What if there is no language lesson? Language can be detected within signals, argued John Elliott and Eric Atwell of Leeds University, even if we cannot read it. They proposed that we respond to an alien message with one of our own that describes us in one of our own languages. The other civilization would recognize the message as language even if they did not understand the content. However, true communication may require a shared code book.[12]

Biological differences may affect our ability to understand alien communications. Human brains seem to have an innate system of grammatical rules that structures all our languages.[13] The brains of extraterrestrials who emerged from separate evolutions are unlikely to be organized the same way.

Philosopher and political theorist John Locke had foreseen in 1689 that our capacity for attaining ideas is limited by our senses; beings on other worlds, assisted by "senses more or perfecter than we have," may develop ideas unavailable to us. Differences in sensory equipment could give aliens a range of sensations totally unexperienced by humans, proposed Ruse; we may have sensations unexperienced by them.[14]

Sagan admitted that it is only a message intended specifically for emerging technical civilizations that we have any good chance of receiving, let alone understanding. With billions of years of independent biological and social evolution, the thought processes and habits of any two communities must differ greatly; electromagnetic communication of programmed learning between two such societies could be a very difficult undertaking.

Drake recognized that we may not be able to communicate with extraterrestrials, even if we are able to detect them. In Stanislav Lem's novel *Solaris*, humans and a planetary intelligence try to converse with each other, but in vain; their minds and their methods of communication are too different.[15]

Perhaps the more advanced sender, aware of the inherent limitations of developing cultures, will compose messages on many intellectual planes. However, our best hope may be to seek out or eavesdrop on civilizations that employ styles of linguistic expression like ours.[16] That may imply societies whose levels of scientific and technological achievement are not very far ahead of our own.

McConnell introduced another dimension: incompleteness. Even under optimal conditions, the listener will not receive the coded message with 100% accuracy. There is a high probability that the signal will be interrupted and that it will be vulnerable to interference, noise, or loss of line of sight between transmitter and receiver. It is extremely unlikely that the listener will intercept this message just as it is beginning; it will not be immediately obvious where the message begins and ends.[17]

What if advanced extraterrestrials have gone through more evolutionary steps, leading to cyborgs and machine intelligence? We might find ourselves communicating with nonbiological beings. If we detected incoming probes, their messages might be intelligent machine language, undecipherable for us.

Computer scientist Michael Arbib challenged the assumption that communicating with artificial intelligences would be more difficult. Divergences in biological evolution may be so great that we would find more in common with an anthropomorphic robot than with many organic beings.[18]

Optimistic speculations about comprehensibility rest on best-case assumptions: The signal we receive will be a structured message; the extraterrestrials want us to understand it and will adjust it to our level; we will grasp the method they choose; we will find enough commonality with our methods of communication to enable translation. We cannot assume that this scenario will prevail. We might be faced with a magnificent puzzle, without all the clues needed to solve it.

They Will Speak Science

Science is the Greek of the interstellar Rosetta Stone.
—Carl Sagan, 1975[19]

Those who are most optimistic about interstellar communication claim that we and the extraterrestrials will have science in common. University of Arizona mathematician Carl De Vito put it this way: We assume that the alien technology that we detect is supported by a reasonably advanced science; that there is an objective reality that is the same throughout the universe; that this reality can be recognized and understood by any intelligent beings; that science is the quantitative study of this reality. Some philosophers would challenge these assumptions; the same reality may be described, even quantitatively, in many ways.[20]

Nature is observed selectively, argued Rescher; we and extraterrestrials might not perceive the universe in the same way. The sameness of the object does not guarantee the sameness of ideas about it. In any case, the language of science is not outside the psychological constraints that determine all other modes of human expression. The differences might be so great, warned Dick, as to prevent the mutual examination of objective knowledge.[21]

When two scientists differ in biology, culture, and history as much as humans and extraterrestrials would differ, Vakoch cautioned, their models of reality may vary considerably. At the core of this problem is the idea that no intelligent species can understand reality without making certain methodological choices. Their metaphors, similes, and other concepts will be quite different. Metaphor plays a very important role in science; to eliminate it would be to alter science drastically.[22]

Western-style science may be critical to the entire process, speculated White. Yet, the development of our kind of science may be a chancy thing.

There is no single-track itinerary of scientific and technological development that different civilizations travel in common.[23]

The scientific revolution on our own planet was not a uniform phenomenon, geographer David Livingstone reminded us, but a complex historical process shaped by geographic conditions. Local knowledge circulated and, by doing so, became universal. What made knowledge universal was standardization, which amounted to the triumph of certain local practices over others. That process took centuries.[24]

One of the prerequisites for the development and growth of modern science on Earth was regular communication among the scientists and interested gentlemen of European scientific societies. That network, McNeill noted, rested on a fertile field of shared culture. It is that common ground that we will lack with extraterrestrial intelligence.[25]

Rescher identified conditions needed for alien science to be functionally equivalent to ours, providing a basis for a meaningful exchange of information. First is formulation: Extraterrestrials must use mathematics like ours. Second is orientation: The aliens must be interested in the same sorts of problems. Third is conceptualization: They must have the same cognitive perspective on nature as we do. Their science will be geared to their sensors, their cultural heritage (which determines what is interesting), and their environmental niche (which determines what is pragmatically useful).

The idea that another civilization is scientifically more advanced requires that they be doing our sort of science. However, cautioned Rescher, natural science as we know it is a man-made creation correlated with our specifically human intelligence. Extraterrestrials, with different needs, senses, and behavior, are unlikely to have any type of science that would be recognizable to us.[26]

Even among humans, the science of one era may be incomprehensible to that of another; the two may not even talk about the same things. Rescher predicted that Earthly science in 100 years would be unintelligible to us today. "Unless the message was specifically tailored to a civilization just emerging into space," Rood and Trefil warned, "an extraterrestrial science book would be as incomprehensible to us as the wiring diagram of a radio would be to an aborigine."[27]

Mind-Stretcher. Musso proposed a different scenario. Scientific progress might not continue indefinitely in any civilization; at some point, it might substantially stop. In that case, even a civilization 1 billion years older than ours might be only two or three centuries ahead of us.[28]

Some see mathematics as a universal. Minsky thought that all intelligent problem-solvers must be subject to the same ultimate constraints: limitations on space, time, and materials. Two principles—sparseness and economics—show that every intelligence will be forced to develop an arithmetic and a language whose structures are rooted in the natures of things.

Because arithmetic is the same everywhere, alien mathematics will be congruent to our own. Because things are, in their most general aspects, the same everywhere, aliens will have evolved thought processes and languages that will match our own to a degree that will enable us to understand them.[29]

Others have challenged this assumption. Mathematics is just another language, argue some linguists, with symbols that are connected to certain ideas only by convention. We can no more assume that extraterrestrials will share mathematics with us than we can assume that they will share English with us. Human mathematics may be only one of several equally valid, physically true languages.[30]

Platonists Versus Anti-Platonists

Some mathematicians have endorsed Plato's concept that numbers and mathematical laws are etherial ideals, existing outside of space and time. Others reject this argument, insisting that mathematics is a human creation like literature, religion, or banking.

If mathematics is universal and eternal, claim the Platonists, aliens will understand concepts like prime numbers and pi. The anti-Platonists dismiss this idea as anthropomorphic; alien brains, responding to different environments, would have radically different mathematics.[31]

As Baird saw it, the messages anticipated by astronomers and engineers read like a shopping list of unsolved scientific problems. This, he declared, is scientific chauvinism.[32]

Many of the modern assumptions about contact reflect the scientific and technological interests of our specific era in history. We emphasize telecommunications, information technology, artificial intelligence, robotics, and genetics. Those interests will change with further scientific and technological advance and with cultural evolution; other civilizations may have moved beyond them.

There is another dimension to this debate: Those who foresee contact as an exchange of scientific information do not represent popular opinion. Surveys show that science is not the only reason, or even the primary reason, for public support of SETI. Most people focus more on other aspects of knowledge, culture, and behavior.

This might be true in the other civilization as well. A society able to afford the enormous effort that interstellar communication requires can hardly be motivated solely by practical considerations, Puccetti argued. They would want to know things that might outweigh in importance further gains in scientific or technological knowledge.[33] Ideas about those subjects may be the most difficult to convey.

Parallelism and Synchrony

The radio search assumes that extraterrestrial civilizations are approximately on the same technological level as terrestrial civilizations, using detection and communication systems similar to our own. Yet, Morrison acknowledged that there is no synchrony anywhere; we are either behind or ahead.[34]

We may be the youngest communicating civilization in the Galaxy, having only just arrived at a stage that allows us to build the powerful transmitters and receivers needed for contact at interstellar distances. However, no stage of communications technology is permanent. Many believe that our present modes of communicating will be short-lived—either because our technologies will change into something quite different, or because we will destroy ourselves.[35] Older technological civilizations also may move on to new communications technologies.

White proposed that the impact of contact would increase as the differences in the levels of development of the two civilizations increases.[36] There may be limits to that paradigm. Civilizations very far advanced beyond our own may not bother with us, and we may not be able to detect or understand them.

Assuming that 2 million intelligent societies have arisen in our Galaxy over the past 5 billion years, Ulmschneider derived an average interval of 2500 years between the births of such civilizations. In this formulation, the societies closest to our present state would be either 2500 years more advanced than we are (4500 A.D.) or 2500 years behind (500 B.C.). The probability that such a civilization is nearby is small. Initial radio "bursts" should be observable from, at most, 30 older societies; the number would be smaller if they spend only a brief time in the radio-emitting phase.[37]

These numbers are statistical artifacts resting on an assumption that intelligent societies have appeared at regular intervals over a 5 billion-year span. The statistics look more promising if there is a flowering of intelligence in our own era. Nonetheless, trying to imagine human society in 4500 A.D. shows us how difficult it is to picture a more advanced civilization.

Mind-Stretcher. SETI conventional wisdom often assumes that older means proportionately more advanced in science and technology. Yet, as we have seen, continuous scientific and technological progress may not be inevitable. A million years older may not mean a million years ahead in knowledge or tools—or in the ability exert influence at a distance.

SETI optimists often assume that more advanced societies will provide us with the information we need to reach their level, homogenizing the Galaxy's technological civilizations. We cannot take it for granted that extraterrestrials wish to accelerate the development of human technological capability. Even if they do help us, they are unlikely to delay their own further development while we catch up. We may always lag behind the leaders.

The idea of interstellar communication actually revives anthropomorphism because of the assumptions we must make, argued Beck. We must believe that a pattern of evolution like ours, from simple organisms to advanced civilization, has been repeated within signaling distance and synchronously with our own development. We must assume that extraterrestrials reciprocate our curiosity and take the same measures that we would take to signal to them, although we also assume that they are sufficiently unlike us to have managed a technological project that probably exceeds our resources of curiosity, patience, and stability. All of these assumptions are highly speculative, but we must make them or else give up the game.[38]

The discovery of aliens will have the profound and weighty consequences it is claimed to have, Regis asserted, only if the extremely improbable occurs: contact with an extraterrestrial culture of just the right degree of similarity to and difference from ourselves. If they are virtually identical to us, we will learn nothing from them; if they are extremely different from us, we will learn nothing from them as well.[39]

These are extremes. We might indeed learn something if we encounter a society different from ours, if the differences are not too radical. We may have to find a civilization that is relatively close to our own level of development; as Bova put it, somewhere between Tarzan and the angels.[40]

They Will Be Generous in Sharing Information

Interstellar radio communication will not be a dialogue. It will be a monologue. The dumb guys hear from the smart guys.

—Carl Sagan, 1973[41]

The classic SETI paradigm is based on the assumption that extraterrestrials will export information by radio. Morrison and others have claimed that, after search and acquisition, communication will consist of massive information transfer, because the purpose of their transmission would be to inform us.[42]

The authors of Project Cyclops proposed that extraterrestrials would use transmissions to ensure the survival of the "galactic heritage" by attracting the attention of young races. Even more optimistically, they assumed that the senders would try to make the job of deciphering and understanding the messages as simple and foolproof as possible.[43]

We cannot assume that exchanging information is the first priority of the other species; nor can we assume that extraterrestrials will want to tell us everything they know. If we are the junior partners, we will have more to learn than to teach. Why would they bother to teach us? Again we encounter the issue of motivation.

When a more powerful society aids a less powerful one through the transfer of knowledge or technology, it loses some of its own power.[44] More advanced civilizations may not want to place state-of-the-art knowledge at the disposal of an alien species, particularly if they see us as ethically underdeveloped or potentially dangerous. They might want to find out first about our own knowledge and capabilities, precisely because knowledge is power. There might be things they would not want to tell us, such as how to achieve interstellar flight or how to build more powerful weapons.

French astronomer Jean Heidmann once suggested that the simplest approach to communicating with another civilization is to transmit our encyclopedias. Would extraterrestrials send theirs? They may send only a sample of what they have to offer, McDonough suggested, because their real goal may be to receive our encyclopedias, the only thing we have to offer a more advanced society.[45] Each side may seek a favorable balance of intellectual trade.

Would we transmit all of our knowledge to an alien species whose capabilities and intentions were unknown to us? Brin speculated that history might speak of no worse traitors to humanity than those who, with the best intentions, cast out to the skies our heritage without asking anything in return.

If their goal is to obtain as much information as possible, they might choose not to communicate at all, as that would disturb our civilization. Instead, they might prefer to observe an undisturbed independent system, delaying contact as long as possible.

We also should recognize that openness among humans—the notion of freely exchanging ideas—is a recent historical development.[46] That practice might not be shared by all technological civilizations.

We cannot assume that alien messages we detect—especially ones not aimed at us—would be rich with information of use to Humankind. Drake admitted that they could be something as trivial as purchase orders.[47]

A Mirror Image. In designing messages or information-rich artifacts, we must try to place ourselves at the other end of the communications process, looking at our symbolic envoy as it might be seen by the recipient. A message or artifact does not simply convey information; it also conveys a state of scientific and technological development, command over energy, and cultural evolution. Rightly or wrongly, the recipient may believe that it also conveys intent.

Everything Will Be Made Public

If and when interspecies contact is made. . . . it may be that we shall encounter ideas, philosophies, ways and means not previously conceived by the minds of men. If this is the case, the present program of research will quickly pass from the domain of scientists to that of powerful men and institutions.

—John C. Lilly, 1961[48]

SETI researchers tend to assume that a detection will become known to the public almost instantly. Horowitz thought it would be impossible to keep any such signal classified, because in the process of verifying it, it is necessary to have scientists at other observatories look at the same place in the sky, just to make sure that you're not seeing an artifact of your own observatory. People who are in on the world's greatest discovery are not going to sit on it.[49]

In fact, scientists have sat on information about important discoveries. In 1967, Cambridge University astronomers detected powerful pulsing signals that may have been the most suggestive of an extraterrestrial intelligent origin that had ever been detected in all the history of radio astronomy. Instead of calling BBC or the *London Times*, the discoverers withheld their results for months while they considered possible explanations, including the idea that this might be evidence of an alien civilization. Their caution proved to be justified; the signals were from a previously unknown type of astronomical object called a pulsar.[50]

Cold War Scenarios

In his novel *The Black Cloud*, Fred Hoyle envisioned that scientists communicating with the alien entity would operate from an estate surrounded by armed guards and cut off from the outside world. This may have been inspired by Bletchley Park, where British code-breakers analyzed German communications during World War II.

The film "2001: A Space Odyssey" (1968) portrayed a fictional situation in which Americans discover an alien artifact while exploring the Moon. The Americans invent a cover story (disease) to prevent access to their base by Soviet personnel. Such a cover story would work only if American authorities also kept their own public in the dark.

As the Annex to this book explains, there was a NASA document providing detailed guidance on how to announce the detection of extraterrestrial intelligence that might be found by the short-lived official search. This tightly controlled procedure, which was designed to prevent announcements that later proved to be mistaken, implied a certain amount of delay.

The research team that found hints of life in a Mars rock deliberately kept news about their work from NASA managers to prevent premature leaks. They commented later that the way to handle a truly exciting discovery is not spelled out in any simple set of rules, so scientists attempt to deal with the situation on a case-by-case basis.[51]

Military or intelligence organizations that serendipitously discover suggestive evidence may hold that information even more tightly. In July 1967, the Department of Defense's Vela 4 satellites detected a brief, intense flash of gamma-ray photons coming not from a nuclear test, but from outer space. The data were not analyzed until March 1969 and were only announced to the astronomical community in 1973—a 6-year delay.[52] Now we call them gamma-ray bursters.

According to writer and editor Randall Fitzgerald, reports of signals from space in the late 1950s were taken seriously by officials of the Central Intelligence Agency because they had been passed on by the National Security Agency. Former CIA officer Victor Marchetti reportedly said that people at the NSA were genuinely puzzled; they thought that the signals were real and intelligent in origin, but did not know what to make of them or what to do with them.[53]

Given this history, we must be realistic. A detection made by persons working for a governmental agency or under a government contract may not be made public for some time. Officials might delay or limit the release of information while the discovery is confirmed, and to allow time for a policy discussion. We cannot assume the inevitability of a leak; some secrets still are kept.

Allen Tough identified reasons why governments may try to keep a detection secret: the belief that people might panic; the fear of a negative impact on religion, science, and other aspects of culture; concern that false alarms may cause embarrassment; the temptation to seek national and individual competitive advantage; avoiding a harmful premature reply; seeking a trade or military advantage; the fear of an extraterrestrial Trojan horse.[54]

In the case of an information-rich contact, political and governmental leaders may think that they need to prepare the public for the news. If they were concerned about the impact on their societies, they might let through only information that they considered safe. Donald Tarter predicted that security agencies would require signal monitoring, information management, and a voice in policy with regard to a reply. Nobelist Wald foresaw a more extreme result: Contact would produce the most highly classified and exploited information in the history of the Earth.[55]

Even Drake, who favors open release of a confirmed detection, recognized that if the signal is information-rich, "you'd better take a close look at the information to see if it would appear threatening to anyone, and make a judgement as to just what you say." Every government will realize that there is possibly very valuable information to be gained, useful for economic, technical, or military reasons. In addition, there is prestige involved; all high-tech countries will pour resources into gleaning alien information.[56]

The detecting nation might choose not to share information with others, at least initially. Later revelation of such decisions could provoke mistrust, encouraging other countries to act independently in communicating with the detected civilization.

Governments are not the only bodies that could be tight with information. A nongovernmental organization that detected extraterrestrial intelligence might not play by the rules followed by most SETI researchers. Such a group could choose to withhold or limit the release of information, considering how it might best be exploited. Those personally involved in the first contact might be possessive about the information and the channel, particularly if they distrusted governments and held a low opinion of the general population.

Entrepreneurs might compete to get first access to alien ideas and to monopolize those with commercial value. As private sources fund more of the search, warned White, we have to ask what happens if an entrepreneur or corporation invests millions of dollars in SETI specifically to market the information generated by a search.[57]

Mind-Stretcher. If the first organization to crack the alien code were a company or profit-minded university, that institution might seek to patent the intellectual property derived from its discovery, charging a fee for access to their findings. The precedent has been set: Government authorities have allowed the patenting of genes found in nature, including 20% of human genes.[58] It could be argued that a signal found through an astronomical search is a comparable discovery.

If a signal is information-rich, receiving, interpreting, and disseminating that information could be a major enterprise, possibly requiring new institutional arrangements. Some believe that decoding a message will be a task of years, decades, even generations. Such lengthy, detailed examinations could mean long delays in the complete release of information to the public. The full contents of the Dead Sea scrolls were not published until 54 years after their discovery.[59]

As control over this information could offer great power as well as high status, there would be a strong temptation to monopolize the channel and limit access by others. Harrison warned of "gate-keepers" who decide if information should be released or suppressed; this could apply at both ends of the communication process.[60]

Our history contains many examples of priesthoods mediating between the heavens and ordinary mortals. In early agricultural societies, knowledge of the movements of the Sun, Moon, stars, and planets provided a basis for determining the coming of the seasons, vital information for timing the crucial activities of the agricultural and pastoral year. In many cases, this knowledge was confined to religious authorities, enhancing their power. The fewer the gate-keepers, the more power they had.[61]

One can imagine scientific priesthoods that decide what the rest of us should know about messages from the skies. They also might send private communications to the other civilization without consulting anyone else.

We cannot assume that the search for extraterrestrials is immune from the ancient motivations of egoism, power, and greed. Decisions that could affect the welfare of the human species might be made by small, non-representative elites.

Their Knowledge Will Solve Our Problems

Into this void of knowledge boldly steps the doctrine of outside assistance, of either a religious or an extraterrestrial flavor.
—John Baird, 1987[62]

Many of those most enthusiastic about SETI hope that contact will bring us solutions to our current problems. Optimists tend to assume universal problems with universal solutions, rather than the unique problems of individual societies and the solutions tailored to fit them.

Most of the problems that we want solved, such as population growth, food shortages, energy supplies, dwindling resources, and environmental degradation, may be peculiar to our level of technological civilization. They also may be peculiar to our period of history. Alien advice is unlikely to meet our unique needs at this precise moment, argued Baird, unless the more advanced civilization sent messages appropriate for multiple levels in a hierarchy.[63]

Some forms of advice that we seek may be specific to our species. Consider one of the most popular categories: medical techniques. Our hope that extraterrestrials will send us cures for human diseases such as cancer rests on unlikely assumptions: either alien biology is enough like ours to make their information useful to us, or they have detailed knowledge of human biology.

There is a practical communications problem as well. In the remote contact scenario, it could take decades or centuries to ask the aliens to clarify their meaning and get a reply. By then, we might have moved far beyond the problems of the moment.

Some SETI scientists have recognized that we should not look to contact as a cosmic cure for the problems of Earth. By implication, we must solve those problems ourselves rather than expect guidance from the sky.

We Will Be Willing to Adopt Their Ideas

Would we be willing to absorb all of forms of alien knowledge, all of their ways of doing things? There might be cultural resistance, particularly outside science and technology. "When scientific revolutions impinge upon metaphysics or social theory," argued astronomer Richard Berendzen, "they are likely to become unusually polemical and possibly unacceptable."[64]

Consider a human case: the impact of the West on other societies. Westernization, which often has provoked social and political instability, has been deeply resented by non-Western cultures. As Huntington put it, what is universalism to the West is imperialism to the rest.

Judging by our own history, religion could provide a rallying point for opposition to alien ideas. The intensified role of non-Western religions is the most powerful manifestation of anti-Westernism in non-Western societies, argued Huntington; it is a declaration of cultural independence. Some analysts see the hostility of Islamic jihadists toward the United States as only the most recent manifestation of a long-running, worldwide reaction to the rise of Western modernity.[65]

Contact Will Unify Humankind

The radio of the future—the central tree of our consciousness—will . . . unite all mankind.
                                   —Russian futurist poet Velimir Khlebnikov, 1921[66]

A remote detection might have a temporary unifying effect, suggested a group of scholars, if political leaders capitalized on the new mood and moved toward greater international cooperation, conciliation, and resolution of differences. However, in the absence of an imminent threat or prospect of immediate gain for humanity, there is little reason to expect that any new sense of shared human destiny would last long enough to cause enduring political change.[67]

Regis, questioning the argument that contact would inspire unity among humans, pointed to an historical parallel. The discovery of the Americas did not have anything like the effect on Europeans that SETI advocates insist that discovery of extraterrestrials will have on us. It did not make differences between Europeans more trivial; it did not serve as an integrating influence among them; it did not make them more tolerant and peace-loving.

What if we find no others? Consistently negative results of SETI programs might reinforce a belief that Humankind is unique. Some believe that this will convince humans of the importance of ending conflict among themselves to preserve their species.

Regis challenged this argument as well. If we examine the claim that failure to find aliens would have a sobering influence on quarreling nation states, we would have to wonder why the pre-Copernican conception of man's specialness failed to have this effect in the past.[68]

Direct contact might have a greater unifying impact, stimulated by fear. If there were a perception of potential threat, nations might be motivated to work together for the common defense.

Contact Will Bring Greater Stability

Optimists tend to see the cultural impact of contact as gradual, because information interpreted and filtered through an elite will be released slowly. If the other civilization were hundreds or thousands of light-years away, there would be no hope of a quick exchange; getting a reply to our own message might take centuries.

We could propose other scenarios. What if the signal is strong and information-rich, and millions of humans have direct access to it through multiple receivers? If the message were relatively easy to decode and interpret, the impact on us could be deep and wide.

What if the signal comes from relatively nearby, say less than 20 light-years? Exchanges of messages could take place within a human lifetime, accelerating the impact. A communicating probe in our solar system could allow exchanges within hours. An information-drenched artifact could have similar effects—if its contents were released to the public.

Even in the case of remote contact, information-rich messages could cause a discontinuity. We might be flooded with new ideas and new ways of doing things; that influx could drive social change. As MacGowan and Ordway put it, new social science and operational science information would accelerate social evolution.[69] By implication, they would be destabilizing.

Alien technologies and ideas about the forms and purposes of economic organization could suggest new opportunities for innovation and growth, or less damaging prosperity. They also could disrupt our economies by undermining independent initiative and the spirit of invention, forcing massive readjustment and unemployment, and threatening existing economic institutions. Rubin foresaw that shortcuts to advanced technology might carry such unintended negative consequences as displaced workers, overpopulation, psychological stress, and social unrest if people came to believe that their governments were powerless or irrelevant.[70]

Major transformations in the nature of work tend to bring wrenching social changes, warned economist Alan Blinder. We would need time to adjust to alien ways of doing things; the evolution of laws, customs, and attitudes that support rather than clash with new technology can take decades.[71]

If the current rate of technological change increases, many humans could find themselves quickly outdated. Fear of such possibilities could provoke a new Luddite movement against alien technologies.

Being in Style

Science and technology are not the only drivers of change. Alien ideas could influence our codes of behavior and styles of social interaction, our arts, and our tastes.

Many humans might emulate alien ways, as we rush to fads and fashions now; this impulse could be stronger if we thought we were imitating superiors. Others would resist the adoption of alien modes, perhaps forcefully.

Even if a more advanced civilization did not engage in physical imperialism, we could be affected profoundly by the imperialism of ideas. "The purpose of any communication," Tipler argued, "is to change the knowledge of the person to whom the message is directed, to colonize a mind with memes (complexes of ideas)." He claimed that meme colonization necessarily extinguishes other memes; it is necessarily imperialistic.[72]

Optimists like Sagan have argued that we are free to ignore the contents of an extraterrestrial message that we do not like; we can select only the information we find acceptable. However, the information might not be ignored if it were made available to the general public. As Jill Tarter put it, we would not be able to put the genie back in the bottle.

We have learned from our own history that a receiving culture cannot take in only those practices it likes from another society; it is affected by the context of those practices, including the broader alien culture. A Syrian scientist, addressing the impact of the West on the Arab world, acknowledged that the arrival of innovations brings with it, directly or indirectly, the lifestyle and socio-cultural values of their innovators.[73]

We might not see the impact coming until it was too late. Historian Bernard Lewis described how the Islamic world was slow to interest itself in the West. By the time it did, Western impact on Muslim societies was irreversible.[74]

Some optimists assume that superior extraterrestrials, knowing what will damage us, would act with restraint. Papagiannis proposed that if a galactic civilization exists, it may know that bringing less advanced cultures into the wider society cannot be rushed. Jill Tarter thought that a "mediated paradigm shift" is less likely to have negative consequences than if it is left in inexperienced human hands.[75] Again, we are shifting the burden of responsibility to the alien civilization.

Bracewell expected that, as the extraterrestrials would anticipate that their probe's message to us would be disturbing, they would prepare that machine with "sociological resourcefulness."[76] One wonders how an alien civilization would program such cultural and political sensitivity about an unknown society into a machine. The first probe to visit may not know what it will encounter.

These speculations about alien behavior assume a degree of omniscience that may not exist. They also assume benign intent.

Their Utopia Will Be Good for Us

Many optimists assume that extraterrestrials will have reached a utopian state that they will eagerly share with us. They will not have succumbed to runaway technology, environmental disasters, or war; they will have full control of their technology, show sensitivity to their environment, and be peaceful.[77] In other words, their society will resemble the ideal toward which we are striving.

Drake foresaw that we would learn what "ultimate social systems" are arrived at in other civilizations. We may discover that evolution leads to a single preferred mode of life. "If this be so," he wrote, "let us know now." Another astronomer, Alistair Cameron, suggested that we might receive valuable lessons in the techniques of a stable world government.[78]

Those who lived through the attempt of national states to impose visions of societal perfection in the twentieth century may have reservations about adopting someone else's perfect social system. Discussing fascism and communism, *New York Times* critic-at-large Edward Rothstein warned that utopias, for all their promises of freedom, turn out to be extraordinarily rigid places, full of rules and demarcations. In practice, that rigidity has turned into cruelty.

The twentieth century was unique not in the kinds of utopia imagined, but in the relentless attempts to bring them into existence and the technology to make them seem possible. The utopian "science" of Marxism and the utopian nationalisms of fascism carried the model to extremes: grand visions of a new age combined with horrific exorcisms and totalitarian control. What goes absolutely wrong, Rothstein warned, is the attempt to make everything absolutely right. Dystopias are failed attempts at utopias.[79]

Political scientist James Scott, examining schemes to improve the human condition through social engineering, found repeated patterns of failure. Soviet collectivization, Mao's Great Leap Forward, the planned city of Brasilia, and compulsory *ujamaa* villages in Tanzania were examples of ambitious projects that extracted a high price from the people they were intended to help. Those people were the victims of what Scott called high modernist ideology—the belief that society can be designed to conform with what are believed to be scientific laws. The imposed ideas of high modernists, who thought that they knew better than ordinary human beings, sometimes had disastrous effects on their peoples.[80]

Reflecting on the terrors of the twentieth century, writer and researcher Robert Conquest concluded that the myth of rationalist politics is based on the frightful idea that some of us know what is best for the rest. The basic attraction is the idea that utopia can be constructed on Earth—the offer of a millenarian solution to all human problems. The conjunction of dreaming and ruling generates tyranny; the dream of salvation will always end in a nightmare.

There is no formula that can give us infallible answers to political, social, economic, ecological, and other human problems. There is no final purpose to history, no perfection, no utopia. Conquest counseled that is is better to stick with the Western liberal culture that implies the absence of absolutes, a disbelief in perfect political wisdom, or in readily predictable futures.[81]

Perfectability

Among the recurrent themes of this long debate are the moral and ethical imperfection of human beings, and the hope that contact will help us to rise to a more perfect level. Kant envisioned a hierarchy of rational beings progressing toward the highest excellence, namely divinity. Tsiolkovskii speculated that we have been set aside in a reserve in order to allow our species to evolve to perfection. Stapledon, too, envisioned our own evolution toward a perfect state.

Is perfection achievable? At what point have we achieved perfection? What are the criteria? Who decides?

Space historian Roger Launius pointed out that the idea of perfect societies also is recurrent in the literature of space advocacy. Arthur Clarke, for one, thought that the exploration of space was a conduit for the improvement of the human race.[82] Although spaceflight has had many important consequences, the social outcomes have been less revolutionary than the most visionary advocates foresaw—and they have not always been positive. Five decades into this effort, we remain far from perfect by anyone's standard.

MacGowan and Ordway offered a utopia ruled by an intelligent machine. They claimed that wars, revolutions, coups, and other forms of major social disturbance would be quickly eliminated if an executive automaton established itself in a position of domination over all segments of a biological society. However, the cost would be high:

A little reflection will show that any significant degree of social progress is inevitably accompanied by the sacrifice of some degree of personal freedom. A significant trend toward freedom is actually a trend toward anarchy and chaos. The ceding of social control to an artificial intelligent automaton would lead to an immediate and undreamed of rate of social progress.... It would mean a fuller and happier life for virtually all members of a biological society. The major social ills such as war, crime, poverty, and injustice would be quickly eliminated. Because of these considerations, a biological society will probably not hesitate to cede its own social control to an intelligent executive automaton.[83]

Some of us disagree.

They Will Be Morally Superior

The grand principles of morality . . . are not to be viewed as confined merely to the inhabitants of our globe, but extend to all intelligent beings.
—Reverend Thomas Dick, 1826[84]

Isn't it axiomatic that any non-human intelligence must be evil?
—Kingsley, a character in *The Black Cloud*[85]

Since the Middle Ages, many who have speculated about intelligent extraterrestrials have argued that they would be morally superior to ourselves. Although this once was meant to inspire human improvement, it has become an assumption that many make when debating the consequences of contact.

Clarke claimed that no culture can advance for more than a few centuries at a time on the scientific and technological fronts alone. "Morals and ethics must not lag behind science," he insisted, "otherwise the social system will breed poisons that will cause its certain destruction. With superhuman knowledge must go equally great compassion and tolerance."[86]

Like Clarke, Grinspoon believed that technical advancement without spiritual progress creates a dangerous and unstable condition that will be selected against. He thought that natural selection on a galactic level will favor those living worlds where technical and spiritual advancement proceed together.[87]

Sagan implicitly assumed that more advanced extraterrestrials would have higher moral standards: while we are asked to imagine enormous progress in their knowledge of the physical sciences, we are also asked to imagine that they are as backward as we are on sociological and ethical questions. Jill Tarter envisioned that there will be a highly established code of ethics among more advanced aliens.[88]

Papagiannis offered a sweeping vision, driven in part by the Limits to Growth theory. Those societies that overcome their innate tendencies toward continuous material growth will be the only ones to survive. As a result, the entire galaxy in a cosmically short time will be populated by stable, highly ethical and spiritual civilizations.[89]

Human history does not support the assertion that social wisdom will accompany scientific and technological progress. Some of the worst horrors in our own history were committed by some of the most scientifically and technologically advanced states, such as Germany in the 1930s and 1940s.

There is no reason to suppose, warned Lamb, that a superintelligence would develop superlevels of compassion and empathy. Shostak acknowledged that technological supremacy is no guarantee of cultural refinement or moral virtuousness.[90]

Burke-Ward proposed a more subtle argument. Extraterrestrial intelligences may be civilized from their own perspective, yet may be dangerous to Humankind. It may not even be a question of intentional malice. According to Pickover, Tsiolkovskii suggested that a superior species would painlessly eliminate animals on other worlds rather than seem them endure the sufferings of evolution.[91]

Ethics and Fishing

The way humans treat dolphins is not reassuring. We enjoy their company, watching them perform like circus acrobats. A few of us try to communicate with them as if they were friends. Yet, we continue to kill them by the thousands when they get in the way of commercial fishing.

Noting that dolphins do not use their great strength against humans despite the outrageous treatment they have received from us, Bruce Fleury speculated that this may indicate an advanced ethical system that might be shared on a planetwide basis. Another intelligent species may not be as quick to forgive.[92]

Others have suggested more restrained claims about morality. To be intelligent beings also means to be moral beings, Musso argued, as that necessarily implies an ability to imagine different possible futures and an ability to choose among them rationally. However, he recognized that to be more advanced does not imply being more moral.[93]

Would aliens have values that we would recognize? Puccetti proposed that no community of intelligent organisms could achieve a technological civilization without certain values: the search for knowledge, the desire for truth, the willingness to subordinate individual interest to social aims for the common benefit. This implies some level of morality.

There might be values shared by a potentially universal community of persons from which we are detached by the accidental dispersion of matter in the cosmos, suggested Puccetti. However, distance would be a major limiting factor; moral relations may not be possible in the case of permanent physical separation.[94]

Mind-Stretcher: Ethicist Michael Gazzaniga proposed that there could be a universal set of biological responses to moral dilemmas—a sort of

ethics—built into human brains.[95] Would aliens who emerged from entirely different evolutions share these responses, or would theirs be different? Would post-biological intelligences lack such responses entirely?

The day that communication is established, predicted dolphin researcher John Lilly, the other species becomes a legal, ethical, moral, and social problem. Do we have any moral obligations to extraterrestrials? Do they have any moral obligations to us? It depends, cautioned Ruse. Are they enough like us that any kind of moral discourse is possible? If the possibility of some sort of reciprocal altruism is there, morality might emerge; otherwise it may not.[96]

Several people have warned that we should not assume that the ethics of extraterrestrials will be like our own; Regis went so far as to argue that aliens may not have any such thing as ethics.[97] Are we prepared to accept all forms of social organization and all forms of behavior as equally worthy of respect? How would we react if we learned that some aspects of an alien society were deeply repugnant?

We know nothing about good and evil in the context of extraterrestrial civilizations. As McKay pointed out, the Copernican Principle is not established with respect to biology, culture, or ethics.[98]

They Will Be Altruistic

No instinct can be shown to have been produced for the good of other animals.

—Charles Darwin[99]

The competition for limited resources is what leads to improved species.
—Frank Drake, 1974[100]

Many contact optimists assume that scientific and technological advance go hand in hand with beneficence. Extraterrestrials are expected to show sympathetic concern for our well-being, even a parental sense of responsibility.

United States Congressman George Brown, a scientifically trained man who was Chairman of the U.S. House of Representatives Committee on Science and Technology, predicted that more advanced aliens would be altruistic teachers. They would look on us as children who need to be encouraged to develop further.[101]

The authors of Project Cyclops were more cautious. We might argue, albeit anthropocentrically, that compassion, empathy, and respect for life correlate positively with intelligence. However, counterexamples are not hard to find.[102]

In the case of our own planet, biologists have documented a basic fact: selfless generosity occurs less often and with decreasing intensity as individuals grow more distantly related. Biological and social evolution have not selected the most altruistic. This fall-off, warned Brin, bodes ill for the likelihood of interstellar altruism. Shostak acknowledged that when it comes to interactions between extraterrestrials and humans, the aliens will have little biological reason to be altruistic, only intellectual ones.[103]

What if more advanced extraterrestrials are postbiological, machine intelligences? Some may have freed themselves entirely from their genetic past. Nonbiological aliens, feeling no kinship with us, might be unconcerned about our survival.

Few of us show altruism toward Humankind's nearest relatives, the chimpanzees. Nor do researchers find much altruism among the chimps themselves; they are indifferent to the welfare of unrelated group members. Cooperative behavior in nonhuman primates is virtually never extended to unfamiliar individuals.[104]

We have elevated altruism from a rare phenomenon to an ideal, argued Brin—something to be striven toward. It is entirely by these recent higher standards that we now project a higher level of altruism upon those we hope to find who are more advanced than ourselves. If we are capable of rationalizing and even exalting brutally unaltruistic behavior, might advanced extraterrestrials be capable of something similar?[105]

Cooperation within a group can make that group more lethally aggressive in its dealings with outsiders, Paul Seabright warned. This is the dark side of reciprocity.[106]

Goodbye, Golden Rule

Even the Golden Rule has been questioned. Lawyer Andrew Haley, in his seminal book *Space Law and Government*, argued that doing to aliens what we would have done to ourselves could be disastrous for the other species.

Consider the sad fate of a captive killer whale returned to the wild. The animal, accustomed to being fed and cared for by its human handlers, stayed close to their installation until it prematurely died. "It is a classic anthropomorphic fallacy," declared psychology professor Clive Wynne, "to believe that an animal's best interests are whatever a human would desire under similar circumstances."[107] An alien species which assumes that it knows what is good for us may be equally wrong.

MacVey cautioned us about seeking altruistic intervention from above. It is not unpleasant to envisage some advanced, humane, and cultured race descending upon our perplexed world and putting it to rights, to contemplate the guiding hand of an elder brother from the stars, a mind knowing all the dangers, all the pitfalls—and all the answers. Would it work out that way? Might not the reality prove hideously and tragically different? Proposing a scenario in which a superior civilization took over the Earth and started to run it for us, MacVey concluded that they probably would run it for themselves. Any benefit accruing to us might be more fortuitous than intentional.[108]

Jill Tarter speculated that more advanced extraterrestrials will act in our best interests, but added an important caveat. Their altruism is likely only if their own longevity and stability do not demand the elimination of our emerging technology.[109]

Perhaps we are thinking on too small a scale; one can imagine altruistic acts that rise far above sending messages packed with useful information. If we were seen as a species endangered by natural threats to our planet, such as collisions with black holes, an altruistic species might intervene to protect us.

We can hardly rely on that expectation. The burden of proof lies on those who think that alien behavior will be more noble than ours.

A Dubious Utopia

Assuming altruistic motivations for the tutelage of less advanced civilizations may be naive. In Arthur Clarke's novel *Childhood's End*, alien Overlords assume control of the Earth. Under their benevolent dictatorship, our planet becomes a scientific and industrial utopia. Only later do the Overlords' intentions become clear: They are evolving humans into a more acceptable species, through human children with exceptional powers. The Overlords depart, taking the gifted children with them and leaving the rest of human civilization behind to disintegrate.[110]

They Speak as One

Most scenarios for contact assume that we are dealing with a united civilization speaking with one voice. We usually picture them as representing their entire species, and we imagine them attempting to communicate with humanity as a whole; communication is between whole civilizations, not between individuals.

We should be wary of treating an advanced civilization as if it were a single individual, with a single set of goals and a single set of moral values. This is not the case in our own civilization and is even less likely to be the case in a more advanced civilization.[111]

Although we value diversity among humans, we tend to envision each type of alien as uniform in its characteristics. This, commented Brin, is the kind of stereotyping that we now try to avoid on Earth. The first exemplars of communicating aliens that we meet may be atypical. Moreover, they may have reasons not to convey this to us.[112]

An extraterrestrial civilization could be made up of many political units, particularly if it has expanded beyond its home planet or its home solar system. We may hear messages from a rebellious political entity or an obscure sect; some of Earth's most powerful radio transmitters have been operated by proselytizing religions. What if we hear from competing groups, each with a different story to tell?

A Mirror Image. Although most of those involved in the recent debate about sending communications to extraterrestrials believe that Humankind should speak as one, others strongly resist this idea.[113] If we are unlikely to speak with one voice, why would extraterrestrials be more likely to do so?

We will want to know to whom we are talking and with whom we should deal. An alien civilization would want to know the same about us.

They Mean What They Say

Optimists tend to assume that an alien message will convey reliable information. To some, it would be a kind of revealed truth from on high. Yet, as space lawyer Ernst Fasan reminded us, the motivations that extraterrestrials have for sending messages may not be easily discerned.[114]

The best test is to ask how honest we would be in telling an alien civilization about ourselves. Would we fully describe Humankind, warts and all? We have not done this in any of the major messages that we have sent out by radio or by plaques on spacecraft. The *Voyager* records portrayed the positive side of the Earth; absent were images of poverty, disease, and nuclear mushroom clouds.[115] A declaration that we are a peaceful species would hardly reflect historical reality. Could we honestly say that we assure social justice among humans?

Some people would want only a censored encyclopedia transmitted, thought McDonough, one that omitted the countless embarrassing and horrifying parts of our history to make us appear more civilized than we really are. Goldsmith foresaw a "lust for censorship."[116] Extraterrestrials also might censor the less attractive facts about their civilizations.

The drafting of a message to aliens would be heavily influenced by the fact that such a transmission also is a message to ourselves. We would be tempted to disguise our problems while inflating our achievements. Extraterrestrials might not be above doing that themselves.

They Will Treat Us Fairly

We men, the creatures who inhabit this earth, must be to them at least as alien and lowly as are the monkeys and lemurs to us.
—H.G. Wells, 1897[117]

A common implicit assumption in optimistic contact scenarios is that a technologically superior civilization will treat us as equals, even if they know that we are far less advanced. Optimists expect extraterrestrials to consider us their wards rather than their inferiors. Yet, a hallmark of complex societies is the inequality of their people.[118] In contact with more advanced extraterrestrials, we might be seen as barbarians.

In our own history, separate codes have governed behavior toward those who are like us and inferiors who are not. Our ancestors defined their own groups as human and ascribed varying degrees of beastliness to those outside them, according to historian Felipe Fernandez-Armesto. Indian, Greek, and Chinese texts mentioned deficient or imperfect categories within our species, including women and barbarians.[119]

According to Plutarch, Aristotle advised Alexander the Great to treat only Greeks as human beings and to look upon all the other peoples he conquered as either animals or plants. Even today, African pygmies need protection from cannibal neighbors who hunt them with impunity because they consider pygmies to be a subhuman species.[120]

A Thought Experiment. **Imagine that some of the species of hominid that once shared the Earth with *Homo sapiens* survived into our own time. If we were in daily contact with Neanderthals or with a population of *Homo habilis*, how would we treat them? As equals, as anthropological subjects, or as apes?**

If many civilizations exist in our galaxy, they are likely to be wildly unequal. Sagan acknowledged that "there is almost no chance that two galactic civilizations will interact at the same level. In any confrontation, one will always utterly dominate the other." Dominance may be a natural—even inevitable—stance of any advanced life-form, Harrison and Dick recognized; it may be a functional necessity for society and culture.[121] Advanced extraterrestrial societies will have some sort of a hierarchy.

Harrison saw positive trends, asserting that human societies have been shifting away from authoritarian forms of government and toward democracies. Others argue that growing disparities in economic power among humans are re-creating dominance in another form. David Christian, in his overview of "big history," concluded that social and economic trends over the past 5000 years offer little hope for a significant reduction in economic and political inequality. On the contrary, they suggest that gradients of wealth will get steeper and that differences between the weakest and most powerful will grow.[122]

The ancient Greek historian Thucydides, commenting on the Peloponnesian Wars, put it bluntly. Right, as the world goes, is only in question between equals in power, while the strong do what they can and the weak suffer what they must.[123]

Technologically Advanced Means Benign

They know of only aggressive, hostile organisms and had never observed
a peaceful, friendly form of life and so could not conceive of one.
—Science fiction writer Edward Grendon, 1951[124]

Contact optimists often assume that more advanced extraterrestrials will treat us benignly. Technologically superior aliens, many argue, will have evolved past the warlike behavior we have seen in our own species.

Sagan and Newman claimed that civilizations that do not self-destruct are "preadapted" to live with other groups in mutual respect. The only societies long-lived enough to perform significant colonization of the Galaxy are precisely those least likely to engage in aggressive galactic imperialism. Any interstellar civilization with a lifetime approaching Galaxy-crossing time will have long before selected itself away from aggressive designs.[125]

"SETI is a screening mechanism," Horowitz asserted. "Civilizations that don't acquire the wisdom to control war will destroy themselves long before they can take to space, so the ones who are trying to contact you will be, by definition, no longer menacing."[126]

Matloff, Schenkel, and Marchan took an even more optimistic position: It should be *assumed* that extraterrestrial intelligence is benign, and that contact would be highly beneficial to Humankind (emphasis added). Mac-Gowan and Ordway were more cautious: "Cooperation and nonviolence rather than competition are *probably* the general mode of extrasolar social life; warfare and violence are *hopefully* unknown"(emphasis added).[127]

The human example provides no support for such optimistic statements. Noting the prediction that a spacefaring society capable of crossing interstellar distances would be comprised of wise and benign beings, Stern commented that the same might have been said of Europeans during the Renaissance period of exploration, when ocean voyages were on par with today's exploration of space.[128] Yet, European conquerors often behaved ruthlessly toward conquered peoples.

Extrasolar intelligent beings, like us, may have had violent pasts, perhaps ascending the slippery slope from barbarism to civilization several times. Their histories may have instilled in them a deep concern for security. A species that had experienced nothing but hostility in its relations with others, perhaps resulting in conflict, would be predisposed to assume the worst.

A Mirror Image. One of the things that we tend to forget in our thinking about contact is how extraterrestrials might react to discovering us. If they have had bad experiences with earlier contacts, they might—at least until they acquired additional information—regard us as a potential threat.

Even if there were no history of conflict, contact might come as an unpleasant surprise to a civilization that had believed itself to be unique, a chosen species. Learning of another civilization could violate the integrity of strongly held beliefs.

The Biological Argument

Some see predation and fear of others as fundamental characteristics of complex animal life. "The most disquieting aspect of natural selection as observed on Earth," commented Easterbrook, "is that it channels intellect to predators."

A necessary precondition for the development of a complex nervous system may be an active, mobile, predatory lifestyle. Any creature we contact, said biologist Michael Archer, will also have had to claw its way up the evolutionary ladder and will be every bit as nasty as we are—an extremely adaptable, extremely aggressive super predator.

Drawing on the assumption that physical and chemical laws are valid throughout the universe, MacVey thought it reasonable to expect that biological laws will be too. In that light, aliens seem more likely to be predatory than benevolent.

Predation and exploitation are not exclusively human traits, physicist George Baldwin warned; they are characteristic of all life, indelible genetic imprints which ensure that some species will survive. He predicted that extraterrestrials will show innate contempt for other beings.

Generations of humans were taught that our early mammalian ancestors were small, meek, retiring creatures that survived the dinosaur age by being inconspicuous, staying out of the way of their dominating rivals. Now we know that at least some early mammals were predators that preyed on small dinosaurs. The remains of one mammal showed that it had swallowed an entire baby saurian.

Mesozoic mammals may have competed with dinosaurs for food and territory. One scientist speculated that rapacious mammals may have driven dinosaurs to get larger, or to get off the ground by becoming avians.[129] It may not be the meek who inherit the Earth—or any other planet.

We are not the only primates to kill our own kind. Research has shown that our closest living relatives, the chimpanzees, do not live in large, peaceful communities as some observers had assumed. Chimps are highly territorial and often violent; males patrol the boundaries between groups, killing rival males from neighboring territories. Their basic goal seems to be eliminating rivals rather than capturing females.[130]

Some optimists have countered this by emphasizing the more peaceful behavior of a related species, the bonobo. Yet, chimpanzee behavior cannot be swept under the ideological rug.

Within the genus *Homo*, as many as 10 recognized species diverged over the past 2 million years, yet only one remains alive today. The absence of any gradations between ourselves and chimpanzees, Bracewell speculated, is due to harsh suppression that occurred at some time in the past when the struggle to determine which intelligent strains would survive was still unsettled.

Humans migrated over the whole Earth and by their presence now preempt the possibility of future evolution in directions that would compete with their supremacy. Signs of intelligence, declared Bracewell, would bring immediate retribution.[131]

Even optimists like Drake acknowledge that among the intelligent species simultaneously existing on land, the one that is most intelligent has annihilated all close competitors.[132] The extinction of rival intelligent species did not require weapons of mass destruction; simpler means sufficed.

Mind-Stretcher. Competition with another humanoid form might have been an important selection feature in a drive toward superior intelligence. Some speculate that such competition may be *necessary* to create our level of intellect.[133]

Deliberate hostility may not be needed, just indifference. Rapidly expanding human populations have devastated chimpanzee and gorilla habitats in Africa; commercial hunting and logging now threaten their last redoubts. Unless they are protected, these other members of our family tree may be pushed to the brink of extinction within the next decade.[134]

Rood thought that most civilizations would harbor as little ill will for us as we do for the snail darter (a small fish whose survival was an environmental issue).[135] Yet, that species survived only because of a decision to grant protection. A species as far advanced beyond us as we are beyond snail darters might or might not decide that we deserve protection from extinction.

A desire for security has been a constant in human social evolution; it is reasonable to assume that extraterrestrials also would make it a primary requirement. An alien society that had experienced conflict within its own species and possibly with others would worry about security and might be primed to make assumptions.[136] One bad experience could be enough, if it lived on in history and legend.

Brin suggested a scenario. An earlier technological species could have unleashed a wave of irresponsible colonization, leaving overexploited worlds and ravaged ecospheres in its wake. Earth might be among the few worlds with life to have escaped. Malevolence is not required, only shortsightedness and unsustainable appetites, traits that are completely consistent with the behavior of the only sapient species we know—ourselves.[137]

Paralyzed by Guilt?

H.G. Wells'novel *The War of the Worlds* included an odd defense of Martian invaders. Before we judge of them too harshly, Wells wrote, we must remember what ruthless and utter destruction our own species has wrought, not only upon animals, but also on its own "inferior races." He cited the extinction of Tasmanians in a war of extermination waged by European immigrants. "Are we such apostles of mercy," he asked, "as to complain if the Martians warred in the same spirit?"

Wells may have intended his novel to be a condemnation of European imperialism, as some suggest.[138] Nonetheless, it may be unwise to let our reactions to contact be conditioned by guilt about our past.

Extrapolating from trends he perceived in recent human history toward democracy, the end of war, and the evolution of supranational systems that impose order on individual nation-states, Harrison concluded that our newfound neighbors will be peaceful. Advanced extraterrestrials will be too rich to be greedy; the vastness of space makes motivations like power and greed meaningless. Very old societies are likely to be democratic as well.

"Belligerent, self-serving states" do not last as long as states that do not initiate war but do enter into defensive alliances, argued Harrison. Some computer models show that societies that refrain from exploiting each other and rush to one another's defense are likely to outlast others; "berserk" or belligerent societies are likely to collapse. These findings, Harrison announced, "free us from the idiosyncracies of world history."[139]

Others challenge such conclusions. Western culture has promoted certain illusions about human nature, observed New York Times columnist William Pfaff, a naïve version of the faith in inevitable human progress that arose during the French Enlightenment. This package of beliefs assumes that everyone is headed not only toward liberal democracy but also toward secularism or religious indifference. Western political and economic values are assumed to be universal, valid for all societies now and in the future; hence the unity of Humankind is only a matter of time. People in the West want to continue believing in these illusions, despite all that history has done to disprove them.[140]

Even if the positive trends were confirmed in the human case, they might not apply to all technological civilizations. We cannot assume that extraterrestrial intelligent beings would follow our political trajectory, leading to rivalry between national states armed with weapons of mass destruction. There could be alternate histories in which a planetary government was achieved before such weapons were invented, particularly if the major land areas on that planet were not separated by oceans. Puccetti suggested that a global community of intelligent organisms could achieve political unity even before the discovery of what we call modern science and technology.[141]

A civilization that had achieved such unity, perhaps as a result of conquest by one society, might well survive the introduction of powerful weapons. Such a civilization would not necessarily welcome the presence of another technological species.

Dyson insisted on objectivity. "Our business as scientists is to search the universe and find out what is there. What is there may conform to our moral sense or it may not. . . . It is just as unscientific to impute to remote intelligences wisdom and serenity as it is to impute to them irrational and murderous impulses. We must be prepared for either possibility and conduct our searches accordingly."[142]

When humans expand away from their home planet, they will take their natures with them; so will intelligent aliens.

War Will Be Obsolete

It takes two to make trade, but only one to make war.
—Murray Leinster, 1945[143]

In the Western world, there has been a widespread belief that human warfare is a modern invention and that prehistorical societies were peaceful. Implicitly, there is a hope that war can be eliminated. Optimistic, progressive-minded English and American readers are not comfortable with military necessity, Barry Gewen observed; they want their historians to explain why warfare is becoming obsolete.[144]

Anthropologist Lawrence Keeley, in his book *War Before Civilization*, pointed out that prehistorians have increasingly pacified the past. Many textbooks ignore the prevalence or significance of warfare.[145] Yet, archaeologists and anthropologists have found evidence of militarism in as much as 95% of the cultures they have examined or unearthed.

Time and again groups that once were lauded as gentle and peace-loving were later exposed as being no less violent than the rest of us. The Maya once were considered a peaceful society led by scholarly priests; that all changed when the texts written by their leaders could be read, revealing a long history of warfare and conquest. Societies waged organized war in the Near East before recorded history; archaeologists have uncovered evidence of a fierce battle fought in northeastern Syria 5500 years ago.

Research has shown that warfare was endemic throughout the entire southwestern United States, with its attendant massacres, population declines, and abandonment of settlements. The prehistoric people who lived in southern California had the highest incidence of warfare deaths known anywhere in the world. All Polynesian societies had warfare.[146]

Attacking what he described as the myth of the peaceful, noble savage, LeBlanc challenged the prevailing scholarly view that warfare was of little social consequence in the past and is relatively unimportant in understanding the human condition. That belief is the result of a universal desire to know that things must have been better. In fact, wars are not an aberration, but a continuation of behavior stretching back into prehistory. Six million years of intergroup conflict might result in a human genetic predisposition for war.

The rise of social complexity results in more organized and intense warfare, observed LeBlanc. Complex societies not only fought among themselves but also attempted to expand into the territories of less complexly organized peoples. Almost all ancient states were involved in enough warfare to be recognized archaeologically.[147]

Greater intelligence did not result in greater peacefulness. Although intelligence alone does not result in war, it seems to be a necessary precondition because it enables the technology and social abilities for more complex warfare. Asimov argued that humans inevitably reach the level of making war not because our species is more violent and wicked than others, but because it is more intelligent.[148]

LeBlanc came to an optimistic but questionable conclusion, arguing that the amount of warfare has declined markedly over the course of human history. Past wars were necessary for survival, he theorized, and therefore were rational. The Industrial Revolution lowered the birth rate and increased available resources; when people no longer have resource stress, they stop fighting.[149] This thesis ignores the fact that the most destructive wars in history have been fought since the Industrial Revolution. Technological advance has greatly increased the killing power of military forces.

Optimism about historical trends is not supported by statistics. A table of war-related deaths from 1500 to 1999 showed a dramatically rising trend both in terms of absolute numbers and in terms of deaths per 1000 people. The twentieth century was by far the bloodiest.[150]

Military historian John Keegan ventured that warfare as we have known it may be drawing to an end, based on a rational calculation that the costs outweigh the benefits. Although this might be true of strategic warfare between nuclear weapons states, that calculation has not prevented frequent smaller conflicts. Between 1900 and the mid-1980s, there were some 275 different wars.[151] Although we have so far avoided nuclear conflict, lesser combats since 1945 have killed millions, the civil wars in Sudan and Cambodia being particularly ugly examples.

Jill Tarter recognized that if the number of civilizations were large and expansion were a natural consequence, competition should be included in calculations about expansion.[152] The closer we are in space and in time, the more intense the competition may be.

What we are likely to detect, Dyson speculated, may be a technology run wild rather than a technology firmly under control and supporting the rational needs of a superior intelligence. Assuming interstellar travel at moderate speeds, such a technological cancer could spread over a galaxy in a few million years. The Great Silence may be a warning that we live in a Darwinian universe, one in which only the stealthiest and most aggressive species survive.[153]

We can refine the issue further by eliminating the requirement that interstellar vehicles transport biological organisms. Intelligent machines not only would suffice, they may be preferred.

What if Tipler is right in foreseeing relentless expansion throughout the Galaxy by machines programmed to act in their own self-interest? Several science fiction authors have portrayed deadly probes either deliberately or accidentally programmed to destructively home in on new civilizations after they become detectable by their radio transmissions. Although such machines may not be likely, they are in no way inconsistent with natural law. They also are quite consistent with the observed state of silence.[154]

Clarke dismissed interplanetary warfare as infinitely improbable; it would only arise in the unlikely event of encountering a civilization at a level of technological development similar to our own. "If ships from Earth ever set out to conquer other worlds," he imagined, "they may find themselves, at the end of their journeys, in the position of painted war canoes drawing slowly into New York Harbor."[155]

The other side of that coin is that we might be unable to defend ourselves against a superior technology. If a race of superbeings moved in, warned philosophy professor Jan Narveson, the survival of mankind on terms at all agreeable to us will be a matter of sheer luck.[156] Resistance indeed might be futile.

This does not mean that conflict is inevitable among technological civilizations. It may or may not occur, depending on which choices are made. Communication between civilizations could influence those choices.

We can hope that their historical experiences have imbued extraterrestrial cultures with the concept of enlightened self-interest, and that interaction at the interstellar level does not rest on social Darwinism. However, hope is not a plan.

Distance Protects Us

I want to show that we need not be afraid of interstellar contact, for unlike the primitive civilizations on Earth that were overpowered by more advanced technological societies, we cannot be exploited or enslaved. . . . they are too far away to pose a threat.

—Frank Drake, 1992[157]

The radar and television announcement of an emerging technical society on Earth may induce a rapid response by nearby civilizations, newly motivated to reach our system directly.

—Carl Sagan and William Newman, 1981[158]

In the remote contact scenario, the impact of contact could be positive, negative, or both. Whatever the cultural consequences of such indirect contact might be, we might feel insulated from danger by distance.

Bernard Oliver, who believed that interstellar flight was effectively impossible, dismissed the idea that "you should keep quiet in the jungle."[159] Yet, as we saw earlier, no law of physics or engineering forbids interstellar travel. The principal issue for a more technologically advanced species would be whether there was sufficient motivation to invest the necessary resources.

Detecting us might provoke a better-equipped species to send out interstellar probes, at least to look us over. Remote contact could lead to direct contact.

Shklovskii and Sagan put this in quasi-diplomatic terms. "If interstellar spaceflight by advanced technical civilizations is commonplace, we may expect an emissary, perhaps in the next several hundred years." A report to the U.S. Congress gave that possibility a more ominous tone. "The receiving civilization might be capable of interstellar flight and dispatch emissaries for further investigation. With no foreknowledge of their character, we might be aiding in our own doom."[160]

Mind-Stretcher. The arrival of a robot interstellar probe from one civilization might lead the other to feel the necessity of developing an interstellar travel capability, suggested the Clarks. This might play a role in a snowballing accumulation of intelligences with interstellar mobility.[161] We could look on this as a parallel to the assumed spread of interstellar communications.

The distance to be traversed depends heavily on the assumptions one makes. Asimov, assuming a uniform distribution of currently existing civilizations in our Galaxy, estimated that the average distance between them may be as little as 40 light-years. Ulmschneider, assuming a lower number of civilizations existing at one time, estimated an average separation of about 1700 light-years.[162]

All estimates of distance between technological civilizations are suspect. The assumption of uniform distribution may be faulty, as some parts of the Galaxy may be much more hospitable to the evolution of intelligence than others. Assumptions about distance also may be wrong if some advanced technological civilizations choose to expand, planting colonies or radiating their presence outward in interstellar arks. The extraterrestrials we detect may be much farther away than the average—or much nearer.

Expansion Will Be Relentless

Two schools of thought imagine that interstellar expansion, once started, will continue until the Galaxy is occupied. The colonization school, often driven by population growth models, foresees humans or their alien counterparts planting settlements around all suitable stars; their colonies will generate more colonies that will continue the expansion. The self-reproducing probe school foresees machines creating artificially intelligent progeny that journey on to other systems, perhaps seeding them with biological life.

Both models are questionable because they assume uniformity and continuity in both purpose and action over millions of years. History tells us that purposes change over much shorter spans of time; policy decisions and the commitment of resources may be even more short term. Waves of colonization may be temporary, for us and for others; they could stall for a variety of reasons. So might expansion by machines.

Tipler and Barrow claimed that there would be no resistance to the expansion of the volume of stars colonized by Von Neumann probes. What happens if the machines of an expanding civilization encounter another with the technological means to resist? The colonists, or the probes, might be defeated.

Let's return to the most disturbing model, Dyson's technological cancer sweeping through the Galaxy. Would those intelligences, perhaps in the form of sentient machines, remain ruthless conquerors through the millennia? Or might they vary their behavior as they evolve? As Sagan and Newman put it, "where are they?" depends powerfully on the politics and ethics of advanced societies.[163]

"Colonizers" Versus "Imperialists"

Tipler and Barrow, who visualized an aggressive interstellar expansion and colonization program, attempted to draw a distinction between that idea and the more pejorative concept of interstellar imperialism. First, they declared that there was no reason to expect imperialism. Then they acknowledged that the existence of "imperialists" would motivate "colonizers" to speed up their occupation of previously unoccupied solar systems, in order to prevent the "imperialists" from seizing them. They cited the rapid conquest of central Africa by European powers as an example of such behavior (African territories were, of course, already inhabited).

Tipler and Barrow seemed not to recognize that a contest between "colonizers" and "imperialists" would be a contest between empires. Nor did they admit that the arrival of a probe from another civilization might be seen as threatening. They even claimed that the colonization by extraterrestrials of all the planets in our solar system other than the Earth would not be imperialism, because those planets are just "dead rock and gas." Yet, they admitted that alien colonization of uninhabited planets would prevent the native intelligent species from eventually colonizing those worlds.[164] Imagine what our reaction would be if we saw extraterrestrials colonizing Mars.

Territory Is the Issue

Human sympathies, moral convictions, political absolutes, philosophical certainties—none...will suborn or suppress the territorial imperative, that biological morality which will still contain the behavior of beings when Homo sapiens is an evolutionary memory.
—Anthropologist and writer Robert Ardrey, 1966[165]

By revealing our existence, we advertise Earth as a habitable planet.
—Project Cyclops, 1972[166]

Many of those foreseeing risks in contact have assumed a territorial motive for extraterrestrial aggression. H.G. Wells had his Martians invade the Earth because they needed our planet to assure their long-term survival. Many other science fiction treatments of contact, including "The X Files," also have assumed that extraterrestrials want the Earth for their own species.

Sir Bernard Lovell, Director of the Jodrell Bank radio observatory in England, issued a warning in an address to the British Association for the Advancement of Science. "We must regard life in outer space as a real and potential danger," he said. "Alien civilizations may be combing the galaxy looking for new resources or a new place to settle."[167]

Although interstellar communication would imply no threat, Von Hoerner thought that interstellar travel would kindle an obsession to annex living space and will lead to the explosive consequences of colonization.

Our own experience suggests that advanced extraterrestrials might still be territorial.[168]

Others believe that territoriality is not a basic biological trait, as many species do not exhibit this drive. Harrison, who seemed to endorse the assumption that rivalry over territory is the cause of war, cited an analysis by Dyson implying that territoriality will become harmless in the vastness of space.[169]

Two expanding civilizations with different planetary requirements might ignore each other, observed Sagan.[170] Others believe that expansive alien civilizations, long accustomed to living away from planets, may have little interest in ours. The Galaxy may be populated with societies that arose on planets but that are comfortable only in the depths of interstellar space.

If territorial motives are lacking, can we therefore dismiss the possibility of conflict? Not if our analysis includes the modern security dilemma of preemptive attack.

During much of the Cold War, the primary worry of each superpower was not invasion, but a fear that it would be preempted by a nuclear strike that would destroy or cripple its ability to retaliate. This led to a hair-trigger confrontation between the United States and the Soviet Union, despite the fact that neither intended to threaten the territorial integrity of the other. More recently, terrorists have reminded us that some major attacks have nothing to do with seizing territory; they are designed either to change behavior or to exact revenge.

Knowing from historical experience that aggressive, powerful civilizations can evolve, technologically advanced species may choose to eliminate potentially dangerous competition. Barrow and Tipler thought that the destruction of other species would be the best strategy for a colonizing species if they believed that the biological "exclusion principle" applied to intelligent beings (that principle states that two species cannot occupy the same ecological niche in the same territory)[171]

Even if a more powerful species has no interest in our solar system, they may see logical reasons to stop us from becoming dangerous. It would be foolish for a more advanced civilization to wait for other life to get to the "star wars" level, O'Neill argued; it makes far more sense to destroy that other life when it is incapable of defending itself. He offered some small comfort: if there were anybody out there hostile to us, we would not have been allowed to get as far along as we are now.[172]

An interstellar probe might be designed to prevent another culture from progressing beyond a certain point. What would be the triggering event? Rood suggested that extraterrestrials might establish "watch stations" throughout the Galaxy to monitor G stars with planets for emerging technology, so that they can intervene between technological emergence and the ability for interstellar travel.[173]

Prudence

In their novel *The Killing Star*, Charles Pelligrino and George Zebrowski imagined that the first civilization in the Galaxy to develop high technology will view newly emerging civilizations as potential threats. This might compel the more advanced civilization to wipe out the emerging technological powers, including those in our solar system. These fictional aliens are not interested in siezing our land or our resources. They simply believe it prudent to destroy us before we have a chance to destroy them.[174]

Even if attack is unlikely, a starfaring civilization may wish to confine other technological species to their home systems. That could close off human expansion, setting a final limit to our growth.

That brings us to an Earthly concept that does have a territorial dimension: defense in depth, controlling buffer zones around your homeland to keep threats farther away. As more advanced civilizations would not want contact with us on Earth, Ulmschneider proposed, they would not want it on their home planets either. They would use their superior technology to deflect us.[175]

There are many conceivable ways in which interstellar conflict could begin without territorial motives. One starfaring species might misunderstand the communications of another. It might overreact to the movements of an alien starship or to the unintentional violation of a buffer zone. A scouting party might be killed by the fearful natives of a visited planet, or its ship might disappear for unknown reasons. Such scenarios would not be confined to first contacts; our own history tells us that societies with prior knowledge of one another, even with a history of negotiation and agreement, can turn to conflict.

The most explicit security threat might be the approach of an alien space vehicle without acceptable assurances of peaceful intentions. If a technological species had sufficient leverage, its authorities might try to place its solar system off limits, perhaps setting up a no-ship sphere 1 light-year out from its star; or it could seek acceptable rules of visitation, but with limits on the number of ships that could be in certain sectors of interstellar space. The possibilities are endless, depending on the relative circumstances of the civilizations in contact.

We cannot assume that the universe is inherently safe because of the distances between the stars. If a security-conscious culture were capable of reaching us with their machines, our survival might depend on their ethics—or our ability to persuade.

The Galactic Club Exists

Although the Galactic Club is a pleasing concept, we have no evidence telling us that it exists. That imagined community is an idealized vision of how international relations should be, rather than how our historical experience tells us they are. It is only one of many possible models of relationships among technological civilizations. Others include isolation, anarchy, centerless cooperation, alliance, federation, dominance, and empire.

Technological civilizations may be so widely separated in their levels of development that they do not ally with each other. Even if something like a Galactic Club does exist, it may be composed of very unequal members. If we were admitted to such an interstellar society, we could find ourselves the most junior partners—the least interesting and the least influential. Would other members of the club treat us as equals, proteges, or inferiors? Societies at very different levels of knowledge and power may not consider themselves even remotely equal.

If there is a Galactic Club, gaining membership may not be as easy as some predict. Evolutionary theorist William Hamilton observed that to protect themselves, cooperative groups evolve strategies to make admission into their ranks difficult. These can take the form of being wary of outsiders, long periods of probation, and initiation ceremonies that are costly to the initiate.[176]

Societies are not independent agents that are always free to pursue idiosyncratic, "egocentric" ends, Harrison argued; they must operate within contexts set by other societies. Who will prosper, he asked, collaborative societies that seek collective security and cooperate in the pursuit of common goals, or "egotistical" societies that aggressively pursue their own ends? This formula assumes a degree of synchrony among civilizations that may not exist.

If the cooperative model prevails, we might hope to be welcomed into a community. Howerer, Harrison recognized that if "self-oriented" states have the highest likelihood of enduring, we will have to be very wary of our new acquaintances.[177]

There Won't Be Any Interstellar Politics

Relations between ourselves and other civilizations might not be binary. If more than one technological species is active on an interstellar scale, contact might draw us into some form of interstellar politics. Discovering that one civilization was capable of transport across interstellar distances would imply that there might be others with similar capabilities.

Politics is inevitably implied by the Galactic Club, a grouping of separate civilizations. Papagiannis speculated that advanced societies would have

The Adjustment of Interests

If human history offers any guide, we can expect interstellar politics to be a process in which societies are continually adjusting their relationships. Other civilizations will have interests, just as we will. Some interests, such as expansion, cannot be maximized by one civilization without ultimately encroaching on the sensitivities of another.

Communicating can be a way of resolving such dilemmas. On the other hand, contact with a much more powerful civilization might give us little with which to negotiate. Although we may wish to be treated as equals, contact is more likely to illustrate the politics of inequality.

The intricacies of galactic geopolitics would be most relevant to us if our contacts were with civilizations at a similar level of technological development. Those may be the civilizations that we are most likely to detect.

divided the Galaxy into jurisdictions, with each civilization supervising its own area and knowing what is happening in it.[178] What if there is more than one "club?" Would they be rivals?

We should be prepared for the possibility that our immersion in interstellar affairs could occur in an almost instant fashion, warned Harrison.[179] How would we relate to multiple alien cultures and political entities? What role could *Homo sapiens* play? As a newcomer, with limited capabilities to affect anything beyond near-Earth space, we might have little influence.

At least at first, we would lack the most significant tool of exerting power beyond our solar system: the technology of interstellar flight. We may be of trifling interest to greater powers as long as our reach is limited to our own solar system.

Bracewell proposed that an alien probe in our solar system would avoid exclusive relations with one power and would not act secretly.[180] That may be our preference, but it is an antipolitical view of the relationship. The probe might find it more advantageous to conduct separate, private dialogues with Earthly powers to extract maximum benefit, which in this case may mean maximum information.

Interstellar Empires Do Not Exist

The popular science fiction stories of interstellar empires and intrigues become pure fantasies, with no basis in reality. . . . the whole concept of interstellar administration is seen as an absurdity.
—Arthur C. Clarke, 1963[181]

The barriers of distance are crumbling; one day we shall meet our equals,
or our masters, among the stars.

—Arthur C. Clarke, 1968[182]

The interstellar empire once was one of the most common themes
in science fiction. Alien empires were depicted either as potentially
dangerous, like "Star Trek"'s Klingons and Romulans, or as ruthlessly
expansionist, like "Star Trek"'s Borg.

In one of the earliest visions, Olaf Stapledon painted a bleak picture:
"By far the commonest type of galactic society was that in which many
systems of worlds had developed independently, come into conflict, slaugh-
tered one another, produced vast federations and empires, plunged again
and again into social chaos, and struggled . . . haltingly toward galactic
utopia."[183]

Asimov gave us a more hopeful vision of human expansion and empire
in his Foundation trilogy, which envisioned a humanized galaxy. As James
Gunn observed, "the pride in being human, the responsibilities of humanity, shone through Asimov's fiction."[184] Other intelligent species might be
equally proud of their empires.

Two Visions

Groff Conklin, writing about science fiction in 1955, proposed that
authors looked at planet Earth in two different ways. In the first, the
Earth is a springboard from which to range over other worlds, a place
of origin from which human explorations begin. In the second, the
Earth is a place to be arrived at by others.

This division may reflect two different human psychologies. "Active
Man has always liked to consider his own adventures among the stars,"
observed Conklin. "Contemplative Man is often entranced by the idea
of alien star adventurers in our midst."[185] In one vision, humans act upon
the universe; in the other, they wait to be acted upon.

Asimov's vision spurred the evolution of what science fiction editor
Donald Wollheim called "the full cosmogony of science-fiction future
history": first, human voyages to the Moon and the planets of our solar
system; second, interstellar flight and human colonies in other star systems;
third, the rise of the human Galactic Empire; fourth, the Galactic Empire
in full bloom; fifth, the decline and fall of the Galactic Empire; sixth, the
Interregnum in which worlds revert to more barbaric conditions; seventh,
the rise of a permanent galactic civilization; eighth, the Challenge to God,
when Humankind's descendants have undreamed-of knowledge and the
power to experiment with Creation.[186]

In more recent times, human interstellar empires have been portrayed
in two very different ways: high minded like "Star Trek"'s Federation (a

de facto empire with a unified military command structure) or threatening and evil like the Empire of "Star Wars." The difference in terminology is significant. When scriptwriters want to describe an empire with bad intentions and brutal methods, they call it by its real name. When those writers want to describe a good empire that behaves toward its members in restrained, cooperative ways, they gave it a less frightening name like Federation—or Galactic Club.

Harrison described the Galactic Club as a supranational system, whose members would accept decisions from an echelon that is higher than that of any individual member state. He speculated that the superior entity would exert control over a limited number of areas; for example, it might withhold information from civilizations that choose not to cooperate. He then admitted a possibility that changes the game: If some members of the Club have overcome the difficulties of interstellar travel, military sanctions could be available.[187]

To apply military sanctions, the Galactic Club—or at least one of its members—would need a superior ability to inflict violence on those who disturb the system. The Earthly analog is not the United Nations, as Harrison suggested, but a multinational alliance like NATO, a concert of powers like that which dominated European politics for much of the nineteenth century, or a multinational empire. The enforcer might be a single "hyperpower" like the present United States.

Harrison argued that the past century of human history has seen a rapid decline—and perhaps the elimination—of empires. His argument may be correct when applied to the European empires created from the sixteenth to nineteenth centuries, or the Soviet empire that collapsed in 1991. However, empires of one type or another have been a recurrent phenomenon in human history. Harrison acknowledged that the recent decline of empires might be a random fluctuation and that *de facto* empires may continue to exist.[188]

Our thinking is heavily influenced by the history of Western imperialism. Before 1500, as Huntington saw it, civilizations on Earth were separated by time and space. During the 400 following years, intercivilizational relations consisted of the subordination of other societies to the West. The immediate source of Western expansion was technological: the invention of the means of ocean navigation for reaching distant peoples and the development of the military capabilities for conquering them. The West won the world not by the superiority of its ideas or values or religion, but by its superiority in applying organized violence.[189]

The history of empires goes back much further; they have been a way of life for most of the peoples of the world, as either conqueror or conquered. For most of human history, the most typical form of government has been an empire.[190] Our own historical experience repeatedly shows an impetus for the outward expansion of powerful societies. Wherever we have civilization, we have imperialism.

Empires do not play by the same rules as ordinary states. They believe they have unique responsibilities and unique rights. They do not want to participate in the international system; they want to be the international system.[191]

Many observers have described a kind of imperial logic that is not confined to any particular era—the logic of extension. As military affairs scholar Eliot Cohen described it, an empire is a multinational or multiethnic state that extends its influence through formal and informal control of other polities. The projection of power results not from the lure of profit or ambition, but from the fear of chaos. To let go never looks safe.[192]

Empires use a range of tools and incentives to maintain their dominance—not just military power, but also political persuasion, economic advantage, and cultural influence. The British and Roman empires survived not just by exerting force, but also by persuading others that it was in their interest to remain part of an empire.

Many human empires were motivated by some kind of ideology; in the case of the Romans, the ideology was "civilization." The Roman Empire constituted not only a state, but a way of living.[193]

Several scholars have argued that empires can have positive effects. Political economist Deepak Lal's survey of ancient and modern empires led him to conclude that they have served as a mechanism for governance for disparate peoples who otherwise would have been trapped in the conflicts and inefficiencies of anarchy.[194] One of the few times of extended peace in Europe was between 100 and 200 A.D., when the Roman Empire exerted centralized control over much of the Western world.

Some empires (particularly the Roman and British) have been seen as progressive historical forces, providing the conditions for prosperity by making international economic order possible. Empires are a time-tested method for imposing order and securing justice, argued historian Niall Ferguson; they have as often been a force for progress as a source of oppression.

Empires spread wealth and technology and allowed the free movement of capital and labor. Historian and author Arthur Herman proposed that the British Navy's dominance of the seas during the era of industrialization enabled the main features of today's global economy—speedy communication and travel across open seas and skies, access to markets, freedom of trade, and an orderly state system that prefers peace to war.[195]

Even those subjected to an empire may see some benefit. Jawaharlal Nehru, who became Prime Minister of India after being a leader in that nation's independence struggle, observed that "a conquest, with all its evils, has one advantage: it widens the mental horizon of the people and compels them to look out of their shells."

Many who write about contact dismiss the idea of interstellar empires because the distances between the stars, and the energy and time needed to traverse them, would make the central administration of such an empire difficult if not impossible. In the human case, the task of governance did become more difficult as empires extended themselves. In order to rule vast and widely separated domains, imperial governments generally found themselves compelled to be broadly tolerant of a diversity of cultures and sometimes of beliefs, as long as those posed no threat to their authority.[196]

In the first human empire on which the sun never set, the Spanish empire of the sixteenth century, conveying messages or people between Spain and far-flung outposts required months of sailing each way. The numbers of Spaniards stationed overseas was, in most cases, small. Yet, the empire held together for centuries, in part because its continued existence was convenient to other powers.[197]

At the high point of their empire, the Romans used a sophisticated system of hegemonic control without occupying all the territories under their sway. They persuaded local rulers that it was in their interest to cooperate. When necessary, they intimidated by example; local rulers knew that rebellion would provoke—sooner or later—a punitive expedition.[198]

Empires do not always have sovereignty over their domains; some have been satisfied with enough preponderance to accomplish their political and economic objectives. Both the Soviet Union and the United States established imperial rule through "hub and spoke" systems of client states and political dependencies. The modern American empire has been seen as innovative because it is not based on the acquisition of territory; it is an empire of bases.[199]

Most empires rely heavily on collaborators, observed historian Paul Kennedy. Imperial governance required proconsuls; around the necessity of delegation grew up cultures of initiative, authority, and responsibility, without which empires would not have survived. David Abernethy noted a behavioral pattern that might be expected when a more technologically advanced species imposes its will on others—the psychology of self-abasement that played into European hands.[200]

We cannot assume that more technologically advanced aliens will reject imperialism as immoral or outmoded. Some human imperialists—notably the British and the French—believed that they were doing good; they thought they had a civilizing mission.

Throughout European history, expansion was generally popular with the majority of the people, so long as it was going well and did not involve too onerous a tax burden. Adventures beyond the boundaries of the homeland were a way for rulers to give their subjects what Machiavelli called "great expectation of themselves."[201]

Interstellar empires may depend on means far more effective than any we now think feasible. Even a very advanced technological species would face an apparent fundamental limitation on exerting physical influence: the light speed limit. Whether civilizations thousands or millions of years more advanced than our own can overcome this barrier is simply unknown.

Before dismissing that idea, we should recall that none of the imperial administrations of the nineteenth century foresaw that a future technology—the airplane—would bring all of the Earth's surface within one day's reach.

Shostak noted another benefit of interstellar empires. They would generate plenty of radio traffic.[202]

Warp Nine

Science fiction has accustomed us to the idea that future starships will overcome the light speed barrier through some imagined technique like warping space. This may be wishful thinking. Those scientists and engineers who have speculated most aggressively about interstellar flight have not yet devised a way to break the light speed limit with any foreseeable propulsion system. Even among more technologically advanced species, starships may not zip across the Galaxy in a matter of days or months.

Although interstellar travel can no longer be considered impossible, declared starflight visionary Robert Forward, it will always be difficult and expensive.[203] Journeys between the stars may never be casual undertakings; even for more technologically advanced civilizations, interstellar voyages may be justified only by major purposes.

The most basic is survival through migration. Encounters with other civilizations may be next on the list.

The Biggest Assumption: Alien Intentions

We cannot and should not endow this animal with human purposes and human ideals. We should not attribute to him kinds of knowledge that belong to human experience and tradition but not to dolphin experience and tradition.

—John Lilly, 1961[204]

Despite the absence of evidence, we imagine that the motives of intelligent extraterrestrials are like our own; aliens with minds and feelings like ours richly populate science fiction and science speculation. Yet, extraterrestrial psychology is, as the Clarks put it, the most conjectural of disciplines.[205]

Harrison and others have warned that we must confront an almost overwhelming tendency to ascribe our own characteristics to aliens. We should not delude ourselves into believing that a nonhumanoid life-form will show humanlike behaviors; human society and consciousness may be products of humanity's unique biological constitution.[206]

A Mirror Image. If we imagine extraterrestrials as projections of ourselves, might aliens do the same, imagining us as projections of themselves? How wrong might they be?

Some SETI astronomers with generally optimistic views about contact have recognized that we cannot assume alien motivations. As Shostak noted, audiences can readily identify with alien impulses that are, in fact, merely transposed human impulses. Yet, we can no better guess their motivations than goldfish can infer ours.[207]

Who knows what values might drive an alien culture, Jill Tarter asked.[208] Would we share common problems, common solutions, common strategies? Extraterrestrials may be developing along a different evolutionary pathway.

Assuming alien motives may be particularly questionable in a post-biological universe; highly intelligent machines may not share our emotions, or our ethics. One solution to the paradox, Dick suggested, is that we live in a universe in which the psychology of biological beings no longer rules.[209]

Our assumptions about alien behavior have not passed the Galilean test. We have no evidence of what motivates extraterrestrials; it may have nothing to do with good or evil as we understand them. The great mystery of contact may be their intent.

Given our ignorance, we must be prudent. If we insist on assigning our best qualities to intelligent extraterrestrials, we also must allow them to have our worst.

The Drake Equation, Take Four

In the best of all worlds, a revised equation would include the most mysterious factors of all—the motivations and intentions of intelligent extraterrestrials. Brin may have come closest to making this a factor in a revised Drake equation through what he called contact cross section, a term derived from nuclear physics. This factor would include the probability of approach or avoidance, which may be different for each culture.

We need to think more broadly and systematically about the motivations that would cause other civilizations to initiate or avoid contact, whether by signals or by interstellar vehicles. We would have to encompass the possibility that a technological civilization might take action against others.

Sentient beings can make choices. We cannot assume that those choices will be the ones we prefer.

Cultural values and policy decisions, as much as physical factors, may determine the fate of intelligence. Success in a search for others may depend on sustained curiosity; positive consequences of contact may depend on ethics.

What Is Missing

Analyses of the Direct Contact Scenario

Astronomers and others who have speculated about the consequences of indirect contact have enjoyed considerable exposure in academic and popular nonfiction literature. The alternative point of view is poorly represented outside science fiction; we lack comparable nonfiction studies of direct contact with extraterrestrial civilizations. It is time to correct this imbalance, drawing on thoughtful speculation as well as on research findings.

A Calculation of Risks and Benefits

We cannot assert that interstellar contact is totally devoid of risk. We can only offer the opinion that, in all probability, the benefits greatly outweigh the risks.

—Project Cyclops, 1972[1]

How can we know which of the possible consequences of contact are most likely? Scientists and others apply probability when they consider the scientific dimensions of finding extraterrestrial intelligence, even when our knowledge of some factors is limited to one example. However, they generally abandon the probabilistic approach when they consider the possible social, cultural, political, and security implications of contact. We get opinions rather than analyses.

Most SETI scientists expect upbeat results. The NASA workshop report of 1977 stated that "the receipt and translation of a radio message from the depths of space seems to pose few dangers to mankind; instead, it holds promise of philosophical and perhaps practical benefits for all of humanity." A *New York Times* editorialist argued that, on balance, the chances of gain from alien intelligence greatly exceed the chances of harm. There is no reason to assume that alien intelligence among the stars *must* be

325

hostile or predatory (emphasis added).[2] The writer's choice of verb is not entirely reassuring.

Others have challenged the optimistic view. The possible downsides of contact are immense and irreversible, argued Brin; given the potentially overwhelming implications, we may be wise to reflect on the full range of possible outcomes, not only those for which we yearn. The risks involved in an encounter with an alien civilization could be real and very great, cautioned MacVey, the chances of benefit remote, perhaps nonexistent. Burke-Ward thought that only a few of the possible outcomes of contact would be neutral or beneficial.[3]

Can we quantify the impact? Ivan Almar and Jill Tarter proposed what is known as the Rio Scale for making an initial judgement about a detection's potential consequences. The factors include the class of phenomenon (e.g., Earth-specific, message vs. leakage radiation), the type of discovery (e.g., result of a SETI program vs. reevaluation of archival data), and distance. Shostak and Almar tested that scale with imaginary scenarios; the implied consequences ranged from minimal to disastrous.[4]

Assuming us to be average has the highest probability of being right, said Von Hoerner.[5] Yet, SETI advocates have tended to shy away from this analogy when it is applied to behavior, presumably because of our unhappy record in dealing with less powerful cultures.

One statistical device used in other fields is the Central Limit Theorem, which states that the sum of a large number of erratic variables tends to follow a normal distribution, assuming the shape of a bell curve.[6] Can we assume that the consequences of contact are most likely to fall somewhere between the extremes foreseen by many commentators? Or would those consequences be among the statistical outliers?

Gaming may help us explore these issues. Simulations are used at the Contact conferences, where teams role-play the first human–extraterrestrial contact. The outcomes have shown that the communication of intentions can be badly misunderstood. In one case, the Human Team found that everything the aliens did was experienced as hostile even though the Alien Team meant those actions to be benign.[7]

It is difficult to play extraterrestrial parts in the absence of hard information; we must rely on assumptions. It would help if those playing alien roles considered only those actions that clearly would be in that civilization's self-interest, rigorously excluding altruistic motivations that might lead to a human-preferred outcome.

Systematic Mirror Images

Trying to understand possible alien behavior by seeing it as a mirror image of our own has been applied very inconsistently, depending on which point an author was trying to make. We use mirror images to show aliens behaving like our most hopeful visions of ourselves, or acting the way we do at our worst.

Consider this example. One of the most frequently suggested legal and ethical concepts in this debate is the Principle of Noninterference, letting other cultures evolve without help or other influences from outsiders ("Star Trek" fans will recognize this as the Prime Directive). To strictly maintain the less powerful civilization's cultural integrity, that society would have to be isolated; the mere knowledge that a more advanced civilization existed would amount to interference.

If we were the "superior" species in a contact situation, what would we do? Dole and Asimov, writing before "Star Trek," argued that contacts with alien intelligence should be made most circumspectly, not only as insurance against unknown factors but also to avoid any disruptive effects on the local population produced by encountering a vastly different cultural system. After prolonged study of the situation, a decision would have to be made whether to make overt contact or to depart without giving the inhabitants any evidence of the visitation.[8]

What would be our duties to aliens? As technologist Robert Freitas asked rhetorically, are we trustees, educators, partners, friends, or colonizers?[9]

Should we avoid interference as much as possible so as to not damage their culture, observing them invisibly from orbit without letting them know of our presence? Or should we intervene to help them progress? At what point do we decide that an "inferior" culture has matured enough to stand the shock of contact with a "superior" one? Is cultural quarantine desirable? Could we assure that all humans would abide by the rules, placing ethics above other motives?

Alien societies might face similar questions. It could be useful to systematically work through reciprocal mirror images about this issue, as well as others related to contact.

The Social Science Dimension

More than 40 years ago, the Brookings study recognized that the consequences of a discovery are presently unpredictable because of our limited knowledge of behavior under even an approximation of such circumstances. That report recommended continuing studies to determine emotional and intellectual understanding and attitudes regarding the consequences of discovering extraterrestrial life. It also called for historical and empirical studies of the behavior of peoples and their leaders when confronted with dramatic and unfamiliar events and social pressures.[10]

Regrettably, this useful advice has not been followed in any systematic way. Physical scientists have continued to dominate discussions about the consequences of contact; few social scientists have ventured into this field.

Physical scientists often are skeptical about social science findings. MacGowan and Ordway declared that "most social science research supports the current local political philosophy ... circumstances usually discourage the investigation of any radical deviations from it." Yet, later in the same book, they commented that both estimates of the prevalence of extrasolar intelligent societies and the immediate future well-being of our present human society hinge on greatly improved understanding of social evolution. This can only be achieved by greater cooperation and mutual support by physical and social scientists.[11]

Five SETI scientists proposed in 1977 that we must begin to explore systematically what the consequences of detection are likely to be in terms of human institutions and thought, drawing on the vision of philosophers, theologians, lawyers, artists, politicians, scientists, and others. Astronomers Black and Stull suggested sweepingly ambitious directions for studies: catalog and classify behavioral patterns and cultural differences; determine how these are related to the environment, physiology, and evolutionary history of each species; determine what traits appear common to all intelligent animals; develop theoretical models that will allow extrapolation to extraterrestrial cultures.[12]

Anthropologist Finney, calling on SETI researchers to draw upon perspectives outside of the biological and physical sciences, speculated that the search could help bridge intellectual gulfs within our own species, particularly between physical scientists and social scientists. This would be useful not only in framing research strategies, but also in developing methods for interpreting any messages received, and for addressing the question of sending a reply.[13]

A serious approach must be interdisciplinary. As NASA scientist David Morrison said about astrobiology research, understanding the consequences will require us to consider economics, environment, health, theology, ethics, quality of life, the sociopolitical realm, and education. Sivier suggested that the consilience offered by sociobiology and a more speculative approach by historians are inducing a cross-fertilization that may give greater insights into both the human condition and the nature and likelihood of extraterrestrial intelligence.[14]

John Billingham, who has argued in favor of social science involvement since the 1980s, wrote in 1994 that the time is ripe to begin a thorough scholarly examination of SETI and society. He identified these issues:

— Historical analogs of detecting extraterrestrial intelligence. These might include the Copernican revolution, the voyages of discovery, and the possible discovery of life on Mars. Researchers might ask how well the predicted consequences equated with what actually happened.
— The immediate responses to detection; planning for activities following detection. This includes such procedures as verification and interpretation.

— The role of journalism and the media. How to deal with media demands for information; how best to transmit information to the public.
— Social, psychological and anthropological implications of a successful search. Convincing analogies will not be easy to find, as these disciplines study contact between members of the same species on Earth.
— Positions and responses of the world's major religions; study and education in schools of divinity and comparative religion, as well as among religions themselves.
— Should we reply? How do we address that question? We might begin by addressing the process for making a decision.
— Political, institutional, international, and legal actions following detection. Perhaps a separate body of law is needed.
— Education about the discovery, and the contribution SETI can make to education. This includes preparing society for the possibility of discovery, which will require a thorough, widespread, and prolonged educational effort.
— Analysis of the cultural aspects of SETI, using modern tools such as systems analysis, utility theory, cost-benefit studies, operations research, and decision theory.[15]

Experts who examined the social consequences of contact recommended directions for social science research—in particular, a thorough examination of published reports on mass psychology, individual and public responses to cultural differences, to strangeness, and to science and technology. They called for cross-cultural studies of popular images of extraterrestrials and saw a need to identify groups with unusual influence.[16]

We might add two themes. First, study societies whose cultures have proved most resilient in similar circumstances. Second, examine our historical experiences in attempting to communicate intent from one society or political entity to another. We might learn useful lessons from the least culture-bound findings about our experiences with negotiation and diplomacy.

The authors of the SETI 2020 report recommended that the SETI Institute expand programs to study the societal implications of detecting signals from an alien civilization, and that it encourage an active interest among scholars in these studies. Yet, the cultural impact of astrobiology was entirely absent from NASA's first biennial conference on the subject. Dick and Strick speculated that this reflected both the difficulty of getting social scientists involved and the lack of encouragement from natural scientists.[17]

Physical scientists and engineers often pigeonhole nonphysical, nontechnological explanations for the behavior of extraterrestrials as "sociological." We could move toward more useful common ground if we called them behavioral.[18] Social scientists try to understand how humans act, and why. Their findings, where thoroughly proven in the real world, might provide a sounder basis for reasoning about alien behavior by analogy.

Harrison and his colleagues proposed ways of getting social scientists more involved in SETI, including conferences, publishing in refereed journals, developing new publication outlets, building a high-profile peer group, training graduate students, and encouraging projects that serve both mainstream disciplinary and SETI interests. All of these would require funding.[19]

Many of the social science research questions have been defined. So far, there has been no systematic effort to commission or fund the research needed to answer them.

To make the best use of social science, people involved in the search must recognize its limitations. Although social science may produce many useful findings, it may not yield laws comparable to those in physical science. Some social scientists believe that, at the broadest level, repeated patterns suggest that there are laws of history; others cringe at the idea.[20]

We can make one prediction now. Social science findings that are driven by ideology or politics will be dismissed by most physical scientists, who will respect only those conclusions based on observable facts. To be credible, social scientists must report human behavior as it is, not as they wish it would be or fear it might be in the future.

Forums for Discussion

One of the biggest gaps is the lack of a journal devoted to the search and the consequences of contact. In 1979, Robert Dixon and John Kraus began publishing a quarterly entitled *Cosmic Search*. Aimed at the educated but nontechnical public, this magazine was one of the best sources for new thinking about the implications of contact. At one point, it had over 3000 subscribers in more than 50 countries. Unfortunately, *Cosmic Search* lost money and went out of business within a few years.

There was a comparable case in the space field, with parallel timing. In the late 1970s, several space advocates founded the Institute for the Social Science Study of Space. That organization—a small invisible college—published one issue of its journal, *The Space Humanization Series*, in 1979, then faded from view as its activists pursued less scholarly interests.[21]

Publishing entrepreneur Carl Helmers started *SETIQuest* in 1994, but stopped publishing that useful periodical 4 years later. At this writing, the SETI League is sponsoring an on line publication called *Contact in Context*.[22]

In the absence of a specialized journal, the primary media for articles on the probability and implications of contact remain the *Journal of the British Interplanetary Society* and *Acta Astronautica*, published by the International Academy of Astronautics.

Those most active in this field also need places for occasional international meetings. Although the annual International Astronautical Congress once provided a useful forum, the extremely high registration fees charged in recent years have placed that venue out of reach for most interested people. The triennial bioastronomy symposia have provided another if less frequent forum, although it is one focused almost enitirely on scientific questions.

Non-Western Perspectives

Speculations about the consequences of contact have been dominated by Westerners, particularly English-speakers from the United States, the United Kingdom, Canada, and Australia, with important contributions from several continental European nations. Although Soviet citizens proposed many early ideas, far less is heard from contemporary Russia.

The debate has lacked significant input from the Earth's most populous continent: Asia. If work is being done in Asian languages, little of it is being translated and made available in the West. This seems doubly unfortunate because of the long historical traditions of Asian religions, philosophies, and worldviews.

Some have defended this Occidental bias. Paolo Musso argued that Western culture is the most likely to experience first the "impact of contact" and to have to make the first decisions about managing it. Historian Joseph Tatarewicz, too, found a Western orientation defensible, but he added that it would be fascinating to explore this issue in other cultures, particularly those that have different metaphysics and epistemologies, and where humans have not traditionally occupied so central a place in the universe.[23]

Intellectual Tolerance

If a tradition of rational thought is to make progress, it is essential that it builds in tolerance. No authority can dictate in advance what can and cannot be believed.

—Charles Freeman, 2002[24]

Twenty years ago, Notre Dame historian Michael Crowe called for greater humility in dealing with the philosophical, religious, and scientific issues central to the extraterrestrial intelligence debate.[25] Jumping to conclusions when facts are so few is plainly premature. We don't know enough to be savaging those who disagree with us about the probability of contact, or its consequences; we are arguing more about possibilities than about truths.

Unless or until contact is actually established, Sagan and Newman observed, this subject is destined to be riddled with uncertainty and honest differences of opinion. No convincing resolution to these issues is likely to come from protracted debates carried on with great passion and sparse data.[26]

Sharp words may reflect impatience with ambiguity. They also may reflect disappointment with the results of our searches so far.

We should not let ourselves be disappointed so easily. Finding evidence of extraterrestrial intelligence may take generations. In the meantime, we should allow speculations—those of others as well as our own—to stretch our minds. There is no place for arrogant assertion when so little is known.

Finding and interacting with extraterrestrial intelligence is one of Humankind's grandest intellectual adventures. Let's enjoy the ride.

Some Conclusions Drawn

The universe isn't obliged to live up to our hopes and expectations.
—Govert Schilling, 1998[1]

Searching

Detecting extraterrestrial intelligence will be more difficult than most searchers had hoped. Finding evidence of alien civilizations may require rigorous and repetitive search and analysis that lasts beyond individual lifetimes. It may require a broader strategy, and a willingness to look in new places. It may require means not yet available to us.

This is not reason to give up. It is reason to be more determined, and more ingenious.

So far, our searches have been driven more by our available means than by a comprehensive approach to the question. They have been very limited in scale and in time.

Ideally, our quest would be continuous; our searches for signals would scan all feasible parts of the electromagnetic spectrum and would be extended deeper into the Galaxy. Thoroughness would require us to encompass the direct contact scenario, recognizing the possibility of alien exploration and expansion not just in our own time but over billions of years in the past. That would mean a more detailed reconnaissance of our solar system, some day including the Kuiper Belt, the Oort Cloud, and the huge volume outside the ecliptic plane.

These explorations always will be constrained by their technologies and their strategies, which are likely to improve, and by their financial resources, which may or may not increase. Our searches also may be limited by our assumptions. Challenges to those assumptions should not discourage searchers, but should drive them to refine their methods.

The search has been described as an example of too much speculation chasing too few facts. That is not an argument for dismissing its conjectures; it indicates a requirement to produce more facts.[2]

Research already has allowed us to better define the ranges of some probabilities, particularly in the astronomical factors. Further work may help us to reduce uncertainty in others, such as the origin of life and the evolution of intelligence.

We are searching for a limited sample of the many extraterrestrial civilizations that may exist: those emitting active or passive signals we can detect, or those that have left their technological traces in our solar system. One detection would imply a galaxy rich with intelligent beings.

No prediction about the timing of contact is reliable, nor can we be sure of its exact circumstances. Our detection of extraterrestrial intelligence could be the serendipitous result of other activities; the first finding may be ambiguous or disputed.

We might approach the timing question more modestly by focusing on interim steps, particularly the detection of a habitable world. The new instruments that are to be deployed during the next decade may well reveal an Earth-sized planet in a possible zone of habitability around another star. We might even detect an atmosphere whose composition suggests the presence of life. That would inspire greater optimism about finding extraterrestrial intelligence and expanded efforts to search for it, although it would not be proof that other minds exist.

Whewell cautioned us more than a century ago that the discussions in which we are engaged belong to the very boundary regions of science—to the frontier where knowledge (at least astronomical knowledge) ends and ignorance begins.[3] The only way to lessen our ignorance is to explore our cosmic surroundings in greater detail.

The fact that this quest is difficult should not deter our best minds; it is a challenge to their abilities. The search is a test of human intelligence.

This quest also requires a generous supply of a currently unfashionable virtue: patience. The search for extraterrestrial intelligence is a project that has to be funded by the century, argued Purcell, not by the fiscal year.[4]

The reward for patience may be great. Some day, there may be another triumph of observation.

The SETI Prize

If credit for a discovery is not enough to motivate new searchers, we might consider more material incentives. We could take a page from the space advocacy's book, encouraging wealthy sponsors to offer a multi-million dollar prize for the first confirmed detection of extraterrestrial intelligence.

There could be interim steps. Chris Boyce suggested an annual prize for the most *responsible* encounter situation presented in a science fiction work—book, story, film, or television show (emphasis added). We might extend that concept to works of scientific speculation.

Going Beyond the Evidence

Science is the best defense against believing what we want to.
—Jack Cohen and Ian Stewart, 2002[5]

Among those who believe that extraterrestrial intelligence exists, and among those who deny it, many have reached conclusions that go well beyond the evidence. In both cases, some of the most extreme spokespersons have damaged the credibility of their arguments by making exaggerated claims.

We may need to remind ourselves of the distinction between what is knowable at any particular moment in our history and what is not. As Crowe put it, the ways of the universe and of God are more difficult to discern than most inhabitants of our planet have been willing to recognize.[6]

We are new arrivals on the interstellar scene. Our knowledge of the cosmos is incomplete; our ability to understand it is limited. We have much to learn. Our current conceptions of the universe will change; our ideas of what to look for will continue to evolve.

There remain many opportunities for innovative work, by social scientists as well as their hard science counterparts. Although the physical and biological sciences are essential to this quest, they cannot answer every question about the probability of success, or its consequences.

Discovery is most likely when we deploy a new capability or employ a new strategy, or invent better ways of analyzing the data we have. Powerful new astronomical instruments, including orbiting observatories, may produce a flurry of astronomical discoveries during the coming decade; some may influence our thinking about extraterrestrial civilizations. However, we may never have comprehensive means of searching; we may never be omniscient.

In the meantime, intellectual honesty requires us to distinguish between what is proven and what is not. We should be particularly suspicious of one-factor analyses and of projections that foresee continued uniform behavior by intelligent beings, such as maintaining permanent beacons or relentlessly expanding through the Galaxy.

We cannot accept the opinion of any distinguished person as the final word; there are no authorities on extraterrestrial intelligence until it is found. There is no place for scientific priesthoods, or any other form of elitism that excludes the rest of us.

Belief, Expectation, and Fact

Faith is the substance of things hoped for, the evidence of things not seen.
—Saint Paul[7]

The idea that we are accompanied by intelligent but invisible others remains powerful. More than ever before, that conception is based on projections from scientific knowledge. Yet, the supposition that we share the universe with other sentient beings remains a belief, not a proven fact. That belief still provokes strenuous resistance from those most attached to anthropocentrism, although their claim that we are alone also lacks proof.

Belief can become expectation. Many humans expect that finding evidence of alien intelligence is just a matter of time and effort. Many expect contact with extraterrestrials to take place in their lifetimes—whether it be through signals or spacecraft.

Perhaps we should be wary of the temptation to believe. We must allow our conceptions to arise from an interpretation of the data rather than trying to make the data fit our conceptions, which may turn out to be wrong. Yet, we should not lightly dismiss belief, which can play an important role in cultural progress. Belief shores up our courage and strengthens our will; it can give us a sense of purpose, and the confidence to attempt more.

The search requires faith, declared the Cyclopeans: faith that the quest is worth the effort, that Humankind will survive to reap the benefits of success, that other intelligences are equally curious and determined to expand their horizons.

Underlying all great civilizations are shared values, shared visions, and confidence in the future. Such civilizations, Achenbach reminded us, are not constructed by the hopeless and the skeptical.[8] Belief can be inspiring, if it does not lead to intolerance.

Martin Rees suggested that agnosticism is the only rational stance on the issue of extraterrestrial intelligence.[9] That is not denial.

The Value of Speculation

In the absence of unassailable evidence, our collective psyche will meanwhile fashion bridges into the imagination, summoning ever deeper realms of possibilities, and excavating ever farther reaches of meaning in the search for our special place in the cosmos.

—Randall Fitzgerald, 1998[10]

There is only one honest answer to the question of what intelligent aliens will be like and how they will behave toward us: No human knows. All of us are speculating. That does not mean that we are being frivolous; speculations can have value.

Our speculations are a way of looking at ourselves, of thinking about Humankind from a novel perspective. The concept of extraterrestrial intelligence encourages us to look again at the meaning of human existence, proposed James Christian, to philosophize again, to theologize again, to take another look at our laws, our ethics, our minds, our knowledge.[11]

Our speculations about other civilizations are a way of looking into our own future. They can suggest either the future we wish for ourselves or the future that we fear will come if we do not change our ways. When we expect extraterrestrials to be morally superior and altruistic, we are hoping that we will be more moral and altruistic in the future. As Achenbach put it, the great moment of contact may simply remind us that what we most want is to find a better version of ourselves.[12]

Our speculations also are warnings. Predicting that other societies may have succumbed to war, runaway population growth, environmental damage, or disastrous experiments is a way of warning ourselves of how we might derail the human adventure. When we imagine aliens to be vicious conquerors, we are projecting fears about human conflicts.

Our speculations are platforms for moral lessons, as they have been for centuries. We still exploit the idea of alien intelligence to advance social, political, and ideological agendas through the imagined views and practices of advanced extraterrestrials. Theories about aliens often are intended to support other beliefs.

Our speculations suggest opportunities for collective self-improvement. Many of us want to change the future, not just let it happen to us. We want to rise above our current condition. The quest for other and better forms of life, society, technology, ethics, and law may not reveal that they are actual elsewhere, suggested Beck, but it may in the long run help us to make some of them actual on Earth.[13]

Imagining other worlds also can be an approach to science. Nobel Prize-winning medical researcher Peter Medawar observed that scientific investigation begins by the invention of a possible world or of a tiny fraction of that world. Another Nobel Prize winner, geneticist Francois Jacob, noted that mystical thought begins the same way.[14] The debate opens our minds to what might be possible.

Our speculations have another practical dimension. They help to prepare us for contact.

A Probable Scenario

We are far more likely to detect evidence of an extraterrestrial civilization than we are to receive long, detailed messages full of valuable information deliberately communicated to us. The most probable scenario, at least initially, is one of low-information contact.

The evidence could take many different forms, from an intercepted message between other civilizations to an eroded chunk of alien metal in our solar system. That evidence may be too exotic to fit within conventional categories. The first detection may not be taken seriously, particularly if it is not made by SETI astronomers. Like Jansky and Reber's findings about our Galaxy's radio emissions, the early data may be ignored or dismissed as tangential to the interests of more established scientists.

SETI advocates have argued for decades that we are most likely to encounter a civilization that is very far in advance of our own, by millions or billions of years. Yet, as we have seen, the most scientifically and technologically advanced civilizations may have little or no interest in us; we may not able to detect them.

This book has presented an alternative: Our search for signals may be a filtering process that makes contact with alien intelligences nearer our scientific and technological level the most probable outcome. Those civilizations may be more common than straight-line statistical projections would suggest. We cannot assume relentless scientific and technological advance in other societies; alien civilizations thousands of years older than ours may not be thousands of years more powerful.

In the remote detection scenario, we are most likely to detect those who use communications technologies similar to ours. Contact may be most likely within a band of scientific and technological progress that extends from our own level to a few thousand years in our future.

If more direct contact takes place, it probably will be with the exploring machines of advanced civilizations, whether or not those devices still function. In this scenario, the time gap that separates our state of scientific and technological development from theirs could be much wider.

A Mirror Image. It is the civilizations closest to us in space and time that are most likely to be provoked by discovering us. Their interest could stimulate a desire to exchange knowledge or engage in cooperative endeavors; it could stir concerns about our capabilities and our intentions; or it might lead in both directions.

Anticipated Consequences

Neither the optimists nor the pessimists have proven their cases about the consequences of contact. Our speculations still rest on analogies with human experience, on our cultural and political contexts, and on our personal biases. They still reflect our hopes and our fears.

Until we have evidence, the anticipated consequences of contact are matters of belief. Those beliefs have led to expectations that may not be supported by fact. Reducing the scientific uncertainties in the Drake equation may do more than refining the probability of discovery; it also could sharpen our speculations about what contact might bring.

Establishing meaningful communication with an extraterrestrial civilization may be far more difficult than many had hoped. We cannot assume massive transfers of useful knowledge; exchanging scientific information may not be their first priority, or ours. If we do receive messages or artifacts, they may not be full of scientific facts. They also may be extremely difficult to understand.

The human hope for intervention from above—the desire to be guided, rescued, or punished by superior beings—remains strong. We cannot assume that voices from the sky will guide us with transforming philosophies or lead us to utopia. However, they may provoke our thinking and broaden our perspective.

Contact could inspire us or demoralize us. We could feel demoted if contact challenges our anthropocentrism and our hubris. We could feel promoted by our entry into a larger community; that might satisfy our desire to be accepted, to be approved.

The biggest change in our way of looking at the consequences of contact is that there are now two established paradigms: remote contact through detection of electromagnetic signals (including the detection of astro-engineering), and direct contact through interstellar travel. The second paradigm implies that the consequences could be much more powerful than the traditional SETI school anticipated.

The more intense forms of contact could provoke cultural shock, testing our ability to adapt. Even if extraterrestrials were well-intentioned, they could damage us. The outcome of such contacts may depend on the motivations and inhibitions of the more powerful species, and on the intellectual and cultural resilience of the less powerful.

Direct contact could happen in different ways, with different implications. Passive direct contact, with an artifact that does not send communications either to us or to its home civilization, would allow us a wide range of choices. Active direct contact, with a machine that communicates or takes other actions toward us, could tilt the range of consequences toward our fears. Even then, worst-case assumptions may not prove true.

The outcome of contact is not a sure thing. Its consequences for Humankind will depend on which scenario prevails, and on the capabilities and intentions of those in contact—the extraterrestrials, and ourselves. We may know our own capabilities and intentions in advance of contact, but not theirs.

The closer we are to the other civilization in time and space, the more potential there is for cooperation—and for conflict. The path chosen will depend on decisions made by intelligent beings. Those decisions could be influenced by the communication of intent, and by actions consistent with those intentions.

The most rational view about the consequences of contact may be ambivalence. Many possible outcomes could be mixtures of the hopeful and the fearful; the effects could be both positive and negative.

We may find beauty; we may find ugliness. If we are intellectually brave, we will face both possibilities, confronting both our angels and our demons.

The Key Factors

The first key factor that would make the difference between controllable and noncontrollable consequences of contact is interstellar flight. Not whether it is possible—as it almost certainly would be for a species more technologically advanced than our own—but whether any civilization actually chooses to do it, whether any society is sufficiently motivated to explore, colonize, or take other action at interstellar distances. We can refine the issue further by eliminating the requirement that interstellar vehicles carry biological organisms. Intelligent machines could do the job.

The second key factor is the choice of actions made by the civilizations in contact—the extraterrestrials, and ourselves.

The Prime Question

The basic aim of CETI is initially to make ourselves known.
—Ian Ridpath, 1978[15]

Given such possibilities, one policy question rises above all others: Should we call attention to ourselves? Should we reply to an alien signal, whether or not it is addressed to us? In the absence of a detection, should we send targeted, high-powered signals in an Active SETI program? Or should we simply leave unchanged what we are doing now, thoughtlessly radiating our normal radar, television, and radio signals out into the cosmos?

We cannot assume that extraterrestrials have detected us or that they will. Many other intelligent beings may believe that their civilizations are unique, as many humans believe their own is. If alien intelligences do search for others, those searches may not be sustained.

If sentient aliens are searching, they may be doing so with technologies that are not optimized for the wavelengths that we emit most strongly. Our normal signals may be below their detection threshold. Our highly detectable phase may be brief. Unless we receive a transmission clearly aimed at us, we won't know if another civilization is aware of us or not.

What if we find alien technology in our solar system? There is no point in remaining silent, Bracewell commented, if a probe has already reached our solar system *and has reported our presence* (emphasis added).[16] The second condition is crucial, as an alien spacecraft might have ceased functioning long ago. If the probe showed no signs of activity, we could not assume that it had informed its sending civilization of our existence. We still might have a choice about calling attention to ourselves.

The underlying issue is whether we think that the effects of contact will be positive or negative. Advocates of Active SETI implicitly assume that alien civilizations are benign, or that distance will insulate us from direct contact if they are not. Those assumptions remain unproven.

We have no basis for assuming that another technological civilization would welcome contact with us; nor do we have any basis for assuming that another civilization would be hostile. We simply don't know how extraterrestrials would react to contact with humans. Do we live in a universe of cooperation and altruism, or in a universe of Darwinian competition? Might this vary from one intelligent species to another?

Some think—or assume—that any extraterrestrial civilization that broadcasts its existence is likely to have peaceful intentions. What if some listeners do not? Our first deliberate message might reveal a great deal about the present level of our science and technology—and about our comparative weakness.

Half a century without a clear signal implies either some degree of scarcity, or else a reticence on the part of aliens to broadcast at full strength. This lack of beacons should be at least somewhat worrisome, Brin concluded, especially to those who feel an urge to shout. The civilizations that survive may be the ones that do not call attention to themselves.[17]

Given our ignorance, a certain degree of prudence may be in order. We have policy choices; we can make a conscious decision about what to do. We could limit Active SETI. We could even reduce the intensity of our normal leakage radiation (this process already is underway in radio and television broadcasting, although not yet in the use of military or planetary radars). Meanwhile, the operators of our noisiest technologies will continue to make the choice for us, without a public policy debate.

Which Signals Could Be Dangerous?

The Earth already radiates extensive signals at certain wavelengths, but most are too weak to be detected at interstellar distances. The brightest are confined to narrow ranges of frequencies. Which are most likely to reveal our presence?

One approach is to classify human-generated signals in three categories: signals not sent in preferred directions, which could be described as including the Earth's normal radio leakage; Active SETI signals sent in preferred directions; transmissions in reply to a signal from an extraterrestrial civilization. Signals that are sent with no spatial or temporal preference are part of Earth's normal background noise and may not be a problem. However, messages that are sent in preferred directions and that are long enough for the other civilization to confirm their origin—such as Active SETI messages—may be dangerous.[18]

The search is science and is properly the domain of scientists. Deliberately calling attention to ourselves is not science, but policy.

Given our lack of knowledge about alien civilizations—and the potential negative consequences of contact—we may need to observe a precautionary principle: Do not call attention to ourselves until we learn something about the civilization that we detect. In practical terms, that means not increasing the power of the Earth's electromagnetic emissions beyond their existing levels.

If we do detect another civilization, we could try to determine its distance and estimate its technological capabilities before deciding whether to make ourselves easier to find. There still would be many unknowns—above all, their intentions. At least we would be making a conscious decision, instead of assigning our future to chance.

Paradigm Shifts

What is important is the attempt to replenish the void, to fill the immense emptiness with meaning, even if this is accomplished by turning emptiness itself into an antagonist that can be confronted in human terms.

—Mark Rose, 1981[1]

Our Place in the Cosmos

Our sense of scale has shifted dramatically within the past century. Astronomer John Kraus, remembering that radio broadcasting was the new sensation in 1920, described how his heart pounded when he first heard a voice from 2000 miles away.[2] Now we are searching for voices from hundreds of light-years away.

Sagan and others have seen the quest for extraterrestrial intelligence as a search for a cosmic context for Humankind, in a universe vaster in both extent and duration than our ancestors ever dreamed.[3] This is not just a scientific issue; many nonscientists are looking for a more appealing, more satisfying, more responsive universe. We want to feel connected with the cosmos rather than isolated and alone.

SETI may be an effort to achieve that connection, working within a framework that is acceptable to the Western scientific model. As White put it, the search might be a way to seek a reintegration that has been shattered by standing apart from the cosmos and examining it as something that is not alive, not intelligent, and separate from ourselves. This emotional dimension does not always manifest itself in scientific terms; it can be expressed as religious feeling, through art and music, and as philosophy and mythology.[4]

Support for the search reflects a desire for community, reassurance, and variety. The universe may not be lonely; it may teem with other lives, other aspirations. No amount of robotic exploration can make us feel emotionally invested in the universe, argue Benjamin and others; only SETI, with its yearning for cosmic encounter, can address this essentially metaphysical need.[5]

What is the place of humans in this vastly expanded space? Many conclude that we are so small and weak as to be inconsequential, and possibly ephemeral. Some scientists seem to be frustrated that their data do not convince everyone of human insignificance.[6]

Others have reacted against the peripheralization of Humankind. We may deserve better.

Our Place in Time

Searching for other civilizations helps us to visualize ourselves in the fourth dimension—time. Contact could bring a Copernican revolution not just in the spatial sense but in the historical sense as well.

The moment of contact is unpredictable; it could range from tomorrow to the end of intelligent life on Earth. Our nearest neighbors may be many light-years away—a measure of time as well as distance. In a radio dialogue across interstellar space, the gap between question and answer could be centuries.

Thinking about our place in time must include a long future. By focusing attention only on the past and the present, argued Tipler, science has ignored almost all of reality.[7]

We ourselves will change over time. If the search goes on long enough, Baird foresaw, the definition of humanness may slowly shift to the extent that the organism that initiated the search in the twentieth century may bear little resemblance to the organism that finally tastes success.[8]

A View from the Millennium

Harrison proposed a perspective from 1000 years in our own future. Initial contact will be part of history; future human attention will be directed somewhere else. Any difficulties or dislocations that occurred during first contact will be long past. Interacting with other civilizations will be no more unusual than interacting with human colonies that will be sprinkled throughout the solar system and beyond. People will be quite different from today, but human interaction with extraterrestrials will account for only some of those differences.[9]

The Copernican revolution and its successors have destroyed the chauvinism of place. The successive demotions of Earth showed us that our planet has no special position in the universe. If we still consider ourselves important, it is not because of our location.

The Copernican Time Principle extends that demotion from space to time. The cosmological epoch in which we humans emerged may have no special place in the vast expanses of cosmic history—if other intelligent beings exist. It may have a very special place if they do not.

Our Place Among Intelligent Beings

Whether the result of the listening effort is positive or not, we will evolve an active idea of our relationship to the universe—either as sharers in . . . or as near unique possessors of intelligence, capable of preserving and diffusing it throughout the universe.

—Five SETI scientists (Black, Tarter, Cuzzi, Connors, and Clark), 1977[10]

After centuries of debate, we still are arguing about our status in the universe. Opposition to the search reflects an underlying fear that success would lower our status and undermine our self-esteem.

The search can be seen as another step in our long journey away from anthropocentrism. Is the continuation of this journey inevitable?

Anthropocentric beliefs may continue to have weight until unambiguous evidence of extraterrestrial intelligence is discovered. Even a confirmed detection might not demolish anthropocentrism. The durability of this idea suggests that contact might drive anthropocentrists to look for another way of arguing that humans are special.

What if our search does not succeed? Nobel Prize-winning biochemist Baruch Blumberg, first head of NASA's Astrobiology Institute, thought that a failure to detect extraterrestrial life would be a step backward from the Copernican revolution. Finding that there are no other intelligent life-forms, White observed, could reinforce religious views of humans as unique.[11]

Asimov speculated that a failure to find other intelligences would cause a new kind of loneliness and desolation, a fear of a vast impersonal universe in which we are lost.[12] That would bring us full circle, back to Blaise Pascal in the seventeenth century.

Until we detect others, we are effectively alone. So far, we have found nothing to exclude that possibility. If we are alone, argued Easterbrook, there are two poles of possibility. One is that life is a fluke without inherent significance. The other is that human life is precious beyond words.[13]

Achenbach went even farther. Until there is reason to believe otherwise, we have the right to view our planet and its inhabitants, and especially its sentient creatures, as the treasures of the known universe.[14]

A Mirror Image. If we can imagine ourselves as the treasures of the known universe, why should we doubt that other intelligent species might come to the same conclusion about themselves?

Regis challenged these views. He found it puzzling to hear human specialness claimed to be a virtue when the pre-Copernican view that mankind is the center of things is supposedly a deadly chauvinism. If *Homo sapiens* becomes more precious if extraterrestrial intelligence is absent, then it must become less so if extraterrestrial intelligence is present.[15]

We may never know for sure that there are other minds. Even if life is a universal, Morris reminded us, we can still be alone. It may be more prudent to assume that we are unique and act accordingly.[16]

The Human Role

As our experience of the destruction of worlds increased, we were increasingly dismayed by the wastefulness and seeming aimlessness of the universe.

—Olaf Stapledon, 1937[1]

The more the universe seems comprehensible, the more it seems pointless.

—Steven Weinberg, 1977[2]

A Search for Purpose

One of the depressing things about the last 300 years of science, Davies lamented, is the way it has tended to marginalize, even trivialize, human beings, thus alienating them from the universe in which they live.[3] Science has convinced many of us that we exist in a cosmos that is not centered on our species, its expectations, or its fate.

To discover an ultimate purpose that links us to the universe, Hoyle and Wickramasinghe observed, humans have traditionally turned to religion. That choice may have become less credible in modern Western society; our emergence from tangible chemical evolution removes the supernatural quality from our origin and our destiny.[4]

The erosion of tradition and the widespread decline of inherited religious belief leaves us without a *telos* (an end to which we strive), argued anthropologist Charles Lindholm. Bereft of a sacred project, a sanctified notion of our potential, we have only a demystified image of a frail and fallible humanity. Surrounded by a disenchanted cosmos, many suffer from what Eric Fromm called The Anxiety of Meaninglessness.[5]

One of the drivers behind our search for other intelligent beings is our desire to find or attribute purpose to our existence. We have an innate yearning to be identified as part of some ill-defined grander scheme of things.[6]

We may ultimately recognize that only intelligent life can provide purpose. Not just superior extraterrestrials, but us.

A Warning. Either there is no general purpose at all behind the cosmos, Stapledon speculated, or that purpose is something wholly unintelligible to human minds. In that case, it is equally false to say either that there is or that there is not a cosmical purpose, as the truth is beyond our comprehension.[7]

A Destiny to Choose

We hold the future still timidly, but perceive it for the first time as a function of our own action.

—J.D. Bernal, 1929[8]

Many have argued that, as long as we are alone, we have an obligation to assure the survival of Humankind as the only known example of creative intelligence. Assuring our survival requires many courses of action: reducing the danger of conflict, controlling the effects of powerful technologies, limiting our impact on our natural environment, and avoiding the exhaustion of essential resources. In the longer run, it also requires expansion beyond the Earth, and ultimately beyond the solar system. Will we take these actions?

There are two futures, Bernal told us, the future of desire and the future of fate.[9] Which will we choose? Our answer will be heavily influenced by our prevailing cultures, our beliefs and our myths, and our opinion of ourselves.

Historian J.M. Roberts identified two central myths at the heart of the Western view of history—myths that enabled the West to dominate the world and to set its agenda for the future. The first is the idea that history is meaningful because it has a direction; it is going somewhere. The other is the idea that humans are able to take charge of their destinies. The most important gift of the West is the belief that humanity might control its fate.[10]

The universe has offered us a chance to define our own significance, declared Seielstad. We are in the right place at the right time to be creatures of consequence.[11]

The test for us will not be just what we are physically able to do, but what we choose to do. Knowledge alone is not enough; we need instrumental means.

Seeding the Galaxy with Life and Mind

If life on Earth is unique, then "Now" is a moment of sharp discontinuity between an equilibrium state in which the universe was virtually devoid of life and one in which life will be profuse—at least if we follow our first steps into space with eventual colonization. . . . Today is the beginning of a bio-eternity.

—George Seielstad, 1989[12]

If we discover no others, if we are effectively alone, the moral task of assuring the survival of intelligence will be ours. That obligation cannot be limited to Earth's biosphere, which ultimately faces extinction. We cannot assume that we have a billion years to go; an astrophysical catastrophe could occur at any time.

Like the nineteenth century's John Chapman, who planted the seeds of apple trees in mid-America, we could seed life throughout the solar system and beyond. We could become agents of the biocosm.

Humankind could initiate a directed panspermia, Mautner and Matloff proposed, propelling radiation-protected microorganisms in lightweight interstellar craft toward stars with potentially hospitable planets. We might be motivated not only to assure the survival of terrestrial life but also to promote the evolutionary trend of expanding life into different and more extensive habitats, and by an intuitive drive to see life affect natural history on a larger scale.[13]

Hoyle predicted that our descendants will realize that life is more important than the manner in which it happens at the moment to be expressed. We are merely convenient packages into which order (or information) has organized itself, proposed Seielstad, so that it can be spread more widely.[14]

Our destiny may be to transmit throughout the Galaxy not only life, but also consciousness. To us may go the task of carrying mind into the mindless void. Just as life seeks not only to survive but to spread, mind may pursue greater control over its surroundings and its future.

In those drives, so unique to living things, is the seed of greatness. Barrow and Tipler foresaw that present-day life would have cosmic significance because of what future life may accomplish.

We are transitions between organisms whose activities are controlled by genes and those whose potential is limitless, argued Seielstad. The gift of creativity, our most precious asset, represents the shrewdest investment life can make to ensure its growth and continuity. Time, nature's ultimate filter of the unfit, will determine whether we are equal to the task.[15]

We represent only the potential for the extension of life and mind into the cosmos, with no guarantee that we will succeed.[16] That future is not inevitable; it requires decision and action. It is time to accept our responsibilities.

Spreading Our Bets

Whatever significance we humans have is that which we make ourselves. . . .
It is very important to realize that many other species have become extinct,
that our survival is not guaranteed, that our future is in our hands, that
some external intervention is unlikely.

—Carl Sagan, 1980[17]

If we come into contact with an extraterrestrial society, we may receive knowledge that could advance our prospects as a species, or we may gain little of practical importance. We might enter into a productive relationship with another civilization, or we might fall before it.

If our search fails—or if we abandon it—none of this may come to pass. There will be no extraterrestrial source of knowledge and inspiration; we will be on our own.

We can spread our bets by investing both in a long-term, patient search for evidence of other minds, and in the slow but determined expansion of human presence and influence into the Galaxy. We can look for help while helping ourselves.

Led by our investigating machines, we can explore, colonize, and establish independent human communities, ultimately beyond our solar system. This expansion would help to assure the long-term survival of our species and its descendants. If we do find other technological civilizations, expansion will place us in a stronger position to deal with them—and to make a contribution to the shared tasks of intelligence.

An Extraterrestrial Paradigm

A dark, illimitable ocean, without bound, without dimension, where length, breadth and height, and time and place, are lost, the universe is mystifying and terrifying. . . . No one knows the terrors of the journey upon which man is about to embark; no one can know until the journey has been made.

—Kenneth Heuer, 1951[18]

Not long ago, space was beyond our familiar world, like the seas of the ancients. The oceanic deep once was a blank slate for the expression of our subconscious fears.[19] So was space until very recent times; we could imagine exotic lands and grotesque monsters, and fear to journey beyond the Pillars of Hercules in our minds. Now we are measuring and demythologizing our larger environment, incorporating its realities into our worldview.

The search for life is one of the forces that draw us outward. Within that broader context lies a special search for extraterrestrial intelligence. We are slowly absorbing the idea that sentient aliens might influence our future, through their transmitted ideas or more direct means. This is part of a larger paradigm shift: accepting the vast realm beyond the Earth as relevant to our future.[20]

Astronomy and solar system exploration have revealed that the external universe is not a neutral background, but an active arena that can affect our lives, our history, and our prospects. The evolution of life on Earth has been influenced by the distant explosions of massive suns; the human body

depends on elements created in the interiors of stars. The impacts of asteroids and comets have radically altered the course of biological evolution
and may do so again.

We are extending our point of view beyond our planet. We have reached
out in a physical sense, using space technology to explore our solar system
in more intimate detail. The surfaces of some other worlds now seem
familiar to us. Many humans aspire to—and expect—a future in which
Humankind expands outward, utilizing larger resources and living space,
escaping the limitations and the ultimate fate of Earth.[21]

These revolutions are dramatically expanding our relevant universe. The
realm of our concern is growing to include worlds far beyond our own. We
are slowly, hesitantly adopting an extraterrestrial paradigm, a new cosmic
context for Humankind.

This is a Copernican-level revolution in the way we look at the human
future. One could see it as making us feel smaller amid the vastness of our
new relevant universe; or we could see ourselves playing on a larger stage.
The outcome may say much about humanity's opinion of itself.

Is our example common, or rare? Are we the rule, or the exception?

Until we detect other civilizations, we have no way of knowing how large
their relevant universes are. Many may never pass beyond the planetary
stage; others might adopt extraplanetary paradigms of their futures.

Those who expand beyond their original biospheres may constitute a
small percentage of all technological civilizations. They may be a very
influential minority.

Reconnaissance

The exploration of space through astronomical instruments and spacecraft,
the search for extraterrestrial intelligence, and human expansion can
be seen as elements of an unarticulated grand strategy for Humankind.
Its basic goal is to improve the prospects for our species in its larger
environment.[22]

The first element of an extraterrestrial strategy is reconnaissance—the
collection of information about our greater surroundings. Astronomy and
solar system exploration are the means. Within that context, SETI is a
reconnaissance of intelligently influenced matter and energy.[23]

This perspective has a clear implication: We should expand the search
with better instruments and more comprehensive strategies, including
spacecraft with enhanced abilities to explore our own solar system. These
systems would have benefits for science even in the absence of contact.

In the longer term, interstellar probes could dramatically improve our
ability to detect extraterrestrial intelligence. No longer would we be limited
to those civilizations that are emitting electromagnetic signals that our
instruments can spot from Earth; we might find intelligent beings that

never had sent such signals, as well as societies that had ceased doing that long ago. Probes also might reveal future habitats for humans—Earth-like planets, or asteroidal and cometary resources that could support artificial worlds.

After the initial detection of another civilization, we would want to gather as much additional information as possible—their proximity, their technological capabilities, their intentions. Only their proximity would be relatively easy to determine. We also would want to know if there are additional extraterrestrial societies.

Other technological civilizations are potential competitors, particularly if they are relatively nearby. They also are potential allies who might help us to improve our prospects for long-term survival.

The more we learn about civilizations that may exist elsewhere in the Galaxy, the better our policy decisions will be. Only fools do not surveil their surroundings, for opportunities as well as for dangers.

Expansion

Exploring the solar system and homesteading other worlds constitutes the beginning, much more than the end, of history.

—Carl Sagan, 1994[24]

Shall we be forever one people, or shall we be a million intelligent species exploring diverse ways of living in a million different places across the galaxy?

—Freeman Dyson, 1979[25]

We humans will use the information we gather in formulating policies for human activities beyond the Earth. One policy decision may be to accelerate the expansion of human presence and influence into the solar system and beyond.

Spaceflight advocates have argued for decades that human expansion is not a frivolous waste of money, but a survival strategy. All life on Earth, including its most intelligent species, will come to an end some day due to changes in the biosphere or the Sun, or an astrophysical accident.

Sagan, who earlier in his career had questioned the utility of sending humans into space, came to advocate expansion and colonization late in his life. Every planetary society is obliged to become spacefaring, he argued, not because of exploratory or romantic zeal, but for the most practical reason imaginable: staying alive.

The spread of Humankind through the solar system would be the best insurance against any fatal catastrophe arising from external or internal threats. The more of us beyond the Earth, the greater the diversity of worlds we inhabit, the more varied the planetary engineering, and the greater the range of societal standards and values, then the safer the human species will be. Peopling other worlds, concluded Sagan, is our new *telos*.[26]

Expansion could give us access to greater resources and larger conceptions of what is possible. It would enlarge our niche in the Galaxy. It would be, quite literally, the expansion of intelligence.

Astronomy may show us how small we are; spaceflight suggests how large we might someday be. Expansion may determine which civilizations have the greatest control over their futures in a universe that is otherwise indifferent to their fate.

Patience, Take Two

The bolder the dream, the more probably must the dreamer leave it to be fulfilled by others.
                              —Science writer Nigel Calder, 1978[27]

Like those who foresaw early detection of alien intelligence, advocates of human expansion have been consistently overoptimistic about how soon each phase will get under way. Visionaries once predicted humans landing on Mars as early as the 1980s; current estimates have slipped to 2030. Further delay would not be surprising.

That is not a reason to give up. Like the search for extraterrestrial intelligence, expansion is a task that extends beyond individual human lifetimes. It, too, implies accepting longer timescales of societal endeavor.

Over the past 500 years, we have changed our understanding of our position in the cosmos. Over the next 50 years, proposed robotics pioneer Rodney Brooks, our new conceptions will empower us to change that position. We are breaking out of our roles as passive observers.[28]

Within a century—a blink of the cosmic eye—our vehicles may carry us to the outer limits of our solar system. Our interstellar probes may be racing outward into the galactic night. This future is not inevitable; it is a matter of choice.

Choosing not to expand beyond the Earth would be a gigantic failure of nerve. It would guarantee the limiting of the human future. By turning away from expansion, we could doom ourselves to being a marginal species, and to our eventual extinction.

Will our descendants be interstellar colonizers? If they are, expansion could lead to the rediversification of humanity. By opening new ecological ranges, it could be a conscious evolutionary step, leading to adaptive radiations and eventually to new species.

We may be on the threshold of changing our role in the universe—but only if we act. If we do not expand our capabilities and our influence, we may be irrelevant to the future of intelligence. We may be ignored as more motivated civilizations go their way.

An Extraterrestrial Ethos

Expansion, with its transcendental implication of infinite wealth and power for Humankind and its progeny, could lead to a new anthropocentrism, a belief that humans and their descendants occupy a special place in the universe. Our descendants could be afflicted with the vice the ancient Greeks called *hubris*.

The time has come to begin formulating an extraterrestrial ethos for Humankind. That ethos must rest on a sense of responsibility to our own species, a commitment to use our growing powers beyond the Earth to better the common lot. At the same time, we may wish to impose ethical limits on our ambition to humanize the nearby cosmos, to make it run on human time.

Preserving the Best

The first test may be our utilization of solar system resources. For decades, visionaries have proposed mining asteroids and comets for raw materials. Should we place limits on that enterprise? We may choose to exploit only those resources clearly needed for our well-being, balancing growth with other measures of value.

We could declare some parts of the solar system off limits. Uhlir and Bishop proposed setting aside space wilderness areas of special interest; Hartmann suggested that we draw up a limited list of sites open to scientific study, but closed to exploitation—a parallel with Antarctica. At the very least, we would want to preserve the most beautiful works of solar system evolution—the rings of Saturn and the oceans of Earth.

The second test would be terraforming, the deliberate altering of planetary environments to make other worlds more habitable by humans. Proposals to remodel other planets—particularly the atmospheres of Mars and Venus—have become increasingly detailed. Speculations about terraforming bring together many strands of science and philosophy in a great thought experiment, proposed Martyn Fogg, one that has Humankind participating in creation as well as preserving it.[28] How far should we go?

What if we find alien life that is not intelligent? Should we protect it, as we protect endangered species on Earth?

Christopher McKay argued that if there is life on Mars, it is not doing well and could use some human help. We would be ethically on good grounds to support it, to encourage it to flourish into a global scale biota like we have on Earth, especially if it were on the verge of extinction. We would not be justified in exterminating the only alien life we know, or in appropriating its environment to our ends. This might inhibit plans to terraform Mars.[29]

Mautner and Matloff looked at this question differently. Although they recognized an ethical objection to interfering with indigenous life-forms, they argued that our "pansperms" would be unlikely to cause damage to indigenous organisms by infection unless the their biology was similar to ours.[30] What if they do? Competition between our pansperms and local biota would be an extension of the evolutionary struggle for the survival of the fittest—a concept that some might find ethically troublesome.

The questions become sharper if we encounter another society of intelligent beings. What will guide our behavior—pure self-interest or a higher ethic? Would we accept limits on human expansion in the regions of interest to the other society?

Other expanding civilizations would face the same questions. The common ground may lie not in science, but in ethics.

The Moment of Truth

Power . . . ingests weaker centers of power or stimulates rival centers to strengthen themselves.

—William H. McNeill, 1963[31]

What is love? It is the ability to confer survival benefits in a creatively enlarging manner upon the other.

—Ashley Montagu, 1972[32]

If some technological species explore or expand over large distances, one may eventually encounter another. That event may be the most crucial point in the evolution of a technological society. In it lies the potential for exchanges of experience and wisdom, and joint efforts in the work of intelligence. In it also lies the seeds of demoralization, cultural imperialism, and destructive conflict.

Would each society see contact as a zero-sum game, a form of interstellar Darwinism in which the less "fit" might be destroyed? Or would they see it as an opportunity for cooperation, even assistance to the less advanced? Ethical considerations and policy decisions, as much as physical factors, may determine the outcome.

Contact could be the moment of truth for technological civilizations, including ours. If that encounter is to have a productive result, a higher ethos, beyond human values alone, might be needed—not fuzzy-minded altruism, but enlightened self-interest in which civilizations help each other to survive.

In the meantime, anticipating contact could motivate us to reduce the disparity between ourselves and a more advanced technology. The greater our own scientific, technological, and natural resources, the more capable our future civilization will be—and the more seriously it will be taken by others. We may not have to resign ourselves to being helpless.

Making such advances would enable our own future even in the absence of contact. It is one part of spreading our bets.

If there is a community of intelligence, our species cannot be part of it unless we search for evidence of its existence and communicate with it. We cannot contribute to its work unless we have the physical means. Technological life, rather than merely intelligent life, will determine membership in a galactic society.

The Triumph of Mind

The same logic that leads to the conclusion that humanity is one brother-hood will . . . lead to the same conclusion about all forms of consciousness in the universe. When this is recognized we will be driven to ask about the common goals of the commonwealth of conscious beings.

—Gerald Feinberg, 1968[33]

Perhaps the optimistic model will prevail. Older intelligences may have polished their skills of communication, diplomacy, and compromise; they may have learned ways to communicate nonaggressive intent that all sentient beings can understand. To succeed, they—and we—must cross not only the barrier of communication but also the barrier of trust.

There could be synergistic effects. A community of intelligence may have perceptions of the universe and its fate beyond the ability of any individual species. Each species may contribute knowledge, insights, skills, and powers, which would interact with those of others to stimulate new syntheses; the whole might be greater than the sum of its parts.

If the optimistic scenario prevails, technological civilizations in contact might move beyond exchanges of information to identify shared interests, develop understandings that would reduce the risk of friction or conflict, and work together on complimentary courses of action. We might see each other as allies against the impersonal forces of nature.

Here is a new opportunity to be part of something much larger than ourselves. Civilizations in contact may conceive of great tasks to assure the survival and further evolution of intelligence. Stapledon foresaw this possibility nearly 60 years ago; the intelligent races he envisioned were engaged upon a vast common enterprise undertaken by the Galactic Society of Worlds.[34] The feasibility of those tasks may depend on how many technological civilizations emerge, survive, and expand beyond their home worlds.

Our cooperative efforts may be limited to those who are not millions or billions of years more advanced than we are. We may never know if the most advanced intelligences are engaged in great endeavors; we may never be participants in their work. They may be driven by millennial purpose, employing their knowledge, skills, and power in tasks so vast as to make planet-crawlers seem like short-lived germs; or they might be so weary of immortality, so inward-directed, so resigned to entropy, that they no longer care.

We may have the opportunity to join in the triumph of intelligent life, or its tragedy. Soft ideas dismissed by hard science—motivation and morale—may determine the outcome.

There is reason for long-term optimism. We may be embedded in a favorable trend: the growing emergence and spread of intelligent life. We may have entered the period of galactic history when mind asserts itself over mindlessness, when intention is imposed on chance.

Annex: Preparing

If extraterrestrial intelligence is abundant, it will be our destiny to interact with that intelligence, whether for good or ill.

—Steven Dick, 2000[1]

Fortune favors the prepared mind.

—Louis Pasteur[2]

Ready or Not

If other technological civilizations exist, we are making contact with them more likely. We have heightened our detectability by radiating radio signals, television carrier waves, and radar pulses out into the Galaxy, making it more probable that they will detect *us*. At the same time, by extending the range, breadth, duration, and sensitivity of our searches, we are making it more likely that we will detect *them*. Some day, we may have to face the consequences of our actions.

As we may be inviting a discontinuity in human history, it seems prudent to prepare as best we can. Preparing could lay the groundwork for an orderly transition into what Billingham called "the rather different universe that may await us."[3]

We do not know which scenario of contact will prevail. Until it occurs, we can only speculate about the capabilities and intentions of the other civilization. Our preparations must be based on possibilities, not certainties.

More than 30 years ago, your present author commented on our psychological readiness for contact:

In our thinking about alien intelligence, we reveal ourselves. We are variously hostile, intolerant, hopeful, naive; influenced by science fiction, we see aliens as implacable, grotesque conquerors, or as benign, altruistic teachers who can save us from ourselves. Usually we think of them as superior to us in some way; either their miraculous but malevolently applied technology must be overcome by simpler virtues, or we must accept them as gods who will raise humanity from its fallen condition. Here we display fear, insecurity, wishful thinking, defeatism, even self-loathing, everything but the calm maturity appropriate for our emergence into the galactic community. We are not ready.[4]

Are we any more prepared now? Space lawyer Steven Doyle, speaking in 1997, thought that the world was probably as ready then for discovering extraterrestrial intelligence as it ever would be.[5] In the sense of popular culture, he may have been right, although the expectations of the general public may not match the reality of contact.

Drake sounded a note of urgency. The sooner contact occurs, the less prepared humanity is likely to be, and the greater the impact. Drake wanted to get thinking adults ready for what he called "the imminent detection of signals from an extraterrestrial civilization."[6]

Once contact happens, warned White, we will be playing "catch-up ball" on an issue that demands the maximum amount of foresight. What would have happened to Native American societies if they could have prepared for the Europeans coming to North America? It might have been far better for both cultures if forethought had informed the process.[7]

We will be well prepared for the initial discovery, Beck believed, because we have to know what it will be in order to know when it has occurred. We are not prepared for the next discovery and the discovery after that; we have no idea what they will be.[8] We will need imagination and flexibility of mind.

Principles, Protocols, and Declarations

Many people dismiss—even ridicule—the idea of preparing for contact with extraterrestrial intelligence. Despite this giggle factor, a small but active invisible college has been making modest efforts to prepare.

As these issues are unlikely to engage the sustained attention of governments in advance of contact, the task has been left to nongovernmental organizations and individuals. Two of the earliest approaches to preparing came from lawyers. Andrew Haley argued in 1956 that we need legal concepts to govern our relations with other intelligences we might encounter. In his 1965 book *Space Law and Government*, he included a chapter on "Metalaw" that suggested broad legal principles for relations among different societies that may exist in our Galaxy. Austrian lawyer Ernst Fasan further developed these concepts in a book published in 1970.[9]

Your present author published an article in 1972 that speculated about how we might conduct relations with an alien civilization. Aimed at foreign policy-makers, that article argued that we could learn much by studying relations among different civilizations on Earth and by considering the lessons of our own history. Although this idea was developed further during the next decade, neither the Department of State nor the world's other foreign ministries showed any institutional interest in the question.[10]

In 1977, the United Nations Committee on the Peaceful Uses of Outer Space held a brief discussion on the scientific search for extraterrestrial intelligence. The Chairman suggested that the COPUOS should not continue to ignore this question, but should give it at least preliminary consideration. Two members of the COPUOS listed the search among the possible future tasks of the committee, and one proposed that the task be assigned to a COPUOS Subcommittee. Although the U.N. Secretariat prepared a brief background paper on the question, no further action was taken. United Nations member states had concluded a treaty in 1967 governing the behavior of states in outer space, but that document did not address contact with extraterrestrial intelligence.

In March 1985, Professor Allan Goodman of Georgetown University began circulating drafts of a paper titled "Diplomatic Implications of Discovering Extraterrestrial Intelligence," which included a proposed international "Code of Conduct" for contact. That code had four principles: (1) Anyone who discovers evidence of extraterrestrial intelligence will publicly report the contact; (2) any response will be formulated by a process of international consultation; (3) visiting extraterrestrials will be regarded as envoys entitled to diplomatic immunity, protection, and aid in the event of accident; (4) in the event that extraterrestrials appear to pose a threat to human health or peace, no nation shall act without first consulting the U.N. Security Council. During the same year, Paul Ney published a brief article proposing an extraterrestrial contact treaty.[11]

Meanwhile, Billingham was stirring up interest in the social and cultural aspects of contact. At the International Astronautical Congress in Stockholm in October 1985, he proposed that a session at the next Congress address the question of international agreements on four points: the need to distribute the details of a discovery to all nations; the establishment of a mechanism to distribute this knowledge; how to determine if a response should be made and who should make the response; how to determine the content of the response.

Goodman and others presented their thoughts on this subject at the next Congress in Innsbruck, Austria in October 1986. This ideation continued at the October 1987 Congress in Brighton, England.

Noticing considerable overlap among these papers, I synthesized the basic elements of various proposals into a single text. As the issues associated with handling a detection appeared to be different from the issues associated with sending a communication, I then separated the relevant language into two draft agreements. When these texts were discussed at an informal meeting of 25 interested people in Brighton, it became clear that it would be much easier to get agreement on how to handle a detection than on how to prepare and send a reply.

Over the next year, I circulated a draft detection agreement to many interested people, making numerous changes in the text in response to their comments. At the 1988 International Astronautical Congress in Bangalore, India, I proposed a course of action for moving this agreement forward.

A finished text was endorsed by the International Academy of Astronautics in April 1989 and by the International Institute of Space Law and other international nongovernment organizations shortly thereafter. At the suggestion of Czech legal scholar Vladimir Kopal, the document was called Declaration of Principles on Activities Following the Detection of Extraterrestrial Intelligence.[12]

This Declaration, more commonly known as the First SETI Protocol, is an informal agreement among searchers, not among governments. Adherence is entirely voluntary; there are no enforcement provisions. The document enunciates three basic principles: verify the nature of any detection in cooperation with other searchers; after verification, publicly announce the discovery; do not send communications to the detected alien intelligence until international consultations have taken place. Other provisions spell out scientific procedures such as recording the evidence and protecting the appropriate wavelengths. Most SETI researchers have adhered to these principles.

Meanwhile, astronomer Jill Tarter and others had been working on a "Signal Detection Protocol" for the NASA SETI program that formally got under way in October 1992. This document went into much more detail about procedures. Four scenarios of detection were outlined, with rules tailored to each case: a confirmed detection of evidence indicating extraterrestrial intelligence, a signal that turns out to be an unsuspected natural phenomenon, an ambiguous signal, and a hoax.

The authors of this protocol were highly sensitive to a premature release of information, stating: "Throughout the verification and confirmation phase, it is imperative to restrict knowledge of the activities to the smallest possible number of people." In the case of a confirmed or ambiguous detection, the protocol required the approval of more senior officials before a press conference could be scheduled by NASA's public affairs office. The NASA Administrator was obligated to inform the President and the National Space Council prior to an announcement. Similarly, NASA's Office of Legislative Affairs was to notify the Chairpersons and ranking minority members of key committees and the majority and minority leaders of both houses of Congress in advance.

After the NASA program was canceled, Tarter and others prepared a similar Signal Detection Protocol for the SETI Institute's Project Phoenix. No member of the Project Phoenix team, SETI Institute management, or the cooperating scientific team was to discuss verification and confirmation activities with anyone outside of the Project. In the case of a confirmed signal, the Project Scientist (Jill Tarter) was to inform the Board of

Directors of the SETI Institute, which, in turn, was to inform the major donors to Project Phoenix. The President of the Board of Directors was then to arrange a press conference. In the meantime, the discovery team was to prepare and submit a scientific paper to a refereed journal.

Ideally, Tarter was quoted as saying, the finding would be kept confidential while radio observatories around the world were given a chance to confirm the observation. The discoverer would notify the world's observatories, using the International Astronomical Union telegram system, in order that the source could be confirmed and monitored continuously. Once the source had been confirmed and clearly identified as artificial and extraterrestrial, it would be time to hold a press conference.[13]

The presence of media people during the discovery of a candidate radio signal in 1997 raised questions about withholding information concerning a detection until it had been confirmed. Tarter recognized the intrinsic conflict between the public interest in the search and the requirement to verify and independently confirm the reality of any candidate detection.[14] What if the detection is a mistake? Media attention to a false positive could do harm to the SETI enterprise.

Some have argued that the Declaration of Principles is focused too narrowly on the detection of an electromagnetic signal and does not apply to other scenarios of contact. Stride thought that we need protocols for both the search for extraterrestrial artifacts (SETA) and the search for extraterrestrial visitation (SETV). These documents would include strict rules for verification, confirmation, even syntax for communication.[15]

As of early 2006, the International Academy of Astronautics was considering a broadened and simplified text of the Declaration of Principles, extending the document's coverage to include other evidence of extraterrestrial intelligence such as the detection of alien artifacts in our solar system. No text had been endorsed by the Academy as of this writing. I published an early version of this revised document in 2005, but that may not precisely reflect the Academy's ultimate position.[16]

Even as revised, this document does not address visits by inhabited spacecraft, a scenario that most people involved in SETI consider highly unlikely. There also is the credibility issue that arises whenever this scenario is proposed, particularly because of its association with UFOs.

Others believe that we should develop a protocol to address the Visit Scenario. Matloff, Schenkel, and Marchan argued that such a visit would be, by far, the most complex and potentially significant form of contact; it might involve the penetration into Earth's atmosphere, even a landing on the Earth's surface, and possibly direct encounters with extraterrestrials. They proposed guiding principles for our own behavior, including offers of assistance if required, proposals for a meeting place, coordination of communication with the visitors, and precautionary measures to prevent undesirable reactions by humans. If the aliens showed hostile intent, the U.N. Security Council would decide on defensive measures. Schenkel already had proposed new declarations of principles, one of which addressed the landing scenario.[17]

Margaret Race of the SETI Institute found that these efforts, modest though they may be, are unusually forward-looking. Within the context of different types of searches for extraterrestrial life, the SETI community was alone in having conducted serious international discussions of how to respond to a detection. Although the Declaration of Principles does not address every issue, it is the only organized attempt to codify guidelines and policies. By contrast, there is no clear guidance on what to do if and when nonintelligent extraterrestrial life is found.[18]

Rating the Impact

False alarms about possible impacts of asteroids on the Earth revealed that no guidelines existed for who should have been informed when, or what emergency measures should have been taken, such as issuing a public warning. This led to the adoption in 1999 of what is known as the Torino Scale for impact risk, based on the probability of impact and the size of the impactor.

Some initially worrisome predictions about close encounters with asteroids raised questions about putting out undigested information. An observer participating in the Lincoln Near-Earth Asteroid Research program posted possible orbits for minor planets without noticing that one of them might have a one in four chance of hitting the Earth within days. A flurry of Internet communications led to further observations showing that this body would miss us. Concerned about future premature announcements of astronomical phenomena, the Central Bureau for Astronomical Telegrams may place more emphasis on confirmation by independent observers.[19] That is likely to mean a delay.

Similar approaches have been suggested for contact with extraterrestrial intelligence. Donald Tarter, assuming the remote detection scenario, suggested that a suspected extraterrestrial signal be examined for five characteristics: certainty of extraterrestrial origin; clarity of signal content; sensory accessibility; activity level demanded of the recipients; intentionality.[20]

In 2000, Ivan Almar and Jill Tarter proposed a scale for classifying the impact of discovering extraterrestrial intelligence. Factors include the class of phenomenon, ranging from traces of astroengineering to an Earth-specific message; the type of discovery—the result of SETI or some other kind of observation, or a re-evaluation of archival data; distance, from extra-galactic to within the solar system (notably, this scheme included the discovery of alien artifacts). The assessment of significance would weigh the level of probable consequences with the credibility of the claim, which can only be estimated subjectively. As this initiative was launched at the International Astronautical Congress in Rio de Janeiro, the method became known as the Rio scale.[21]

Releasing the News

The First SETI Protocol was designed to build a consensus favoring open sharing of information after a detection is confirmed. However, that protocol is a nongovernmental document without the force of law or regulation. Government and private organizations can ignore it if they wish.

David Morrison, a senior scientist with NASA's Astrobiology Institute, noted that many people distrust the way governments and scientists handle information about impacts.[22] That problem may be even more acute when the issue is the detection of extraterrestrial intelligence. The best policy, as Morrison suggested for impacts, is openness together with an effort to educate the media.

This suggests a need for a policy decision requiring public release of information about a detection made by a public sector organization or by a nongovernmental organization working under a government contract. That, in turn, implies a need to get relevant government agencies to adopt something like the main principles of the First SETI Protocol. In the case of the United States, obvious candidates are NASA and the National Science Foundation, which support space-oriented research. Hopefully, efforts to assure an open information policy would be extended to the international level.

Governmental organizations are unlikely to give this question high priority in advance of a detection. Getting defense and intelligence agencies to adopt such principles may be even more problematic; the whole effort may depend on a few enlightened officials. In the meantime, it may be useful to encourage such agencies to inform the civilian scientific community of unexplained phenomena that pose no evident threat to national security.

Several authors have examined the question of how best to communicate with the public if extraterrestrial intelligence is detected. The social implications group recommended steps that might be taken in advance of detection with educational institutions, news media, and entertainment media.[23] As institutional memories are short, those steps would have to be repeated at frequent intervals.

Sending Communications

It is when we respond *to such signals that we assume any risks that may exist. Before we make such a response or decide to radiate a long-range beacon, we feel the question of the potential risks should be debated and resolved at a national or international level.*

—Project Cyclops, 1972[24]

With the first SETI Protocol on its way to public release, some people began addressing the question of a response to an extraterrestrial signal. In 1989, your present author suggested basic principles for an international agreement on communicating with extraterrestrial intelligence. In a joint paper presented a year later, Billingham, Jill Tarter, and I proposed that a collective message be sent on behalf of all Humankind. "The time to begin our studies on a reply from Earth," we wrote, "is now." Meanwhile, Goldsmith had proposed that the International Astronautical Federation and the International Astronomical Union create a joint committee that would encourage consultation on a worldwide reply.[25]

Subsequent discussions within the International Academy of Astronautics produced a Position Paper proposing a decision process for sending communications to other civilizations. Under this plan, the United Nations Committee on the Peaceful Uses of Outer Space (COPUOS) would consider a draft declaration of principles that, if approved by member governments, would become a statement of international policy. The draft Declaration (an attachment to the Position Paper) is known informally as the Second SETI Protocol.[26]

In 1996, IAA President George Mueller (a former senior official of NASA) sent the Position Paper and the draft Declaration of Principles to the foreign ministers of nations that are members of COPUOS, suggesting that they consider placing this issue on the committee's agenda. Australia responded that it would support this idea if another nation took the lead, but none did.

Jill Tarter, then Chairperson of the IAA SETI Committee, presented a briefing on these documents to COPUOS in 2000. The U.N. General Assembly formally noted the Academy's presentation, but took no action beyond filing the documents for future reference.

The authors of this proposal had no illusions about getting a formal agreement among nations in advance of a detection. They did hope to set up a process that would require governments to address the issue.

What would stir the United Nations to action, other than a confirmed detection? Related discoveries such as fossils on Mars or an Earth-like planet near another star might provoke a discussion, possibly leading to consideration of something like the proposed Declaration of Principles on communicating with extraterrestrial civilizations. Although the United

Nations is unlikely to take policy action in advance of contact, its members could call for a study of that event's implications.

To U.N. or Not U.N.

We must recognize the limitations of pursuing this idea through the United Nations. Although the U.N. is the most comprehensively inclusive intergovernmental body, that organization's operating style has disadvantages. The United Nations, not known for its efficiency, may be slow to react to contact.

Doyle, assuming the remote contact scenario, described the probable U.N. process. That organization eventually would convene a meeting of the General Assembly to hear general statements from national representatives. The issue would be referred to COPUOS and possibly to other organizations such as the International Council of Scientific Unions. These bodies would conduct assessment studies that could take years. In Doyle's view, there would be no reason to hurry.[27]

COPUOS is in session for only a few weeks a year. As the committee operates on the basis of consensus, a single nation could block the approval of a proposed policy statement.

If the U.N. process proves too slow and unwieldy, some governments may turn to other options. There may be pressure for a quicker response, particularly if we find alien technology in our own solar system. This has led to proposals that the matter be referred to the U.N. Security Council, which is always in session. Referring contact with extraterrestrials to that body makes some people uncomfortable because it implies that they are a threat to human security.

What if there is no agreed international procedure when contact occurs? Nations with the necessary technical capabilities might act preemptively in an uncoordinated way—for example, by sending separate messages to the detected civilization. One or more governments could head this off by quickly proposing a coordinated set of actions, within or outside of the U.N. system.

Preparing Governments for Contact

Policy-makers should not dismiss the search for extraterrestrial intelligence as just an exotic scientific enterprise. Contact will raise policy questions; governments will be forced to respond. Without preparation, they may respond in ways that are not in the best interest of their nations or of Humankind.

Government and political leaders respond best when the event is of a familiar type: an economic crisis, a natural disaster, even a war. They may have had personal experience in reacting to such events; they may remember what has happened previously. Officials and politicians are least well prepared for events that are unfamiliar and that seem unlikely to occur.

Contact with extraterrestrial intelligence is an excellent example. Because there is as yet no confirmed evidence, most government and political leaders do not regard the issue as one deserving their attention. The few who think about it may fear that they will be ridiculed if they discuss preparations for contact.

If government officials or politicians are informed that an extraterrestrial civilization has been detected, the first thing they will want to know is whether the detection has been confirmed. No government or political leader will want to risk the embarrassment of making public statements about a detection that turns out to be false. In some cases, this confirmation can be done within a day; in others, the nature of the detection might be ambiguous and confirmation might take longer. Officials and politicians will be wise to say "let's wait until we're sure."

The next policy question is: What should the government say to the public, and when? Reporters and others will ask officials not only what they know but also what they plan to do. Many media people think that the credibility of government announcements on this subject would be low, implying a need for careful preparation.

Who should speak for the government? Who is in charge of the issue? Should the executive branch try to maintain exclusive control? How soon should they consult with lawmakers and political party leaders? Legislators might take action on their own, passing laws that direct the way that contact should be handled.

Should foreign governments be fully informed from the beginning? Should they be invited to join postcontact processes?

The longer-term policy issues concerning contact with ETI will depend on the scenario of detection. In a remote detection, one issue will be whether to send a communication to the detected civilization.

To many people, the answer seems obvious. Of course we should! Some attempts already have been made. However, questions remain. Do the extraterrestrials know that we exist? If they do not know of our existence, should we call attention to ourselves? Is there any possible danger in advertising our location? If we do decide to send a message, should Humankind speak with one voice or with many? What should the message say? Is there some information that we should withhold? Who should decide?

Governments have not yet addressed these questions. They may have to.

The second long-term issue is whether releasing information from extraterrestrials would threaten human cultures. The type and degree of impact will depend on the nature and comprehensibility of the signal or artifact. A message that is rich in information will have a far greater impact than a signal that is little more than a carrier wave or radar pulse—if we can understand it.

The direct contact scenario would raise other questions. Do we need to take defensive measures? If so, who is in charge?

In 1994, the social implications group proposed a number of policy initiatives to prepare governments and international organizations in advance of contact. These focused on educating officials most likely to be involved in responding to contact, and on setting up procedures within their organizations to make Humankind's responses more effective.[28]

At the national level, searchers and their allies could brief government and political leaders more pointedly, informing officials and politicians of the possible cultural, social, economic, and political implications of contact. Such potential consequences are more likely to get the attention of public officials than purely scientific findings. (In this context, it would be useful to periodically remind national security agencies that they might unintentionally detect evidence of extraterrestrial intelligence.) Briefings could be extended to the international level, including both foreign governments and international organizations. Officials should be made aware of the full range of possible consequences, not just those we prefer.

Harrison proposed his own set of steps, with the United States his apparent focus: acknowledge the possibility of contact; develop a workable organizational or committee structure; develop different contact scenarios, then rehearse appropriate responses; improve credibility by resolving jurisdictional issues; be realistic about how quickly a coherent and credible report could be issued; resist groupthink. Harrison recognized that these efforts would be hampered by public perceptions of low credibility and competing organizational priorities.[29]

Who Speaks for the Earth?

... a small group of desperate men, who to gratify insatiable ambitions had allied themselves with the thing in the sky, men who were guilty of treason against the entire human species.

—Fred Hoyle, *The Black Cloud*[30]

Many advocates of the radio detection scenario have foreseen no urgent need to respond to an alien signal. According to Morrison, Sagan, and others, we will have time to analyze the incoming message and to frame an appropriate reply; radio astronomers generally have seen no technical reason to rush. Vakoch questioned that assumption, arguing that the rules of the game might change once we know that intelligent life is out there.[31]

If we extend our thinking beyond the astronomical community, we find that the desire to send immediate communications to detected aliens is nearly overwhelming. People with access to transmitters might fire a barrage of signals into space. Whether their messages would reach their destination would depend on the power of the transmitters, on the distance of the other civilization, and on the sensitivity of its receivers. A remote civilization may be far beyond the range of most radio operators.

The real issue may be whether superior transmitters such as large radio telescopes would be used to send messages. Nearly all of those facilities are funded by governments. Depending on the nature of the contact, policy-makers might have an opportunity to make conscious decisions about sending powerful signals from Earth.

Should Humankind speak with one voice, or many? Most people involved in this debate have supported a collective message, believing that the human end of the communications process should speak for all Humankind, not a particular political or occupational subdivision. Drake, for one, believed that any reply should be crafted on a worldwide basis.[32]

Others argue that anyone with access to a transmitter should have the right to send separate messages. Some see this as freedom of speech; others believe that many individual messages would more correctly reflect human diversity.

A Mirror Image. Having Humankind speak with many voices may be congruent with individual rights and cultural diversity, but may be bad policy. Imagine yourself in the place of extraterrestrials who receive thousands of uncoordinated messages from Earth. How could you conduct a rational dialogue with such mixed signals? Which ones matter the most? Which most accurately reflect human policy? Who would you believe, those humans who seek an exchange of scientific information, those who want to convert you to the true faith, or those who announce their intent to exterminate you?

Donald Tarter proposed that we send a brief initial reply that says "we have received your message and will communicate with you in the near future." The intent would be to establish an "official" channel of communication. If this signal were sent with the most powerful transmitter available, the power and priority of this transmission would differentiate it from others.[33]

Communication is not the only barrier to non-zero-sum interactions between civilizations; the other barrier is trust.[34] Using a preferred channel could help to establish greater mutual confidence.

In the absence of an agreed process, our response is likely to be ad hoc. Nations with the needed technical capabilities might act in an uncoordinated way, sending different messages.

If Humankind chooses to send a collective response, the best way to assure acceptance is to make the process as inclusive as possible, both among nations and within them. That process may be laborious and slow; impatient organizations or groups might short-circuit the procedure.

Active SETI

Some of those involved in the interstellar communication debate disagree with the requirement for international consultations before we send signals intended to attract the attention of an alien civilization.

As most of us recognize, there is not much point in expecting consultations about the radio, television, and radar signals that our civilization emits every day. The focal point of the debate is the exceptionally powerful signals like the one sent from Arecibo in 1974, signals that dramatically increase our detectability. One suggested approach is to set a quantitative threshold for such consultations—for example, by requiring them for signals that exceed the power and duration of normal pulses from military or planetary radars.

Almar proposed what he called the San Marino Scale, intended to quantify the potential hazard of transmitting messages into space. The main factors are the signal strength in relation to Earth's normal background radiation and the characteristics of the transmission such as information content, direction, and duration. He included the most subjective factor of all: intentions.[35]

What Should We Say?

What would be our purpose in sending a communication to an alien civilization: to exchange scientific information, to describe ourselves, to propagate our values and cultures, to propose some course of action, or all of these? What does the human species want or hope for from contact? What do we wish to convey? We would need a clear vision of our own purposes.

Physical scientists have tended to focus on sending and receiving scientific information. Nonscientists have other priorities. They want to describe human history, cultures, religions, values, and ways of organizing societies, as well as policy issues that we humans currently face. They want to ask the other civilization about the same things.

As our message will be received by a society very unlike ours, it may be interpreted in ways quite different from what we intend. Rather than simply accepting our transmission as a vehicle for scientific information, the receiving party may exhaustively analyze the message from many points of view. What does the message (and the medium through which it was sent) imply about the knowledge and capabilities of the senders? Why did they send it? What are their intentions toward us?

Many civilizations may be surprised by contact. Extraterrestrials, like us, might have both hopes and fears about this event.

There is no perfect solution to this problem, as the knowledge, culture, and politics of the receiving society will be unknown to us. We might try to reduce the risks involved by drawing on our most universal, least culture-bound findings about the psychology and behavior of intelligent beings. Perhaps there are universal principles recognized by all intelligent species from their experiences with their own evolution, with their environments, and with each other.

Several authors have proposed approaches to drafting interstellar radio messages. Vakoch suggested three: first, construct a message that avoids areas of disagreement; second, allow multiple compositions, with each reply highlighting a different perspective; third, draft a coherent, unified message that includes perspectives that may be irreconcilable with each other (this seems likely to produce a very long message!) Vakoch cited linguist Kirsten Refsing's view that we should send out as many different kinds of message as we are able to compose, in the hope that at least one of them would be understood.[36]

Your present author and others have proposed that we try to draft an outgoing message in advance of a detection, as a way of focusing our thinking. That message would be reviewed by an international body that would decide whether to send it. This exercise would have implications reaching beyond the immediate issue of message content.[37]

Vakoch thought that public support for SETI could be promoted through a widespread discussion of the contents of a reply message. He gave us several other reasons for drafting a message in advance: first, it is only when we begin composing a reply message that we will begin to understand that range of issues involved in deciding the content; second, this effort may facilitate the initial decryption of messages from extraterrestrials; third, we can begin to see interstellar compositions as a form of art; fourth, it may be advantageous to have a standard reply message in place; fifth, drafting such a message would provide a source of intellectual stimulation and a concrete sense of accomplishment to SETI researchers; sixth, this effort would prepare us for an active search strategy, in which we transmit messages to attract the attention of others.[38]

At the same time, we must recognize that no stock reply could match more than one of the infinite possibilities for the alien message's content. What we say in response depends on the nature of the alien signal and what it is telling us. It depends on distance, whether we can understand the message, and on the world's reaction.[39]

The practice involved in drafting detailed messages would leave us better prepared for contact. It also would offer opportunities for insight into ourselves; by explaining to another species what it means to be human, we may gain a clearer understanding. Building a global consensus on how to represent Humankind would have a significance reaching far beyond the immediate issues of contact.[40]

The drafting of a message to extraterrestrials would be heavily influenced by the fact that it also is a message to our own civilization. Many humans would be highly sensitive to its content. Should we withhold some information because it advertises our vulnerabilities? If some humans transmit such information, would that be treason?

The first message would be crucial. It could set the tone of the entire communications exchange, even determine if there will be an exchange. It could initiate a highly productive relationship, or it might generate suspicion and hostility where none was intended.

There could be more subtle reactions. A message wholly made up of scientific and mathematical information could convey a one-dimensional impression of our culture as coldly technological, without other values. An overly simple message might even insult the intelligence of more sapient beings.

How Should We Say It?

In science fiction, intelligent aliens speak Earth languages or become easily intelligible when their statements pass through translating machines. Real contact is not likely to be that convenient. One might ask how many human languages extraterrestrials would know and use: one, such as English; several, such as the official languages of the United Nations; or all of the hundreds that exist on our planet?

Several people have proposed languages for interstellar messages, nearly all of them resting on basic concepts in science and mathematics. The most general approach is to present the basics of a communications language at the beginning of a message and develop it further as the transmission proceeds. Although often clever, these languages still spring from our own cultural assumptions.

More than a century ago, statistician and meteorologist Francis Galton suggested a language using dots, dashes, and lines. H.W. Nieman and C. Wells Nieman proposed a mathematical approach to building up a common language in their 1920 *Scientific American* article "What Shall We Say to Mars?" Mathematician Lancelot Hogben outlined an "Astroglossa" in 1952, basing his proposed communication language on the number concept and knowledge of celestial events; we would begin with symbols for mass, temperature, and distance, working toward more complex concepts.[41]

The best known example may be mathematician Hans Freudenthal's 1960 proposal for a mathematics-based Lingua Cosmica, using radio signals of different lengths and wavelengths. A year later, mathematician Solomon Golomb challenged earlier proposals for cosmic languages because they all are based on terrestrial logic. He recommended prime number sequences or arithmetic progressions.[42]

By the time of the 1974 Arecibo message, thinking had migrated toward sending pictures that could be converted into two-dimensional images—if the recipient discerned the principle. The trouble with pictures, computer scientist Michael Arbib wrote 5 years later, is that they are too literal to communicate general truths. He thought that our communications should contain a combination of symbolic messages and pictures, although no single message or small set of messages can reliably convey the meaning of a given set of symbols.[43]

Carl DeVito revived the idea of languages based on science in 1992. More recently, Canadians Yvan Dutil and Stephane Dumas, drawing on Freudenthal's work, developed what they called an "Interstellar Rosetta Stone," a set of symbols that can be thought of as a mathematical language.[44]

Vakoch proposed that in addition to pictorial representations and three-dimensional images, we should draw on semiotics (the theory of signs). He suggested transmissions that simulate natural phenomena as icons. Given that we are not sure which sensory modality will be primary for extraterrestrials, symbols that are not reliant on any particular modality would be preferable.[45]

Given our ignorance of the other civilization, there is no perfect method of communicating meaning, particularly intent. Misunderstandings may be inevitable.

Managing the Relationship

If we do begin exchanging communications with an extraterrestrial civilization, we will be entering into a long-term relationship. We will need to think beyond our immediate reaction or our first message.

If that civilization is hundreds or thousands of light-years away, we will face long delays between an outgoing message and a reply. Human cultures and human politics could change in ways that affect the dialogue or even bring its continuation into question.

Should Earth's nations communicate with extraterrestrials through an international institution? Does the United Nations provide an adequate framework, or would we need a new body? This is not a minor bureaucratic issue. The communicating organization would become *Homo sapiens'* foreign ministry, where human interests are aggregated and expressed to nonhumans.

Long-term relations require a strategy and a continuous calculation of Humankind's collective interests. Building global consensus on the human response may be the ultimate test of coalition-building. Achieving it could greatly extend the concept of enlightened self-interest.

Harrison proposed that a civilization belonging to the Galactic Club would know how to develop working relationships, guarantee mutual security, and explore the possibilities of trade and commerce.[46] This is the essence of diplomacy, a practice with which we have some experience.

A dialogue with another civilization would be a project for many generations in succession, demanding a continuity of purpose that human societies rarely attain. Swenson speculated that it may require an enduring organization based on immutable dogma, like one of the world's major religions.[47]

We will need a clear idea of what we want. We also will need a clear understanding of what we are prepared to give in return.

Our foundation must be the enlightened self-interest of the human species. We may be able to move beyond that to find common ground, to identify actions that could be seen as positive by both civilizations.

Toward a New Legal Order

If two intelligent races—both of which have special rules of behavior— come into contact with each other, the basic understanding of such mutual rules will lead to a kind of code of conduct.

—Ernst Fasan, 1998[48]

As far back as Immanuel Kant, some have speculated about a legal system that would apply to all intelligences in the universe. What would be its basic principles?

Half a century ago, space lawyer Andrew Haley proposed what he called The Great Rule of Metalaw: Do unto others as they would have you do unto them.[49] It is not clear how we could observe this principle in the absence of extensive knowledge about the other civilization. We may need detailed, sophisticated communication to find out.

Fasan expanded on Haley's concepts, proposing basic metalaws applicable to all sentient beings:

1. No partner of Metalaw may demand an impossibility.
2. No rule of Metalaw must be complied with when compliance would result in the suicide of the obligated species.
3. All intelligent species have in principle equal rights and values.
4. Every partner of Metalaw has the right to self-determination.
5. Any act which causes harm to another species must be avoided.
6. Every species is entitled to its own living space.
7. Every species has the right to defend itself against any harmful act performed by another species.
8. The principle of preserving one's species has priority over the development of another species.
9. In case of damage, the damager must restore the integrity of the damaged party.
10. Metalegal agreements must be kept.
11. To help the other species by one's own activities is an ethical principle, not a legal one.[50]

Interpretation of such high-minded principles could be problematic. Principles 7 and 8, for example, could be seen as authorizing preemptive attacks on incoming alien spacecraft. Fasan explicitly stated that the right of self-defense is a legal consequence of the right of freedom of will. What one species sees as self-defense may be seen by another as aggression.

Harrison and Dick suggested that if we encounter civilizations that already have formed a Galactic Club, they may offer us an interstellar legal framework. Would we accept it? Establishing a metalegal order might require significant changes in our own laws, which, as Freitas noted, do not include nonhumans as legal persons.[51]

Planetary Defense

If we continue to survive and if our technological civilization continues to advance, we will become progressively more capable of defending ourselves against outsiders.

—Isaac Asimov, 1979[52]

The threat of asteroid and comet impacts has provoked some humans to think in terms of planetary defense. Telescopes already scan the skies for Near-Earth Objects, some of which might intersect our planet's path. We are extending our concept of security beyond individual nations to the entire Earth.

Scientists and engineers have proposed technological means for dealing with any bodies that look threatening, ranging from propulsion systems that would deflect an asteroid or comet nucleus from a collision course to nuclear weapons that would demolish the threatening body. The European Space Agency's Don Quijote mission, planned for 2011, would practice steering asteroids away from Earth. Maccone has even proposed stationing weapons in space that would deflect incoming asteroids; this might violate the Outer Space Treaty and arms control agreements.[53]

Current search efforts focus on large asteroids, although smaller ones could do serious damage to our civilization. A NASA committee concluded in 2003 that the search for smaller objects could be done for less than $400 million, but no funding agency has volunteered to pay. Asteroid hunting does not fulfill an obvious scientific mission, observed Grant Stokes of the MIT Lincoln Laboratory; it is more like a public service.[54]

As a Planetary Society fund-raising letter pointed out, there is still no concrete plan in place for humanity's response if we discover an asteroid heading our way. Some analyses indicate that we can divert Earth-crossing asteroids and comets only if we reach them years before their projected impact. For an asteroid 200 meters in diameter, we would need roughly 20 years; for a larger asteroid, the lead time would be longer.[55]

To starflight theorist Gregory Matloff, that meant building an infrastructure in the outer solar system—at a minimum, the lookout posts that would watch for incoming bodies.[56] That capability also could give us the means for spotting other potentially dangerous intruders—the probes or inhabited vehicles of another civilization. Earth security would be extended to solar system security.

Planetary defense can be seen as a rehearsal for direct contact. It provides one model of preparing ourselves to deal with the exploring machines of a more advanced technology. Whether we could defend ourselves would depend on the relative capabilities of the two civilizations. Whether we would need to would depend on the intentions of the more powerful one.

SETI conventional wisdom assumes that because we will be much less technologically advanced than any other civilization that we contact, we would be helpless if the extraterrestrials were hostile. This disparity may turn out to be true, but it remains unproven. To assume our weakness in advance would be preemptive capitulation.

Educating Ourselves for Contact

The social implications group proposed ways of informing educational institutions, the media, and the public about the search, although all were based on the remote contact scenario. These steps included listing "reliable" SETI books for librarians, informing professional organizations of teachers, and establishing liaison with the entertainment industry.[57]

The search for extraterrestrial intelligence, with its unique ability to provoke interest in science, already is used in school curricula. Materials prepared by the SETI Institute for primary and secondary schools (initially based on the Drake equation) have been supported by federal grants.

Schenkel recommended that all colleges and universities include courses on the possibilities and potential benefits of encounters with other civilizations.[58] He did not mention potential risks. A more balanced educational effort would have to address both.

About the Author

Michael Michaud has been a leading figure in preparations for possible future contact with extraterrestrial intelligence. As chairperson of International Academy of Astronautics working groups that considered this subject, he coordinated the drafting of the Declaration of Principles Concerning Activities Following the Detection of Extraterrestrial Intelligence, also known as the First SETI Protocol. Michaud also is a member of the International Institute of Space Law, the American Institute of Aeronautics and Astronautics, and the American Astronautical Society, and is a Fellow of the British Interplanetary Society. He has published more than thirty articles and papers on the implications of contact, as well as more than sixty articles on other subjects. He is the author of a previous book entitled Reaching for the High Frontier: The American Pro-Space Movement, 1972–1984.

During his career with the U.S. Department of State, Michaud served as Director of the Office of Advanced Technology and as Counselor for Science, Technology, and Environment at the American embassies in Paris and Tokyo. He led U.S. delegations in the successful negotiation of international science and technology agreements. He played an active role in reviving U.S.-Soviet space cooperation and in initiating U.S.-Soviet talks on outer space arms control. He represented the State Department in interagency space policy discussions, and testified before the U.S. Congress four times on space-related issues. He presently lives in Europe.

References

JBIS = Journal of the British Interplanetary Society
QJRAS = Quarterly Journal of the Royal Astronomical Society

Introduction

1. Arthur C. Clarke, The Exploration of Space, New York, Harper, 1951, 194.
2. Quoted in Roger D. Launius and Howard E. McCurdy, Imagining Space: Achievements, Predictions, Possibilities, 1950–2050, San Francisco, Chronicle Books, 2001, 39. Steven Dick described the concept of extraterrestrial life as the biological dimension to the debate over the status of humanity in the universe. "The Search for Extraterrestrial Intelligence and the NASA High Resolution Microwave Survey: Historical Perspectives," Space Science Reviews, Vol. 64 (1993), 93–139.
3. Stephen Webb, Where is Everybody? Fifty Solutions to the Fermi Paradox and the Problem of Extraterrestrial Life, New York, Copernicus, 2002, 4.
4. Quoted by David W. Swift in SETI Pioneers: Scientists Talk About Their Search for Extraterrestrial Intelligence, Tucson, University of Arizona Press, 1990, 219.
5. David Lamb, The Search for Extraterrestrial Intelligence: A Philosophical Inquiry, London, Routledge, 2001, 196.
6. Karl Guthke, The Last Frontier: Imagining Other Worlds from the Copernican Revolution to Modern Science Fiction, translated by Helen Atkins, Ithaca, NY, Cornell University Press, 1990, originally published in German as Der Mythos der Neuzeit, Bern, A. Francke AG Verlag, 1983; Michael J. Crowe, The Extraterrestrial Life Debate 1750–1900, Cambridge, Cambridge University Press, 1986, republished by Dover in 1999; Steven J. Dick, The Biological Universe: The Twentieth-Century Extraterrestrial Life Debate and the Limits of Science, Cambridge, Cambridge University Press, 1996.
7. Arthur C. Clarke, Greetings, Carbon Based Bipeds!: Collected Essays, 1934–1998, New York, St. Martin's Press, 1999, 229. The concept of disrupting the mundane is from science fiction author Bruce Sterling, "A Century of Science Fiction," Time, 29 March 1999.
8. Quoted in Analog Science Fiction/Science Fact, December 1985, 121.
9. Crowe, 548.

10. Edwin Hubble, The Realm of the Nebulae, New Haven, CT, Yale University Press, 1936 and 1982, 201–202.
11. David Grinspoon, Lonely Planets: The Natural Philosophy of Alien Life, New York, Ecco (Harper Collins), 2003, 294.
12. Iosif Shklovskii and Carl Sagan, Intelligent Life in the Universe, New York, Dell, 1966, 22. Originally published by Holden-Day, 1964.
13. S.A. Kaplan, editor, Extraterrestrial Civilizations: Problems of Interstellar Communication, Washington, DC, NASA, 1971, translated from Russian by the Israel Program for Scientific Translations, 1.
14. John Billingham, et al., editors, Social Implications of the Detection of an Extraterrestrial Civilization, Mountain View, CA, SETI Press, 1994, 53.
15. Glen David Brin, "The Great Silence: The Controversy Concerning Extraterrestrial Intelligent Life," QJRAS, Vol. 24 (1983), 283–309; Carl Sagan and William I. Newman, "The Solipsist Approach to Extraterrestrial Intelligence," in Edward Regis, Jr., editor, Extraterrestrials: Science and Alien Intelligence, Cambridge, Cambridge University Press, 1985, 151–161.
16. Christof Koch, "Thinking about the Conscious Mind" (review of Mind, by John R. Searle), Science, Vol. 306 (5 November 2004), 979–980; Jung is quoted in John Casti, Paradigms Lost: Tackling the Unanswered Mysteries of Modern Science, New York, Avon, 1989, 411.
17. Giuseppe Cocconi and Philip Morrison, "Searching for Interstellar Communications," Nature, Vol. 184 (1959), 844–846, republished in Donald Goldsmith, editor, The Quest for Extraterrestrial Life, University Science Books, 1980, 102–104; Frank D. Drake, "Project Ozma," Physics Today, Vol. 14 (1961), 40. The Drake equation is described in Frank Drake and Dava Sobel, Is Anyone Out There?, London, Souvenir, 1993, 52; first published by Delacorte in 1992.
18. Philip Morrison, "Viking, Life, and a Search for Signals from the Stars," Astronautics and Aeronautics, July–August 1976, 65–67.
19. Quoted by Eliot A. Cohen in "History and the Hyperpower," Foreign Affairs, Vol. 83, Number 4 (July/August 2004), 49–63.

A Belief in Other Minds

1. Stanislav Lem, Summa Technologiae, quoted in Billingham, et al., editors, Social Implications, 55.
2. Edward Harrison, Masks of the Universe, New York, Macmillan, 1985, 3; Dick, the Biological Universe, 12.
3. Guthke, 36–37.
4. Crowe, 3; Walter Sullivan, We Are Not Alone (revised edition), New York, Plume, 1994, 2–4.
5. Sullivan, 4.
6. Albert A. Harrison, After Contact: The Human Response to Extraterrestrial Life, New York, Plenum, 1997, 173; Angus Armitage, Copernicus and Modern Astronomy, Mineola, NY, Dover, 2004, 40; originally published as Copernicus: The Founder of Modern Astronomy, New York, Yoseloff, 1957.

7. Fred Hoyle and Chandra Wickramasinghe, Our Place in the Cosmos, London, Dent, 1993, 182; Roland Puccetti, Persons: A Study of Possible Moral Agents in the Universe, London, Macmillan, 1968, 121.

8. Frank J. Tipler, "A Brief History of the Extraterrestrial Intelligence Concept," QJRAS, Vol. 22 (1981), 133–145; Sullivan, 6; Grinspoon, Lonely Planets, 8.

9. Jennifer Birriel, "Struggling to Know the Universe," (review of Measuring the Cosmos), Sky and Telescope, February 2005, 110–112.

10. Charles Freeman, The Closing of the Western Mind: The Rise of Faith and the Fall of Reason, New York, Knopf, 2003, xviii, 311. Lucio Rosso described how Hellenistic science was lost to the Western world for more than a thousand years before it was rediscovered. The Forgotten Revolution: How Science Was Born in 300 B.C. and Why It Had To Be Reborn (translated by Silvio Levy), New York, Springer, 2004.

11. J.M. Roberts, The Triumph of the West: The Origin, Rise, and Legacy of Western Civilization, London, Phoenix Press, 2001, 71 (first published by the British Broadcasting Corporation in 1985); Edward Harrison, 4.

12. Lewis White Beck, "Extraterrestrial Intelligent Life," in Regis, editor, 3–15; Guthke, 41.

13. Guthke, 38–39; Grinspoon, 15–16; Dick, 39.

14. Quoted in Ronald Bracewell, The Galactic Club: Intelligent Life in Outer Space, San Francisco, Freeman, 1974, 127.

15. Roger D. Launius and Howard E. McCurdy, Imagining Space, San Francisco, Chronicle, 2001, 38.

16. Guthke, 47; J.H. Parry, The Age of Reconnaissance: Discovery, Exploration, and Settlement, 1450–1650, London, Phoenix, 1963, 2. Copernicus wrote of heliocentrism in an unpublished work in 1514.

17. Carl Sagan, Cosmos, New York, Random House, 1980, 146; Armitage, 89–90.

18. Arthur O. Lovejoy, The Great Chain of Being: A Study of the History of an Idea, Cambridge, MA, Harvard University Press, 1936, 102–103, reprinted by Harper Torchbooks in 1965.

19. Lovejoy may have been the first to use this term, in The Great Chain of Being. Also see Guthke, 39, 41 and Dick, 12–13.

20. Lovejoy, 116; Guthke, 67–76, 86; Michael White, The Pope and the Heretic, New York, Perennial, 2003, 71–72; Joel Achenbach, Captured by Aliens: The Search for Life and Truth in a Very Large Universe, New York, Simon and Schuster, 1999, 45; John C. Baird, The Inner Limits of Outer Space, Hanover, NH, University Press of New England, 1987, 119; Armitage, 171–172; Paolo Musso, "Philosophical and Religious Implications of Extraterrestrial Intelligent Life," paper presented at the 2004 International Astronautical Congress (IAC-04-IAA-.1.1.2.11). A selection from Bruno's work is included in Goldsmith, editor, 5–7.

21. Anthony Aveni, Conversing with the Planets: How Science and Myth Invented the Cosmos, Boulder, CO, University of Colorado Press, 2002, 179, 191.

22. Dick 17; Guthke, 96–97.

23. Guthke, 50, 110–111.

24. Guthke, 80, 95, 105.

25. Guthke, 156.

26. Guthke, 149–50; Sullivan, 18; John D. Barrow and Frank J. Tipler, The Anthropic Cosmological Principle, Oxford, Oxford University Press, 1986, 576.
27. Guthke, 122.
28. For other thoughts on the importance of the Moon, see Isaac Asimov, Extraterrestrial Civilizations, New York, Crown, 1979, 21–24.
29. Guthke, 186; Dick 17–18.
30. Guthke, 187, 197.
31. Guthke, 227–239; Grinspoon, 25.
32. Guthke, 239–243. A sample of Huygens' thinking is in Goldsmith, editor, 10–16.
33. Guthke, 57, 245.
34. Edward Harrison, 101.
35. William H. McNeill, The Rise of the West: A History of the Human Community, New York, Mentor, 1963, 744.
36. Guthke, 213.
37. Guthke, 220–221; Billingham, et al., editors, Social Implications, 54.
38. Guthke, 35, 210.
39. Guthke 212; Ernan McMullin, "Life and Intelligence Far from Earth," in Steven J. Dick, editor, Many Worlds, Philadelphia, Templeton, 2000, 151–173.
40. Crowe, 161, 167.
41. Dick, The Biological Universe, 23–26; Guthke, 331–335.
42. Crowe, 315.
43. McNeill, 829.
44. Antonio Lazcano, "Just How Pregnant is the Universe?" (review of de Duve's Life Evolving), Science, Vol. 299 (17 January 2003), 347–348; Steven J. Dick, "The Search for Extraterrestrial Intelligence and the NASA High Resolution Microwave Survey: Historical Perspectives," Space Science Reviews, Vol. 64 (1993), 93–139.
45. Crowe, 360; Alan W. Hirshfeld, "Starlight Detectives: The Birth of Celestial Spectroscopy," Sky and Telescope, August 2004, 44–49; Guthke, 349.
46. Asimov, Extraterrestrial Civilizations, 2, 5.
47. Percival Lowell, Mars as the Abode of Life, New York, Macmillan, 1908, 188. A facsimile edition was published by Brohan of Waterbury, CT in 2000.
48. Crowe, 485–486, 496. The term canali was first used in an 1859 publication by Angelo Secchi (Crowe, 486).
49. Lowell's story is told in William Graves Holt, Lowell and Mars, Tucson, University of Arizona Press, 1976, and in a more sociological way by David Strauss, Percival Lowell: The Culture and Science of a Boston Brahmin, Cambridge, MA, Harvard University Press, 2001.
50. Crowe, 515.
51. Guthke, 357–358; Percival Lowell, "Proofs of Life on Mars," Century, Vol. 76, Number 2, 1908, 292–303.
52. Strauss, Percival Lowell, 203.
53. Kurd Lasswitz, Auf Zwei Planeten (On Two Planets), 1897, translated by Hans H. Rudnick, Carbondale, IL, Southern Illinois University Press, 1971; Guthke, 383–385.

54. H.G. Wells, The War of the Worlds, serialized in 1897, published in book form in 1898; for a more recent reprint, see The Complete Science Fiction Treasury of H.G. Wells, New York, Avenel Books, 1978, 265–388. Also see Guthke 386–388.
55. Dick, The Biological Universe, 52.
56. Dick, The Biological Universe, 52–53; Nicolas Cusanus had foreseen this in 1440, when he wrote that "the fabric of the world has its center everywhere and its circumference nowhere." Edward Harrison, Masks of the Universe, 174–175.
57. Dick, The Biological Universe, 37–38.
58. McNeill, 830–831.
59. Dick, The Biological Universe, 53. Jeans argued in 1929 that the odds were 100 thousand to 1 against a star being surrounded by planets. Life, he believed, is limited to an almost inconceivably small corner of the universe ("Life in the Universe," in The Universe Around Us, New York, McMillan, 1929). Jeans had changed his opinion by 1942, when he wrote that there may be 2 billion planetary systems and that life-evolving processes may be at work on a great many planets. Smithsonian Institution Annual Report, 1942, 145–150.
60. Dick, The Biological Universe, 93–94; Crowe, 539; Frank D. Drake, "Reflections on the Modern History of SETI," Acta Astronautica, Vol. 26 (1992), 143–144. The eminent astronomer E.E. Barnard had made drawings of Martian features he observed in 1892 and 1894, including mountain ranges and craters. Barnard kept those drawings to himself because he thought that other astronomers would not believe them and might ridicule them. See Rodger W. Gordon and William Sheehan, "Solved: The Mars-Crater Mystery," Sky and Telescope, November 2005, 64–67.
61. Dick, The Biological Universe, 222–223; Guthke, 391.
62. Ben Bova, Faint Echoes, Distant Stars: The Science and Politics of Finding Life Beyond Earth, New York, Morrow, 2002, 131.
63. Steven J. Dick and James E. Strick, The Living Universe: NASA and the Development of Astrobiology, Brunswick, NJ, Rutgers University Press, 2004, 12; Edward Edelson, "Who Goes There?", Garden City, NY, Doubleday, 1979, 19.
64. Freeman Dyson, "Dynamic Universe," Nature, Vol. 435 (23 June 2005), 1033.
65. Percival Lowell, Mars as the Abode of Life, 69.
66. Dick, The Biological Universe, 189, 195–196.
67. For an overview of Oparin's ideas in English, see A.I. Oparin, The Origin of Life on the Earth, New York, Academic Press, 1957. Haldane's work is described in Dick, The Biological Universe, 341–342.
68. S. I. Miller, "A Production of Amino Acids Under Primitive Earth Conditions," Science, Vol. 117 (15 May 1953), 528–529; S.I. Miller, "Production of Some Organic Compounds Under Possible Primitive Earth Conditions," Journal of the American Chemical Society, Vol. 77, Number 9 (12 May 1955), 2351–2361; Jeffrey L. Bada and Antonio Lazcano, "Prebiotic Soup—Revisiting the Miller Experiment," Science, Vol. 300 (2 May 2003), 745–746; Dick and Strick, 16.

69. Otto Struve, "Life on Other Worlds," Sky and Telescope, Vol. 14, Number 4 (February 1955), 137–140, 146.
70. Harlow Shapley, Of Stars and Men: Human Response to an Expanding Universe, Boston, Beacon Press, 1958, 74, 85, 112–113.
71. Shapley, 108.

A New Era

1. Paul Davies, "A Quantum Recipe for Life," Nature, Vol. 437 (6 October 2005), 819.
2. Dick, The Biological Universe, 141–142; Dick and Strick, 18.
3. Luann Becker, et al., "Higher Fullerenes in the Allende Meteorite," Nature, Vol. 400 (15 July 1999), 227–228; Dick, The Biological Universe, 372; David Darling, Life Everywhere: The Maverick Science of Astrobiology, New York, Basic, 2001, 49.
4. Bova, 91; Dick and Strick, 78.
5. Darling, 58; Dick and Strick, 80.
6. Sean C. Solomon, et al., "New Perspectives on Ancient Mars," Science, Vol. 307 (25 February 2005), 1214–1219; Richard A. Kerr, "On Mars, A Second Chance for Life," Science, Vol. 306 (17 December 2004), 2010–2012; Richard A. Kerr, "And Now, the Younger, Dry Side of Mars is Coming Out," Science, Vol. 307 (18 February 2005), 1025–1026.
7. H. Strughold, "Is Mars Covered with Ice?" International Science and Technology, Number 30, June 1964, 11–12; Philip R. Christensen, "The Many Faces of Mars," Scientific American, July 2005, 32–39.
8. G.L. Chaikin, "A Transitional Hypothesis Concerning Life on Interstellar Bodies," Popular Astronomy, Vol. 59, Number 1 (January 1951), 50–51.
9. Kenneth Chang, "Life on Mars? Could Be, But How Will They Tell?" The New York Times, 29 March 2005; Michael Carroll, "Europa: Distant Ocean, Hidden Life?", Sky and Telescope, December 1997, 50–55.
10. Christopher P. McKay, in Dick, editor, Many Worlds, 45–56.
11. James Lovelock, "The Living Earth," Nature, Vol. 426 (18/25 December 2003), 769–770; James E. Lovelock, "The Recognition of Alien Biospheres," Cosmic Search, Year End 1980, 2–4, 18; E.G. Nisbet and N.H. Sleep, "The Habitat and Nature of Early Life," Nature, Vol. 409 (22 February 2001), 1083–1100; George Gaylord Simpson, "The Nonprevalence of Humanoids," in Goldsmith, editor, 214–221, originally published in Science, Vol. 143 (1964), 769–775.
12. See, among others, Luis P. Villarreal, "Are Viruses Alive?" Scientific American, December 2004, 101–105.
13. Philip Ball, "Seeking the Solution," Nature, Vol. 436 (25 August 2005), 1084–1085.
14. George A. Seielstad, At the Heart of the Web: The Inevitable Genesis of Intelligent Life, Boston, Harcourt, Brace, Jovanovitch, 1989, 238.
15. Steven J. Dick, "Extraterrestrial Life and Our World View at the Turn of the Millennium" (Dibner Library Lecture), Washington, DC, Smithsonian Institution, 2000, 31–32.
16. Darling, xiii, 169.

17. Tony Reichhardt, "NASA Beefs up Life Science Involvement," Nature, Vol. 399 (27 May 1999), 293; Christopher P. McKay, "All Alone After All?" (review of Rare Earth), Science, Vol. 288 (28 April 2000), 625.
18. Jack Cohen and Ian Stewart, "Where Are the Dolphins?" Nature, Vol. 409 (22 February 2001), 1119–1122; Dick, Dibner Lecture, 34–36.
19. Gerald Feinberg and Robert Shapiro, Life Beyond Earth: The Intelligent Earthling's Guide to Life in the Universe, New York, Morrow, 1980, 415.
20. Svante Arrhenius, "The Propagation of Life in Space," in Goldsmith, editor, 32–33.
21. F.H.C. Crick and L.E. Orgel, "Directed Panspermia," in Goldsmith, editor, 34–37, originally published in Icarus, Vol. 19 (1973), 341–346; David Warmflash and Benjamin Weiss, "Did Life Come from Another World?", Scientific American, November 2005, 64–71. Also see M. Mautner and G.L. Matloff, "Directed Panspermia: A Technical and Ethical Evaluation of Seeding Nearby Solar Systems," JBIS, Vol. 32 (1979), 419–423.
22. A.D. Taylor, et al., "Discovery of Interstellar Dust Entering the Earth's Atmosphere," Nature, Vol. 380 (28 March 1996), 323–325; G.J. Flynn, "Collecting Interstellar Dust Grains," Nature, Vol. 387 (15 May 1997), 248; G.M. Muñoz Caro, et al., "Amino Acids from Ultraviolet Irradiation of Interstellar Ice Analogues," Nature, Vol. 416 (28 March 2002), 403–405; Robert Irion, "The Science of Astrobiology Takes Shape," Science, Vol. 288 (28 April 2000), 603–605; Kathy Sawyer, "Far-Out Building Blocks," International Herald Tribune, 31 January 2001; Darling, 45; Max P. Bernstein, Scott A. Sandford, and Louis J. Allamandola, "Life's Far-Flung Raw Materials," Scientific American, July 1999, 42–49. Studies of interstellar dust suggest that old stars and ultraviolet light can create a pervasive prebiotic haze. "Astronomers Sweep Space for the Sources of Cosmic Dust," Science, Vol. 310 (28 October 2005), 614–615.
23. Fred Hoyle and Chandra Wickramasinghe, Life Cloud: The Origin of Life in the Universe, London, Sphere, 1979 (first published by Dent in 1978); Fred Hoyle and Chandra Wickramasinghe, Diseases from Space, New York, Harper and Row, 1980, originally published in the United Kingdom by Dent in 1978; Fred Hoyle and Chandra Wickramasinghe, Cosmic Life Force, New York, Paragon House, 1990, originally published in the United Kingdom in 1988; Fred Hoyle and Chandra Wickramasinghe, Our Place in the Cosmos: The Unfinished Revolution, London, Dent, 1993; Fred Hoyle, The Intelligent Universe, London, Michael Joseph, 1983; Fred Hoyle, The Black Cloud, New York, Penguin, 1957, originally published in London by Heinemann.
24. N. Chandra Wickramasinghe, "Why Alien Intelligence May Not Be So Alien," in SearchLites (published by the SETI League), Vol. 8, Number 4 (Autumn 2002), 3–7.
25. Irion, "The Science of Astrobiology Takes Shape," 605.
26. Crowe, 403; Dick, The Biological Universe, 326–327.
27. Kenneth Chang, "Martians Landing on Earth? If You Mean Bacteria, Maybe," The New York Times, 31 October 2000.
28. David S. McKay, et al., "Search for Past Life on Mars: Possible Relic Biogenic Activity in Martian Meteorite LH84001," Science, Vol. 273 (16 August 1996), 924–930. For a brief overview, see Oliver Morton, Mapping Mars, New York, Picador USA, 2002, 252–257.

29. John Noble Wilford, "Replying to Skeptics, NASA Defends Claims About Mars," The New York Times, 8 August 1996; Dick and Strick, 179–201. For a defense of the McKay team's position, see Everett K. Gibson, Jr., et al., "The Case for Relic Life on Mars," Scientific American, December 1997, 58–65.

30. Jeffrey J. Walker, et al., "Geobiology of a Microbial Endolithic Community in the Yellowstone Geothermal Environment," Nature, Vol. 434 (21 April 2005), 1011–1014.

31. Morton, Mapping Mars, 313.

32. Paul C.W. Davies, "Biological Determinism, Information Theory, and the Origin of Life," in Steven J. Dick, editor, Many Worlds, Philadelphia, Templeton Foundation Press, 2000, 15–26; "No Panspermia Between Stars," Sky and Telescope, May 2000, 23; Sun Kwok, "The Synthesis of Organic and Inorganic Compounds in Evolved Stars," Nature, Vol. 430 (26 August 2004), 985–990.

33. Bova, 199.

34. Freeman J. Dyson, "Looking for Life in Unlikely Places," in Yoji Kondo, et al., editors, Interstellar Travel and Multi-Generation Space Ships, Burlington, Ontario, Apogee, 2003, 105–119.

Searching for Intelligence

1. E.W. Barnes, Nature, Vol. 128 (Suppl) (1931), 719–722; Dick, The Biological Universe, 414.

2. Carl Sagan, editor, CETI: Communication with Extraterrestrial Intelligence, Cambridge, MA, MIT Press, ix.

3. Andrew J.H. Clark and David H. Clark, Aliens: Can We Make Contact with Extraterrestrial Intelligence?, New York, Fromm, 1999, 115.

4. Isaac Asimov, "Terrestrial Intelligence," in Ben Bova and Byron Preiss, editors, First Contact: The Search for Extraterrestrial Intelligence, New York, NAL Books, 1990, 21–30.

5. David Lamb, The Search for Extraterrestrial Intelligence: A Philosophical Inquiry, London, Routledge, 2001, 53.

6. Steven J. Dick, "Back to the Future: SETI Before the Space Age," paper presented at the International Astronautical Congress in 1993 (IAA.9.2-93-790); Louis Berman, "What it Was Like Before Ozma," Cosmic Search, Vol. 1, Number 4 (Fall 1979), 17–20; Frank Drake and Dava Sobel, Is Anyone Out There? The Scientific Search for Extraterrestrial Intelligence, London, Souvenir, 1993, 172, first published in New York by Delacorte (Bantam Doubleday Dell) in 1992; Dick, The Biological Universe, 401; Sullivan, 159; Grinspoon, 291–292; "No Mars Messages Yet," The New York Times, 16 June 1922, in Goldsmith, editor, 80.

7. Dick, The Biological Universe, 406–408; Bova, 133; Thomas H. White, "Extraterrestrial DX. Circa 1924: Will We Talk to Mars in August?" SearchLites, Vol. 9 (Summer 2003), 3–4.

8. Swift, SETI Pioneers, 6–7; Dick, The Biological Universe, 410; Sullivan, 163–164; Jeffrey M. Lichtman, "Exploring the Radio Sky," Sky and Telescope, January 2005, 127–131.

9. Sullivan, 164; Lichtman, 127.

10. "50, 100 & 150 Years Ago," Scientific American, December 2003, 22.

11. Giuseppe Cocconi and Philip Morrison, "Searching for Interstellar Communications," Nature, Vol. 184 (1959), 844–846. Reprinted in Goldsmith, editor, 102–104.

12. Frank D. Drake, "How Can We Detect Radio Transmissions from Distant Planetary Systems?" Sky and Telescope, 19 (1959), 140 (also in A.G.W. Cameron, editor, Interstellar Communication, New York, Benjamin, 1963, 165–174); Frank D. Drake, "Project Ozma," Physics Today, Vol. 14 (April 1961), 40–46. Also see Drake and Sobel, passim.

13. Robert T. Rood and James S. Trefil, Are We Alone? The Possibility of Extraterrestrial Civilizations, New York, Scribner's, 1981, 2.

14. Ronald N. Bracewell, "Communications from Superior Galactic Communities," Nature, Vol. 186 (1960), 670–672, reprinted in Goldsmith, editor, 105–107.

15. Dick, The Biological Universe, 436; Clark and Clark, 68–69; Swift, 188; Carl Sagan, "On the Detectivity of Advanced Galactic Civilizations," Icarus, Vol. 19 (1973), 350–352; reprinted in Sagan, editor, CETI, 365–370, and in Goldsmith, editor, 140–141.

16. Clark and Clark, 70–71; Sullivan, 207; Marina Benjamin, Rocket Dreams: How the Space Age Shaped Our Vision of a World Beyond, New York, Free Press, 2003, 196–198.

17. Sagan, editor, CETI, 350.

18. I.S. Shklovskii and Carl Sagan, Intelligent Life in the Universe, New York, Dell, 1966.

19. Roger A. MacGowan and Frederick I. Ordway, III, Intelligence in the Universe, Engelwood Cliffs, NJ, Prentice-Hall, 1966.

20. Project Cyclops: A Design Study of a System for Detecting Extraterrestrial Intelligent Life, NASA Ames Research Center, 1972, revised edition 1973; Sullivan, 188.

21. Swift, 253.

22. The text is reproduced in the preface to James Gunn's novel The Listeners, New York, Scribner's, 1972.

23. Carl Sagan, The Cosmic Connection, New York, Anchor Press (Doubleday), 1973.

24. Philip Morrison, John Billingham, and John Wolfe, editors, The Search for Extraterrestrial Intelligence, NASA SP-419, 1977, 13, 14; Jill C. Tarter and Christopher F. Chyba, "Is There Life Elsewhere in the Universe?" Scientific American, December 1999, 118–123.

25. Linda Billings, "From the Observatory to Capitol Hill," in Bova and Preiss, editors, 223–239; Carl Sagan, Cosmos, Random House, 1980.

26. The petition appeared in the issue of Science dated 29 October 1982, page 426. It was republished in The Planetary Report of May–June 1996.

27. Billingham, et al., editors, "Social Implications," 47.

28. Dick and Strick, 148. Over the years 1975 to 1992, Congress appropriated $78 million for SETI. Ronald D. Ekers, et al., editors, SETI 2020: A Roadmap for the Search for Extraterrestrial Intelligence, Mountain View, CA, SETI Press, 2002, xlvii.

29. Stephen J. Garber, "Searching for Good Science: The Cancellation of NASA's SETI Program," JBIS, Vol. 52 (1999), 3–12.

30. Dick and Strick, 153; Grinspoon, 306.

31. Richard A. Kerr, "No Din of Alien Chatter in Our Neighborhood," Science, Vol. 303 (20 February 2004), 1133. For descriptions of SETI Institute activities, see the website at www.seti.org.

32. Seth Shostak, "The Future of SETI," Sky and Telescope, April 2001, 42–53.

33. Guillermo A. Lemarchand, "SETI Technology: Possible Scenarios for the Detectability of Extraterrestrial Intelligence," paper presented at the International Astronautical Congress, 1997 (IAC-97-IAA.7.1.02); Shklovskii and Sagan, 422. Von Hoerner was quoted in Alfred Roulet, The Search for Intelligent Life in Outer Space, New York, Berkley, 1977, 135. (originally published in French by Julliard in 1973).

34. "Microsoft Co-Founder Dishes Out Millions to Search for Aliens," Nature, Vol. 428 (25 March 2004, 358; Jill Tarter, "For the ATA, Life Begins at 32," SETI Institute News, Third Quarter 2003, 1; Leo Blitz, "Exciting Astronomy with the ATA," SETI Institute News, Second Quarter 2003, 2; SETI Institute fund-raising letter, 23 June 2005; David Schlom, "The Allen Telescope Array," Ad Astra, Spring 2006, Vol. 18, Number 1, 19–20.

35. Ekers, et al., 12.

36. See Gale E. Christianson, Edwin Hubble: Mariner of the Nebulae, Chicago, University of Chicago Press, 1995, and Ronald Florence, The Perfect Machine: Building the Palomar Telescope, New York, Harper Perennial, 1995.

37. The Planetary Society web site is at www.planetary.org. For the early history of the Planetary Society, see Michael A.G. Michaud, Reaching for the High Frontier: The American Pro-Space Movement, 1972–1984, New York, Praeger, 1986, 207–213.

38. M. Lampton, et al., "The SERENDIP Piggyback SETI Project," Acta Astronautica, Vol. 26 (1992), 189–192; David Koerner and Simon LeVay, Here Be Dragons: The Scientific Quest for Extraterrestrial Life, Oxford, Oxford University Press, 2000, 172.

39. Alan M. MacRobert and Andrew J. LePage, "SETI@home: Catching the Wave," Sky and Telescope, October 1999, 68–72; Benjamin, 150, 166.

40. Clark and Clark, 74; Tom Van Horne, "The Future of SETI," in Bova and Preiss, editors, 159–164.

41. The SETI League Web Site is at www.setileague.org;. Also see Kent Cullers and William R. Alschuler, "Individual Involvement," in Bova and Preiss, editors, 281–301.

42. Clark and Clark, 75, 110; Seth Shostak and Alex Barnett, Cosmic Company: The Search for Life in the Universe, Cambridge, 2003, 108–109; Donald E. Tarter, "Treading on the Edge: Practicing Safe Science with SETI," The Skeptical Inquirer, Vol.17 (Spring 1993), 288–296; Drake, "Promising New Approaches."

43. Chris Salter, "Arecibo's Bright Future," SETI Institute News, Vol. 12 (Third Quarter 2003), 10; Robert Irion, "Tuning in the Radio Sky," Science, Vol. 296 (3 May 2002), 830–833; Sagan, editor, CETI, 250; Carolyn Collins Petersen, "LOFAR: A Deep Look Through a Neglected Window," Sky and Telescope, July 2003, 22.

44. Ekers, et al., SETI 2020, xliv–xlvii, 229.

45. Kerr, "No Din of Alien Chatter."

46. Quoted in Michael A.G. Michaud, "SETI Hearings in Washington," JBIS, Vol. 32 (1979), 116–118; Christopher F. Chyba, "Life Beyond Mars," Nature, Vol. 382 (15 August 1996), 576–577.

47. Michael J. Klein, "Where and What Can We See?", in Bova and Preiss, editors, 143–158; SETI Fund letter, January 2005.

48. Robert L. Forward, "Technological Limits to Space Exploration," in Valerie Neal, editor, Where Next, Columbus? The Future of Space Exploration, New York, Oxford University Press, 1994, 171–193.

49. Diane Richards, Interview with Dr. Frank Drake, SETI Institute News, First Quarter 2003, 5–6. This is explained concisely in Donald Goldsmith and Tobias Owen, The Search for Life in the Universe, Menlo Park, CA, Benjamin/Cummings, 1980, 386–387.

50. In Sagan, editor, CETI, 232; David Black, et al., "Searching for Extraterrestrial Intelligence: The Ultimate Exploration," Mercury, July–August 1977, 3–7.

51. David H. Smith, "From Cambridge to Cosmology" (interview with Martin Rees), Sky and Telescope, August 1983, 107–109; Clark and Clark, 97; John Billingham, editor, Life in the Universe, Cambridge, MA, MIT Press, 1981, 333.

52. R.N. Schwartz and C.H. Townes, "Interstellar and Interplanetary Communication by Optical Masers," Nature, Vol. 190 (1961), 205–208, reprinted in Cameron, editor, 223–230, and in Goldsmith, editor, 110–113; Sullivan, 182.

53. Ekers, et al., 135; "Searching for Alien Rays," Science, Vol. 291 (2 February 2001), 823.

54. Bova, 252; Seth Shostak, "The Future of SETI," Sky and Telescope, April 2001, 42–53; "Two Telescopes Join Hunt for ET," Nature, Volume 440 (13 April 2006), 853.

55. Philip Morrison, "Interstellar Communication," in Cameron, editor, 249–271; George Musser, "All Screwed Up," Scientific American, November 2003, 22; "A New Way to Find ET," Sky and Telescope, July 2005, 22.

56. Ekers, et al., editors, 61, 98, 229.

57. Ekers, et al., editors, 54, 56, 134. Also see the special section entitled "NASA's Other Space Telescopes," Sky and Telescope, January 2006, 34–50.

58. S. Yu. Sazonov, et al., "An Apparently Normal Gamma Ray Burst with an Unusually Low Luminosity," Nature, Vol. 430 (5 August 2004), 646–647; Andrew MacFadyen, "Long Gamma Ray Bursts," Science, Vol. 303 (2 January 2004), 45–47.

59. Webster Cash, et al., "Laboratory Detection of X-ray Fringes with a Grazing-Incidence Interferometer," Nature, Vol. 407 (14 September 2000), 160–161; Dennis Normile, "Improved X-Ray Telescope Takes Flight," Science, Vol. 309 (15 July 2005), 363; Charles Seife, "Neutrino Hunters Borrow Military Ears—and Eyes," Science, Vol. 298 (4 October 2002), 43–45; Jay M. Pasachoff and Marc L. Kutner, "Neutrinos for Interstellar Communication," Cosmic Search, Vol. 1, Number 3 (Summer 1979), 2–8; Robert Irion, "A Positron Map of the Sky," Science, Vol. 305 (24 September 2004), 1899.

60. Peter G. Friedman and D. Christopher Martin warned that we will soon lose our view of the ultraviolet sky unless we preserve or replace the few remaining UV space missions. "Galex and UV Observations," Science, Vol. 306 (1 October 2004), 54.

61. Lowell, Mars as the Abode of Life, 211.

62. Grinspoon, 405.

63. Swift, 318–324.

64. In Sagan, editor, CETI, 190.

65. Freeman Dyson, "Search for Artificial Sources of Infrared Radiation," Science, Vol. 131 (1960), 1667–1668, reprinted in Cameron, editor, 111–114, and in Goldsmith, editor, 108–109; Andrew J. LePage, "Where They Could Hide," Scientific American, July 2000, 40–41; Grinspoon, 406.

66. Ben Ianotta, "Longer View for a New Space Telescope," Aerospace America, November 2005, 26–31.

67. See, for example, Jean Heidmann, "SETI From the Moon: An Invitation to COSPAR," COSPAR paper F3.1-0008, 1996; Jean Heidmann, "Recent Progress on the Lunar Farside Crater Saha Proposal," paper presented at the 1997 International Astronautical Congress (IAA-97-IAA.9.1.05), and Jean Heidmann, "Sharing the Moon by Thirds: An Extended Saha Crater Proposal," Advances in Space Research, Vol. 26 (2000), 371. Jill C. Tarter and Christopher F. Chyba briefly discussed the issue in "Is There Life Elsewhere in the Universe?" Scientific American, December 1999, 118–123. For a dissenting view, see Dan Lester, "Here's Dirt in Your Eye," Sky and Telescope, April 2006, 110. For a more general perspective, see Michael A.G. Michaud, "Farside," Future Life, May, 1980, 31; Claudio Maccone, "NASA Gateways at L-1 and L-2 and the Radio-Quiet Moon Farside Initiative," paper presented at the 2004 International Astronautical Congress (IAA.1.1.1.08); Peter G. Stanley, "Prospects for Extraterrestrial Life," JBIS, Vol. 50 (1997), 243–248.

68. Freeman J. Dyson, "Gravitational Machines," in Cameron, editor, 115–120; John Kraus, "ABCs of Space," Cosmic Search, Vol. 2, Number 1 (Winter 1980), 28–29.

69. Lemarchand, "SETI Technology."

70. William A. Reupke, "Efficiently Coded Messages Can Transmit the Information Content of a Human Across Interstellar Space," Acta Astronautica, Vol. 26 (1992), 273–276.

71. Richard A. Kerr, "No Din of Alien Chatter in Our Neighborhood," Science, Vol. 303 (20 February 2004), 1133.

Sending Our Own Signals

1. Quoted in Frank White, The SETI Factor: How the Search for Extraterrestrial Intelligence is Changing Our View of the Universe and Ourselves, Walker, 1990, 17.

2. Crowe 205, 207, 394, 398–399; MacGowan and Ordway, 327–329; Drake and Sobel, 170–174; Sullivan, 158–159; Swift, 6; Grinspoon, 324. Douglas Vakoch reviewed this history in "Pictorial Messages to Extraterrestrials, Part 1," SETIQuest, Vol. 4 (First Quarter 1998), 8–10.

3. See W. Bernard Carlson, "Inventor of Dreams," Scientific American, March 2005, 79–85. In 1931, Tesla told Time magazine about his plans to signal the stars with his "Teslascope"—a giant radio transmitter. Nothing seems to have come of this project.

4. MacGowan and Ordway, 331.

5. Vakoch, "Pictorial Messages to Extraterrestrials;" Carl Sagan and Frank Drake, "The Search for Extraterrestrial Intelligence," Scientific American, May 1975, 80–89. Sagan, Drake, and Linda Salzman Sagan described the Pioneer plaques in "A Message From Earth," Science, Vol. 175 (1972), 881, republished in Goldsmith, editor, 274–277. Also see Carl Sagan, Murmurs of Earth: The Voyager Interstellar Record, New York, Random House, 1978, and George Abell, "The Search for Life Beyond Earth: A Scientific Update," in Christian, editor, 53–71.

6. The Staff at the Arecibo National Astronomy and Ionosphere Center, "The Arecibo Message of November, 1974," in Goldsmith, editor, 293–296; Paul Davies, Are We Alone?, 56; Carl Sagan, Cosmos, New York, Random House, 1980, 297; Drake and Sobel, 183; Stephen Webb, Where is Everybody? Fifty Solutions to the Fermi Paradox and the Problem of Extraterrestrial Life, New York, Copernicus, 2002, 111.

7. Donald Goldsmith, Voyage to the Milky Way: The Future of Space Exploration, New York, TV Books, 1999, 223.

8. Koerner and LeVay, 174; Seth Shostak, Sharing the Universe: Perspectives on Extraterrestrial Life, Berkeley, CA, Berkeley Hills Books, 1998, 149–150. Also see Michael A.G. Michaud, "Ten Decisions that Could Shake the World," Space Policy, Vol. 19 (May 2003), 131–136.

9. Carl Sagan, "The Quest for Intelligent Life in Space Is Just Beginning," Smithsonian, May 1978, 38–47; Ridpath, 130. In his 1973 book The Cosmic Connection (page 195), Sagan wrote that the Arecibo radio telescope could communicate with an identical copy of itself anywhere in our galaxy, implying a range of at least 70,000 light years. Seven years later, Sagan wrote in Cosmos (page 297) that the Arecibo observatory could communicate with an identical radio telescope 15,000 light years away, about 20% of the previous distance and only halfway to the center of the Milky Way. For Sagan's change of view, see John Casti, Paradigms Lost: Tackling the Unanswered Mysteries of Modern Science, New York, Bard (Avon), 1989, 384.

10. Philip Morrison, et al., editors, SETI: The Search for Extraterrestrial Intelligence, NASA SP-419, 1977, 14; Ekers, et al., 5.

11. Ekers, et al., eds, xxxi, 5, 235.

12. White, 68; Morrison, et al., editors, 1977, 8, 14.

13. Frank Drake, "Methods of Communication," in Cyril Ponnamperuma and A.G.W. Cameron, editors, Interstellar Communication: Scientific Perspectives, Boston, Houghton-Mifflin, 1974, 118–139.

14. Ekers, et al., 5, 229.

15. MacGowan and Ordway, 330, 355.

16. The quote is from Ekers, et al., 244. Also see Douglas A. Vakoch, "The Dialogic Model: Representing Human Diversity in Messages to Extraterrestrials," Acta Astronautica, Vol. 42 (1998), 705–710; Douglas A. Vakoch, "Message Policy for Active SETI," paper presented at the International Conference on SETI in the 21st Century, Sydney, Australia, 1998.

17. MacGowan and Ordway, 330.
18. Vakoch, "The Dialogic Model," 709.
19. Team Encounter website (www.teamencounter.com), September 2003; Lamb, 33.
20. Michaud, "Ten Decisions"; Ekers, 244; David Brin, "A Contrarian Perspective on Altruism: The Dangers of First Contact," 30 September 2002, available at www.davidbrin.com.

Probabilities

1. Jacques Monod, Chance and Necessity, translated into English by A. Wainhouse, New York, Knopf, 1971, 180, also published by Vintage, 1972.
2. Sullivan, 200.
3. Lamb, 41–42.
4. Paul Davies, Are We Alone?, 64.
5. Drake and Sobel, 52; Swift, 81.
6. Govert Schilling, "The Chance of Finding Aliens: Reevaluating the Drake Equation," Sky and Telescope, December 1998, 36–42.
7. Shostak, Sharing the Universe, 179–183.
8. Ekers, et al., xxxv.
9. Donald Goldsmith, "Optimists and Pessimists," in Goldsmith, editor, 186–187.
10. Drake and Sobel, 62; White, 196.
11. Isaac Asimov, A Choice of Catastrophes, New York, Fawcett, 1979, 265; MacGowan and Ordway, 366.
12. Clark and Clark, 93, 95.
13. Quoted in Randall Fitzgerald, Cosmic Test Tube: Extraterrestrial Contact, Theories and Evidence, Los Angeles, Moon Lake Media, 1998, 360–361.
14. William C. Burger, Perfect Planet, Clever Species: How Unique Are We?, Amherst, New York, Prometheus, 2003, 282.
15. Brin, "A Contrarian Perspective."
16. Robert T. Rood and James S. Trefil, Are We Alone? The Possibility of Extraterrestrial Civilizations, New York, Scribners, 1981, 7, 60, 120.

Probabilities: The Astronomical Factors

1. Guthke, 7.
2. Krafft A. Ehricke, "A Long-Range Perspective and Some Fundamental Aspects of Interstellar Evolution," JBIS, Vol. 28 (1975), 713–734.
3. G. Mark Voit, "The Rise and Fall of Quasars," Sky and Telescope, May 1999, 40–46; Alan Heavens, et al., "The Star-Formation History of the Universe from the Stellar Populations of Nearby Galaxies," Nature, Vol. 428 (8 April 2004), 625–626; "The Fading Firmament," Science, Vol. 301 (29 August 2003), 1159.
4. John Cowan, "Elements of Surprise," Nature, Vol. 423 (1 May 2003), 29; "Earliest Star-Forming Region," Sky and Telescope, November 2003, 24;

Charles H. Lineweaver, et al., "The Galactic Habitable Zone and the Age Distribution of Complex Life in the Milky Way," Science, Vol. 303 (2 January 2004), 59–62; Robert Irion, "Are Most Life-Friendly Stars Older Than the Sun?" Science, Vol. 303 (2 January 2004), 27.

5. Martin Rees, Our Final Hour: A Scientist's Warning, New York, Basic Books, 2003, 5.

6. Robert Irion, "Astronomers Shine a Light Upon Dim Nearby Stars," Science, Vol. 304 (11 June 2004), 1587–1589; Margaret Turnbull, "SETI and the Smallest Stars," SETI Institute News, Vol. 12, No. 3 (Third Quarter 2003), 12–13.

7. Govert Schilling, "Getting to Know Our Stellar Neighbors," Sky and Telescope, April 2004, 21. For a concise description, see "The Spectral Types on the Main Sequence," Mercury, January–February 2005, 28.

8. Seth Shostak, "From the Science Editor's Desk," SETI Institute News, Vol. 12 (Third Quarter 2003), 6; Diane Richards, "Interview with Dr. Frank Drake," SETI Institute News, Vol. 12 (First Quarter 2003), 5–6.

9. Schilling, "Getting to Know Our Stellar Neighbors;" Paul Kalas, et al., "Discovery of a Large Dust Disk Around the Nearby Star U Microscopii," Science, Vol. 303 (26 March 2004), 1990–1992.

10. Robert Naeye, "Extrasolar Planets: Pictured at Last?" Sky and Telescope, August 2005, 39–42; Gibor Basri, "A Decade of Brown Dwarfs," Sky and Telescope, May 2005, 34–40; Jill C. Tarter, "Brown Dwarfs and How They Grow Old," unpublished Ph.D. thesis, University of California at Berkeley, 1975.

11. Virginia Trimble, "Nucleosynthesis and Galactic Evolution," in Hart and Zuckerman, editors, 135–140.

12. Quoted in Eugene F. Mallove and Gregory L. Matloff, The Starflight Handbook: A Pioneer's Guide to Interstellar Travel, New York, Wiley, 1989, 217.

13. Stephen H. Dole and Isaac Asimov, Planets for Man, New York, Random House, 1964, 139–140, 171–172, 214, 225–226. The original study was published as Stephen H. Dole, Habitable Planets for Man, New York, Blaisdell, 1964.

14. Paul Kalas, et al., "Discovery of a Large Dust Disk Around the Nearby Star 'U Microscopii,'" Science, Vol. 303 (26 March 2004), 1990–1992; Paul Kalas, et al., "A Planetary System as the Origin of Structure in Fomalhaut's Dust Belt," Nature, Vol. 435 (23 June 2005), 1067–1069.

15. David R. Ardila, "The Hidden Members of Planetary Systems," Scientific American, April 2004, 63–69; Jane S. Greaves, "Disks Around Stars and the Growth of Planetary Systems," Science, Vol. 307 (7 January 2005), 68–70; John Noble Wilford, "A New Look at How Planets Are Formed," The New York Times, 19 October 2004.

16. Robert Naeye, "A Bonanza of Exoplanet Discoveries," Sky and Telescope, May 2005, 19; Ray Jayawardhana, "Unraveling Brown Dwarf Origins," Science, Vol. 303 (16 January 2004), 322–323; Katharina Lodders, "Brown Dwarfs—Faint at Heart, Rich in Chemistry," Science, Vol. 303 (16 January 2004), 323–324; James Liebert and William B. Hubbard, "Big Planets and Little Stars," Nature, Vol. 400 (22 July 1999), 316–317.

17. A. Wolszczan and D.A. Frail, "A Planetary System Around the Millisecond Pulsar PSR 1257 + 12," Nature, Vol. 355 (1992), 145.

18. Michel Mayor and Didier Queloz, "A Jupiter-Mass Companion to a Solar-Type Star," Nature, Vol. 378 (23 November 1995), 355–359.

19. John MacVey, Interstellar Travel, Past, Present, and Future, New York, Stein and Day, 1977, 12.

20. Tim Appenzeller, "The Search for Other Earths," National Geographic, December 2004, 73–95; "Everywhere You Turn," Science, Vol. 307 (7 January 2005), 63; Naeye, "A Bonanza of Exoplanet Discoveries;" Stuart Ross Taylor, "Why Can't Planets be Like Stars?" Nature, Vol. 430 (29 July 2004), 509. For a periodic update, check the Jet Propulsion Laboratory website called PlanetQuest (http://planetquest.jpl.nasa.gov) the California and Carnegie Planet Search at http://exoplanets.org or The Extrasolar Planets Encyclopedia at www.obspm.fr/planets. You can participate in one form of planetary discovery through a distributed computer program known as Planet Quest, at www.planetquest.org.

21. Kenneth Chang, "Star Is a Near Twin of the Sun, Scientists Say," The New York Times, 7 January 2004; "Close to Home," Scientific American, September 2003, 36; Ben Bova, Faint Echoes, Distant Stars: The Science and Politics of Finding Life Beyond Earth, New York, Morrow, 2004, 214–228.

22. Ekers, et al., editors, 50; Bova, 220–221, Darling, 107.

23. Dennis Overbye, "It Orbits a Star, but Does It Qualify for Planethood?" The New York Times, 5 April 2005.

24. Dennis Overbye, "Found: Earth's Distant Cousin (About 15 Light-Years Away)," The New York Times, 14 June 2005; Guy Gugliotta, "Rocky Earth-like Planet Discovered," The Washington Post, 14 June 2005; Alan MacRobert, "A Low-Mass Exoplanet," Sky and Telescope, September 2005, 19; J.-P. Beaulieu, et al., "Discovery of a Cool Planet of 5.5 Earth Masses Through Gravitational Microlensing," Nature, Vol. 439 (26 January 2006), 437–440; Didier Queloz, "Light Through a Gravitational Lens," Nature, Vol. 439 (26 January 2006), 400–401.

25. Artie P. Hatzes and Gunther Wuchterl, "Giant Planet Seeks Nursery Place," Nature, Vol. 436 (14 July 2005), 182–183.

26. Stephen J. Garber, "Searching for Good Science: The Cancellation of NASA's SETI Program," JBIS, Vol. 52 (1999), 3–12.

27. Sullivan, 65.

28. Joanne Baker, "Planetary Orbits Do the Time Warp," Nature, Vol. 425 (4 September 2003), 34; Govert Schilling, "The Chance of Finding Aliens: Reevaluating the Drake Equation," Sky and Telescope, December 1998, 36–42.

29. Dennis Overbye, "Hunting for Life in Specks of Cosmic Dust," The New York Times, 19 July 2005.

30. Charley Lineweaver, "Finding Earths, Life and Intelligence in the Universe," SETI Institute Explorer, Special Edition 2006, 7, 36

31. David A. Hardy provided a good sampling of artists' imaginary depictions of alien worlds in Visions of Space, New York, Paper Tiger (Dragon's World), 1989. Also see Frederick C. Durant and Ron Miller, Worlds Beyond: The Art of Chesley Bonestell, Norfolk, VA, Donning, 1983.

32. P.E. Cleator, "Extraterrestrial Life," JBIS, Vol. 1 (October 1934), 3–4; Arthur C. Clarke, By Space Possessed, London, Gollancz, 1993, 139–145.

33. Darren M. Williams, et al., "Habitable Moons Around Other Planets," Nature, Vol. 385 (16 January 1997), 234–236; Christopher F. Chyba, "Life on Other Moons," Nature, Vol. 385 (16 January 1997), 201; Sullivan, 44; Darling, 89. Our solar system contains one moon with an atmosphere—Saturn's large satellite Titan. Conditions on that body, particularly its intense cold, may be too extreme to allow life to evolve.

34. David J. Stevenson, "Life-Sustaining Planets in Interstellar Space?" Nature, Vol. 400 (1 July 1999), 32; Richard L.S. Taylor, "Planets Without Stars," JBIS, Vol. 54 (2001), 19–26.

35. J. Kelly Beatty, "Distant Planetoid Sedna Baffles Astronomers," Sky and Telescope, June 2004, 14–15; John Noble Wilford, "Scientists Find an Icy World Beyond Pluto, and the Solar System Suddenly Seems Bigger," The New York Times, 16 March 2004; Brett Gladman, "The Kuiper Belt and the Solar System's Comet Disk," Science, Vol. 307 (7 January 2005), 71–75; David Tyrell, "The New Kings of the Kuiper Belt," Sky and Telescope, October 2005, 28–31; Scott S. Shepard, "A Planet More, a Planet Less?" Nature, Vol. 439 (2 February 2006), 541–542; F. Bertoldi, et al., "The Trans-Neptunian Object UB 313 is Larger Than Pluto," Nature, Vol. 439 (2 February 2006), 563–564.

36. Scott J. Kenyon and Benjamin C. Bromley, "Stellar Encounters as the Origin of Distant Solar System Objects in Highly Eccentric Orbits," Nature, Vol. 432 (2 December 2004), 598–600; Dennis Overbye, "Sun Might Have Exchanged Hangers-On with Rival Star," The New York Times, 2 December 2004; "Sedna's Dark Origin," Sky and Telescope, December 2004, 20; Kenneth Chang and Dennis Overbye, "Planet or Not, Pluto Now Has Far-Out Rival," The New York Times, 30 July 2005; Kenneth Chang, "10 Planets? Why Not 11?", The New York Times, 23 August 2005.

37. David Shiga, "Imaging Exoplanets," Sky and Telescope, April 2004, 44–52.

38. Robert Irion, "The Search for Pale Blue Dots," Science, Vol. 303 (2 January 2004), 30–32.

39. Darling, 174; Ben Zuckerman, "Searches for Electromagnetic Signals from Extraterrestrial Beings," in Hart and Zuckerman, editors, 9–17. Astronomers detected hydrogen from the atmosphere of an extrasolar planet in 2003. See David Charbonneau, "Atmosphere Out of that World," Nature, Vol. 422 (13 March 2003), 124–125; James F. Kasting, "When Methane Made Climate," Scientific American, July 2004, 78–85.

40. "Boiling Planets," Science, Vol. 303 (19 March 2004).

41. Giovanna Tinetti, "Detecting Biosignatures in Extrasolar Terrestrial Planets," SETI Institute Explorer, 2006, 12–13, 37; Clark and Clark, 226.

42. Rood and Trefil, 124–125; Peter D. Ward and Donald Brownlee, Rare Earth: Why Complex Life is Uncommon in the Universe, NY, Copernicus, 2000, 250–251, 257–275; William C. Burger, Perfect Planet, Clever Species: How Unique Are We?, Amherst, NY, Prometheus, 2003, 282; Grinspoon, 91; Isaac Asimov, Extraterrestrial Civilizations, 21–24; Isaac Asimov, "The Triumph of the Moon," Fantasy and Science Fiction, July 1973, 137–147; Darling, 97.

43. H. Strughold, "The Ecosphere in the Solar Planetary System," paper presented at the International Astronautical Congress in September 1956.

44. Philip Ball, "Water, Water, Everywhere?" Nature, Vol. 427 (1 January 2004), 19–20.

45. Rood and Trefil, 58.
46. Guillermo Gonzales, Donald Brownlee, and Peter D. Ward, "Refuges for Life in a Hostile Universe," Scientific American, October 2001, 62–67.
47. Charles H. Lineweaver, et al., "The Galactic Habitable Zone and the Age Distribution of Complex Life in the Milky Way," Science, Vol. 303 (2 January 2004), 59–62.
48. John Noble Wilford, "Sun's Cruise Through Space Is About to Hit Bumpy Patch," The New York Times, 18 June 1996.

Probabilities: Life

1. In Goldsmith, editor, 27.
2. Davies, Are We Alone?, 22–37. Nearly forty years earlier, N.W. Pirie had listed theories about the origin of life that were presented at a symposium: life has always pervaded space and is transferred from place to place; life was created by divine intervention; evolution on Earth through inevitable, normal processes; life appeared whenever there was a suitable environment. N.W. Pirie, "The Origins of Life," Nature, Vol. 180 (2 November 1957), 886–888.
3. Stephen Jay Gould, "The Evolution of Life on the Earth," Scientific American, October 1994, 85–91; Bor Luen Tang, "Many Possibilities for Life's Emergence;" JBIS, Vol. 58 (2005), 218–222; Koerner and LeVay, 242; Dick and Strick, 111.
4. Darling, 25; Nicholas Beukes, "Early Options in Photosynthesis," Nature, Vol. 431 (30 September 2004), 522–523.
5. Darling, 28; Grinspoon, 92.
6. Kenneth M. Towe, "Biochemical Keys to the Emergence of Complex Life," in John Billingham, editor, Life in the Universe, Cambridge, MA, The MIT Press, 1981, 297–303; Linda C. Kah, et al., "Low Marine Sulphate and Protracted Oxygenation of the Proterozoic Biosphere," Nature, Vol. 431 (14 October 2004), 834–837.
7. Philip Morrison, "Viking, Life, and a Search for Signals from the Stars," Astronautics and Aeronautics, July–August 1976, 65–67; Kenneth Chang, "Surprising Footprints in Old Sand," The New York Times, 1 March 2005.
8. Darling, xii; Shapley, 15.
9. "Microbes Made to Order," Science, Vol. 303 (9 January 2004), 158–161.
10. Quoted in Darling, 92.
11. Jacques Monod, Chance and Necessity, translated by A. Wainhouse, New York, Viking, 1971; Arthur Peacocke, "The Challenge and Stimulus of the Epic of Evolution to Theology," in Dick, editor, Many Worlds, 89–115; Bernd-Olaf Kuppers, "The World of Biological Complexity: Origin and Evolution of Life," in Dick, editor, Many Worlds, 31–43.
12. Leonard Ornstein, "A Biologist Looks at the Numbers," Physics Today, March 1982, 27–31.
13. Francois Jacob, "Evolution and Tinkering," Science, Vol. 196 (10 June 1977), 1161–1168.
14. Seielstad, 214.
15. Mark Ridley, The Cooperative Gene, New York, Simon and Schuster, 2001; Andrew Berry's review in Nature, Vol. 412 (26 July 2001), 379–380; Heather

L. True, et al., "Epigenetic Regulation of Translation Reveals Hidden Genetic Variation to Produce Complex Traits," Nature, Vol. 431 (9 September 2004), 184–187; Sagan, editor, CETI, 86.

16. Simon Conway Morris, Life's Solution: Inevitable Humans in a Lonely Universe, xii–xiii, 8–9, 298; Simon Conway Morris, "Not So Alien," SETI Institute News, Vol. 12 (First Quarter 2003), 10–11.

17. Morris, 297–298; Douglas H. Erwin's review in Science, Vol. 302 (5 December 2003), 1682–1683.

18. Stephen J. Gould, The Hedgehog, the Fox, and the Magister's Pox, New York, Harmony, 2003; see Robert N. Proctor's review in Science, Vol. 302 (31 October 2003), 785.

19. Sean B. Carroll, "Chance and Necessity: The Evolution of Morphological Complexity and Diversity," Nature, Vol. 409 (22 February 2001), 1102–1109.

20. Darling, 138.

21. Ernst Mayr, "The Probability of Extraterrestrial Life," in Regis, editor, 23–29.

22. Nicholas Wade, "How Did Life Begin?" The New York Times, 11 November 2003, and Carol Kaesuk Yoon, "Is Evolution Truly Random?", The New York Times, 11 November 2003. For an example of recent findings, see Pamela F. Colosimo, et al., "Widespread Parallel Evolution in Sticklebacks by Repeated Fixation of Ectodysplasin Alleles," Science, Vol. 307 (25 March 2005), 1928–1936.

23. Armand Delsemme, Our Cosmic Origins: From the Big Bang to the Emergence of Intelligence, Cambridge, Cambridge University Press, 2001, quoted in Bova, 87.

24. Paul Davies, "Biological Determinism, Information Theory, and the Origin of Life," in Dick, editor, Many Worlds, 15–27; Darling, 173; Feinberg and Shapiro, in Hart and Zuckerman, editors, Where Are They, 116.

25. Duncan Steel, Rogue Asteroids and Doomsday Comets, New York, Wiley, 1995, 107. For an overview of the end Permian extinction, see Michael J. Benton, When Life Nearly Died: The Greatest Mass Extinction of All Time, London, Thames and Hudson, 2003.

26. Christian de Duve, "The Onset of Selection," Nature, Vol. 433 (10 February 2005), 581–582.

27. Quoted in Peter Ulmschneider, Intelligent Life in the Universe, New York, Springer, 2003, 145.

28. Paul Davies, The Fifth Miracle: The Search for the Origin and Meaning of Life, New York, Simon and Schuster, 1999, 252.

29. Davies, Are We Alone?, xii, 24–37, 58, 77, 80–87; Davies, The Fifth Miracle, 265.

30. Christian de Duve, Life Evolving, Oxford, Oxford University Press, 2002; Christian de Duve, "Lessons of Life," in Dick, editor, Many Worlds, 3–13.

31. Sullivan, 307.

32. Grinspoon, 270–272.

33. Feinberg and Shapiro, 435; Grinspoon, 140, 263; Feinberg and Shapiro, in Hart and Zuckerman, editors, 114–115; Rood and Trefil, 84; Carl Sagan, Pale Blue Dot: A Vision of the Human Future in Space, New York, Random

House, 1994, 36; Clifford Pickover, The Science of Aliens, NY, Basic Books, 1998, 101.

34. Mayr, in Regis, editor, 24.

35. James Lovelock, "The Living Earth," Nature, Vol. 426 (18/25 December 2003), 769–770; Michael M. Tice and Donald R. Lowe, "Photosynthetic Microbial Mats in the 3,416-Myr-old Ocean," Nature, Vol. 431 (30 September 2004), 549–552.

36. Grinspoon, 273–274.

37. Pace, quoted in John Noble Wilford, "Come Out, Come Out, Wherever You Are," The New York Times, 1 January 2000; Singer, in Hart and Zuckerman, editors, 86.

38. Douglas H. Erwin, "Seeds of Diversity," Science, Vol. 308 (17 June 2005), 1752–1753; Dan Jones, "Personal Effects," Nature, Vol. 438 (3 November 2005), 14–16.

39. Michael B. Kastan and Jiri Bartek, "Cell-Cycle Checkpoints and Cancer," Nature, Vol. 432 (18 November 2004), 316–322; Lynn J. Rothschild and Rocco L. Mancinelli, "Life in Extreme Environments," Nature, Vol. 409 (22 February 2001), 1092–1100; William J. Broad, "Deep Under the Sea, Boiling Founts of Life Itself," The New York Times, 9 September 2003; D.N. Thomas and G.S. Dieckmann, "Antarctic Sea Ice: A Habitat for Extremophiles," Science, Vol. 295 (25 January 2002), 641–644. For nontechnical overviews, see Seth Shostak, "Extremophiles: Not So Extreme?," SETI Institute Explorer (First Quarter 2005), 6–9, and Lynn Rothschild, "Life in Extreme Environments," Ad Astra, Vol. 14, Number 1 (January–February–March 2002), 33–40; Bova, 38–39.

40. Ian M. Head, et al., "Biological Activity in the Deep Subsurface and the Origin of Heavy Oil," Nature, Vol. 426 (20 November 2003), 344–351; Steven D'Hondt, et al., "Distributions of Microbial Activities in Deep Subseafloor Sediments," Science, Vol. 306 (24 December 2004), 2216–2220.

41. Von R. Eshleman, "SETI and UFO Investigations Compared," in Peter A. Sturrock, The UFO Enigma: A New Review of the Physical Evidence, New York, Warner, 1999, 146–149; Paul Davies, "Goodbye Mars, Hello Earth," The New York Times, 10 April 2005.

42. Paul Davies, "New Hope for Life Beyond Earth," Sky and Telescope, June 2004, 40–45; Anurag Sharma, et al., "Microbial Activity at Gigapascal Pressures," Science, Vol. 295 (22 February 2002), 1514–1516; Darling, 174.

43. Bor Luen Tang, "Many Possibilities for Life's Emergence." JBIS, Vol. 58 (2005), 218–221.

44. Dick and Strick, 130; William J. Broad, "Scientists Widen the Hunt for Alien Life," The New York Times, 6 May 1997.

Probabilities: Intelligence

1. From "The War of the Worlds," in The Complete Science Fiction Treasury of H.G. Wells, New York, Avenel, 266.

2. Schilling, "The Chance of Finding Aliens," 41; Michael D. Papagiannis, "Colonies in the Asteroid Belt, or A Missing Term in the Drake Equation," in Hart and Zuckerman, editors, 77–86.

3. Carl Sagan and Frank Drake, "The Search for Extraterrestrial Intelligence," Scientific American, May 1975, 80–90; Casti, 403–404. Carter produced a statistical argument showing that the existence of other intelligences is highly improbable because evolutions to intelligence are likely to take longer than the lifetime of a host star on the main sequence.

4. Morris, Life's Solution, xii, 298–299; Brian McConnell, Beyond Contact, Sebastopol, CA, O'Reilly, 2001, 71.

5. George Gaylord Simpson, "The Nonprevalence of Humanoids," Science, Vol. 143 (1964), 769, reprinted in Goldsmith, editor, 214–221.

6. Project Cyclops, 24.

7. Stephen Jay Gould, "The Wisdom of Casey Stengel," Discover, March 1983, 62–65.

8. Stephen Jay Gould, "The Evolution of Life on the Earth," Scientific American, October 1994, 85–91.

9. Clarke, in Bova and Preiss, editors, 310–311.

10. Davies, Are We Alone?, 103, 106–107, 128; Billingham, et al., editors, Social Implications, 45.

11. Sagan, editor, CETI, 87–91, 105.

12. Thore J. Bergman, et al., "Hierarchical Classification by Rank and Kinship in Baboons," Science, Vol. 302 (14 November 2003), 1234–1236; Rood and Trefil, 89.

13. Gary F. Marcus, "Before the Word," Nature, Vol. 431 (14 October 2004), 745.

14. Ernst Mayr, Letters column, Science, Vol. 259 (12 March 1993), 1522; Letters column, Science, Vol. 260 (23 April 1993), 474–475; Mayr, in Regis, editor, 24. The texts of the Mayr-Sagan debate were published in The Planetary Society's Bioastronomy News, Vol. 7 (Fourth Quarter 1995).

15. Frank Drake, "Methods of Communication," in Cyril Ponnamperuma and A.G.W. Cameron, editors, Interstellar Communication: Scientific Perspectives, Boston, Houghton-Mifflin, 1974, 118–139.

16. Frank Drake, "On Intelligence: A Wave of the Past and Future," Cosmic Search, Vol. 3, Number 2 (Spring 1981), 5.

17. Dale A. Russell, "Speculations on the Evolution of Intelligence in Multicellular Organisms," in Billingham, editor, Life in the Universe, 259–270; Richard G. Klein with Blake Edgar, The Dawn of Human Culture, New York, Wiley, 2002, 22; Oliver Curry, "A Change of Mind?" (review of Buller's Adapting Minds), Nature, Vol. 435 (26 May 2005), 425–426; Lovejoy, in Billingham, editor, Life in the Universe, 326; Hansell H. Stedman, et al., "Myosin Gene Mutation Correlates with Anatomical Changes in the Human Lineage," Nature, Vol. 428 (25 March 2004) 415–418; Pete Currie, "Muscling in on Hominid Evolution," Nature, Vol. 428 (25 March 2004), 373–374.

18. Klein and Edgar, 8, 252; Jean-Jaques Hublin, "An Evolutionary Odyssey" (review of Klein's book The Human Career), Nature, Vol. 403 (27 January 2000, 363; Carel Van Schaik, "Why Are Some Animals So Smart?", Scientific American, April 2006, 65–71.

19. Ann Gibbons, "Bone Sizes Trace the Decline of Man (and Woman), Science, Vol. 276 (9 May 1997) 896–897; C. Owen Lovejoy, "Evolution of Man and its Implications," in Billingham, editor, Life in the Universe, 317–327.

20. "Big Brains Rule the Roost," Science, Vol. 307 (25 March 2005), 1867; Fred Hoyle, The Intelligent Universe, 229.
21. Lovejoy, in Billingham, editor, Life in the Universe, 326.
22. William H. Calvin, "The Emergence of Intelligence," Scientific American, October 1994, 101–107; Morris, 246; Stephen Mithen, The Prehistory of the Mind: The Cognitive Origins of Art and Science, London, Thames and Hudson, 1996, 209–210.
23. Christian de Duve, "Lessons of Life," in Dick, editor, Many Worlds, 3–13; John H. Mauldin, Prospects for Interstellar Travel, San Diego, American Astronautical Society, 1992, 279.
24. William H. Calvin, A Brain for All Seasons: Human Evolution and Abrupt Climate Change, Chicago, University of Chicago Press, 2002; Emma Bakes, "DNA as a Mechanism for SETI," SETI Institute News, Vol. 11 (Fourth Quarter 2002), 1–3.
25. Darling, 100.
26. Seth Shostak and Alex Barnett, Cosmic Company, New York, Oxford University Press, 2003, 63.
27. Ian Tattersall, "Once We Were Not Alone," Scientific American, January 2000, 56–62; Ian Tattersall and Jeffrey Schwartz, Extinct Humans, Boulder, CO, Westview Press, 2000; Ann Gibbons, "Homo erectus in Java: A 250,000 Year Anachronism," Science, Vol. 274 (13 December 1996), 1841–1842; John Noble Wilford, "3 Human Species Coexisted On Earth, New Data Suggest," The New York Times, 3 December 1996.
28. M.J. Morwood, et al., "Archaeology and Age of a New Hominin from Flores in Eastern Indonesia," Nature, Vol. 431 (28 October 2004), 1087–1091; P. Brown, et al., "A New Small-Bodied Hominin from the Late Pleistocene of Flores, Indonesia," Nature, Vol. 431 (28 October 2004), 1055–1061; Marta Mirazon Lahr and Robert Foley, "Human Evolution Writ Small," Nature, Vol. 431 (28 October 2004), 1043–1044; Kate Wong, The Littlest Human, Scientific American, February 2005, 56–65; Nicholas Wade, "Miniature People Add Extra Pieces to Evolutionary Puzzle," The New York Times, 9 November 2004.
29. Michel Balter, "Small But Smart? Flores Hominid Shows Signs of Advanced Brain," Science, Vol. 307 (4 March 2005), 1386–1389; Ann Gibbons, "Bone Sizes Trace the Decline of Man (and Woman), Nature, Vol 387 (8 May 1997), 896–897; John Kappelman, "They Might be Giants," Nature, Vol. 387 (8 May 1997), 126–127; Felipe Fernandez-Armesto, Humankind: A Brief History, New York, Oxford University Press, 2004, 82.
30. Ann Gibbons, "The Mystery of Humanity's Missing Mutations," Science, Vol. 267 (6 January 1995), 35–36; "Stone Age Sophistication," Scientific American, June 2005, 89; Jorge A. Vazquez and Mary R. Reid, "Probing the Accumulation History of the Voluminous Toba Magma," Science, Vol. 305 (13 August 2004), 991–993.
31. John C. Lilly, Man and Dolphin, New York, Pyramid, 1961, 154.
32. Lilly, 7.
33. Morris, 247–260. Also see Diana Reiss, "The Dolphin: An Alien Intelligence," in Bova and Preiss, editors, 31–40.
34. Gregory Benford, "Alien Technology," in Bova and Preiss, editors, 165–177.

35. Morris, 247–248; Seth Shostak, "The Drake Equation Revisited," SETI Institute Explorer, Special Edition, January 2005, 4.

36. Nathan J. Emery and Nicola S. Clayton, "The Mentality of Crows: Convergent Evolution of Intelligence in Corvids and Apes," Science, Vol. 306 (10 December 2004), 1903–1907.

37. In Bova and Preiss, editors, 39.

38. Mark Derr, "Brainy Dolphins Pass the Human Mirror Test," The New York Times, 1 May 2001; Morris, 258; Ian L. Boyd, review of Hal Whitehead's book Sperm Whales: Social Evolution in the Ocean, Science, Vol. 302 (7 November 2003), 990–991; Clive D. L. Wynne, Do Animals Think?, Princeton, NJ, Princeton University Press, 2004, and Sara Shettleworth's review in Nature, Vol. 430 (8 July 2004), 148; Roulet, 161–162. Other species besides humans and dolphins engage in vocal imitation—other primates, some birds, bats, and elephants. Researchers speculate that mimicry helps maintain social bonds in fluid societies. Joyce H. Poole, et al., "Elephants Are Capable of Vocal Learning," Nature, Vol. 434 (24 March 2005), 455–456.

39. Morris, 260.

40. Jill Tarter, "Searching for Extraterrestrials," in Regis, editor, 171.

41. Norman H. Sleep, "Out, Out, Brief Candle?" (book review), Nature, Vol. 422 (17 April 2003), 663–664.

42. Bruce E. Fleury, "The Aliens in Our Oceans: Dolphins as Analogs," Cosmic Search, Vol. 2, Number 2 (Spring 1980), 2–5.

43. Fleury, "The Aliens in Our Oceans;" N.J. Berrill, Worlds Without End, New York, Macmillan, 1964, 207; Lilly, 128.

44. William L. Abler, Letter to the Editor, Nature, Vol. 438 (24 November 2005), 422.

45. Ze Cheng, et al., "A Genome-Wide Comparison of Recent Chimpanzee and Human Segmental Duplications," Nature, Vol. 437 (1 September 2005), 88–92.

46. Duane M. Rumbaugh and David A. Washburn, Intelligence of Apes and Other Rational Beings, New Haven, CT, Yale University Press, 2003. Also see Andrew Whiten's review in Nature, Vol. 425 (2 October 2003), 454.

47. David Premack, "Is Language the Key to Human Intelligence?," Science, Vol. 303 (16 January 2004), 318–320; Amy Poremba, et al., "Species-Specific Calls Evoke Asymmetric Activity in the Monkey's Temporal Poles," Nature, Vol. 427 (29 January 2004), 448–450.

48. Craig Stanford, Significant Others: The Ape-Human Continuum and the Quest for Human Nature, New York, Basic Books, 2002; circular letter from the Jane Goodall Institute, April 2004; Peter D. Walsh, et al., "Catastrophic Ape Decline in Western Equatorial Africa," Nature, Vol. 422 (10 April 2003), 611–613.

49. Edward Harrison, Masks of the Universe, 115.

50. De Duve, in Dick, editor, Many Worlds, 10–11; Morris, Life's Solution, 253.

51. "Why Life May Be Common—and Intelligence Rare," Sky and Telescope, February 2003, 26; MacGowan and Ordway, 189; David Brin, "Xenology: The New Science of Asking Who's Out There," Analog Science Fiction/Science Fact, May 1983, 64–82.

52. Drake and Sobel, 209.

53. Dale A. Russell and R. Seguin, "Reconstruction of the Small Cretaceous Therapod *Stenonychsaurus inequalis* and a Hypothetical Dinosaurid," Syllogeus, Vol. 37 (1982), 1–43; David M. Raup, "ETI Without Intelligence," in Regis, editor, 31–42; Darling, 144. For excellent pictures of this reconstruction, see Terence Dickinson and Adolf Schaller, Extraterrestrials: A Field Guide for Earthlings, Camden, Ontario, Camden House, 1994, 32–33.
54. Mark Greene, et al., "Moral Issues of Human-Non-Human Primate Neural Grafting," Science, Vol. 309 (15 July 2005), 385–386.
55. Doris and David Jonas, Other Senses, Other Worlds, New York, Stein and Day, 1978.

Probabilities: Civilization, Technology, and Science

1. Anthony Padgen, Peoples and Empires, New York, Modern Library, 2003, 26–27.
2. K. Flannery, in Sagan, editor, CETI, 94–95.
3. T.M. Lenton, et al., "Climbing the Co-Evolution Ladder," Nature, Vol. 431 (21 October 2004), 913.
4. On convergence, see Robert Wright, Non-Zero: The Logic of Human Destiny, New York, Vintage (Random House), 2000. A condensation of Volumes I to VI of Toynbee's A Study of History was prepared by D.C. Somervell, New York, Oxford University Press, 1947.
5. Nigel Davies, The Toltec Heritage: From the Fall of Tula to the Rise of Tenochtitlan, Norman, OK, University of Oklahoma Press, 1980, 320.
6. Ben Finney, "SETI, Consilience, and the Unity of Knowledge," in Tough, editor, 139–142; Steven J. Dick, "They Aren't Who You Think," Mercury, November–December, Vol. 32, Number 6 (2003), 17–26.
7. Wright, Non-Zero, 108.
8. White, The SETI Factor, 84.
9. Kaplan, editor, Extraterrestrial Civilizations, 4, 214–215; Project Cyclops, 180.
10. Examples include Robert Crowley, editor, What If?: The World's Foremost Military Historians Imagine What Might Have Been, New York, G. Putnam's Sons, 1999, and What If? 2: Eminent Historians Imagine What Might Have Been, New York, G. Putnam's Sons, 2001. A particularly well thought out alternative history can be found in David Downing, The Moscow Option: An Alternative Second World War, London, Greenhill Books, 1979, revised 2001.
11. Greg Miller, "Tool Study Supports Chimp Culture," Science, Vol. 309 (26 August 2005), 1311; Andrew Whiten, et al., "Conformity to Cultural Norms of Tool Use in Chimpanzees," Nature, Vol. 437 (29 September 2005), 737–740; Carel P. van Schaik, et al., "Orangutan Cultures and the Evolution of Material Culture," Science, Vol. 299 (3 January 2003), 102–105.
12. Shklovskii and Sagan, 411.
13. MacGowan and Ordway, 182; McConnell, 62.
14. Morrison, et al., editors, SETI, 51–52.
15. Bernard Campbell, "Evolution of Technological Species," in Billingham, editor, Life in the Universe, 277–285; Morris, Life's Solution, 262, 268.

16. Davies, Are We Alone?, 49.
17. Felipe Fernandez-Armesto, Civilizations: Culture, Ambition, and the Trans-formation of Nature, New York, Macmillan, 2001.
18. K. Flannery, in Sagan, editor, CETI, 96; Jared Diamond, Guns, Germs, and Steel: The Fates of Human Societies, New York, Norton, 1997.
19. Richard Burke-Ward, "Possible Existence of Extraterrestrial Technology in the Solar System," JBIS, Vol. 53 (2000), 2–12.
20. The first thought is from Ronald Puccetti, Persons, 101, the second from Carl De Vito, "Languages Based on Science," Acta Astronautica, Vol. 26 (1992), 267–271.
21. Jack Cohen and Ian Stewart, What Does a Martian Look Like?, Hoboken, NJ, Wiley, 2002, 297.
22. David Sivier, "SETI and the Historian," JBIS, Vol. 53 (2000), 23–25.

Probabilities: Longevity

1. David Schwartzman and Lee J. Rickard, "Being Optimistic about the Search for Extraterrestrial Intelligence," American Scientist, Vol. 76 (1988), 364–368.
2. Sagan, editor, CETI, 6.
3. Sagan, Cosmos, 41.
4. Ekers et al., 22, 115; Carl Sagan, "On the Detectivity of Advanced Galactic Civilizations," Icarus, Vol. 19 (1973), 350–351, reprinted in Goldsmith, editor, 140–141; Sagan, editor, CETI, 366.
5. Shapley, 155.
6. Jared Diamond, Collapse: How Societies Choose to Fail or Succeed, New York, Viking/Allen Lane, 2005.
7. White, 86; Swift, 367; David Brin, "Mystery of the Great Silence," in Bova and Preiss, editors, 118–139.
8. Sagan, "The Quest for Intelligent Life;" Drake and Sobel, 211.
9. Carl Sagan and William I. Newman, "The Solipsist Approach to Extrater-restrial Intelligence," in Regis, editor, 151–161; Sebastian von Hoerner, "The Likelihood of Interstellar Colonization, and the Absence of its Evidence," in Hart and Zuckerman, editors, 29–33; "von Hoerner on SETI" (interview), Cosmic Search, Vol. 1, Number 1 (January 1979), 40–45.
10. Swift, 113; Bracewell, in Cameron, editor, 242; Goldsmith and Owen, 357.
11. Sullivan, 296; Bracewell, in Cameron, editor, 242.
12. Alfred Roulet, The Search for Intelligent Life in Outer Space, originally published in French in 1973, translated and edited by William A. Packer, New York, Berkley, 1977, 164; Frank Drake, "Methods of Communication," in Cyril Ponnamperuma and A.G.W. Cameron, editors, Interstellar Communi-cation: Scientific Perspectives, Boston, Houghton-Mifflin, 1974, 118–139; Swift, 367.
13. Swift, 293.
14. Sullivan, xii.
15. Burger, 294.
16. Burger, 289.

17. Dick Taverne, review of Red Sky at Morning, Nature, Vol. 432 (25 November 2004), 443–444.
18. Gregg Easterbrook, review of Jared Diamond's book Collapse, The New York Times, 30 January 2005.
19. Platt, in Sagan, editor, CETI, 153.
20. Christian de Duve, Life Evolving, 261.
21. Rees, Our Final Hour, 3, 8, 139, 185–186.
22. Steen Rasumussen, et al., "Transitions from Nonliving to Living Matter," Science, Vol. 303 (13 February 2004), 963–965; Gretchen Vogel, "How Can a Skin Cell Become a Nerve Cell?", Science, Vol. 309 (1 July 2005), 85.
23. "Microbes Made to Order," Science, Vol. 303 (9 January 2004), 158–161; W. Wayt Gibbs, "Synthetic Life," Scientific American, May 2004, 75–81; Philip Ball, "Starting from Scratch," Nature, Vol. 431 (7 October 2004), 624–626; "Synthetic Biologists Face Up to Security Issues," Nature, Vol. 436 (18 August 2005), 894–895; "Futures of Artificial Life," Nature, Vol. 431 (7 October 2004), 613; Andreas von Bubnoff, "The 1918 flu virus is resurrected," Nature, Vol. 437 (6 October 2005), 794–795.
24. James Glanz, "Experiments on Dense Matter Evoke Big Bang," The New York Times, 16 January 2001.
25. Sheldon L. Glashow and Richard Wilson, "Taking Serious Risks Seriously," Nature, Vol. 402 (9 December 1999), 596–597; Charlie LeDuff, "When Worlds, and Fears, Collide," The New York Times, 6 April 1999.
26. David Brin, Earth, New York, Bantam, 1990; Bernard J. Carr and Steven B. Giddings, "Quantum Black Holes," Scientific American, May 2005, 48–55.
27. Rees, Our Final Hour, 116, 125.
28. Ulmschneider, 200–204.
29. Tenzin Gyatso, "Our Faith in Science," The New York Times, 12 November 2005.
30. Steven J. Dick, "Cultural Evolution, the Postbiological Universe and SETI," International Journal of Astrobiology, Vol. 2 (2003), 65–74.
31. Sivier, "SETI and the Historian;" MacGowan and Ordway, 248.
32. Clark and Clark, 216–217; Shostak, 104.
33. David Brin, "Mystery of the Great Silence," in Bova and Preiss, editors, 118–139; Shostak, 104.
34. Peter Schenkel, "The Nature of ETI, Its Longevity and Likely Interest in Mankind: The Human Analogy Re-Examined," JBIS, Vol. 52 (1999), 13–18.
35. Arthur C. Clarke, Foreword to Michael Benson's Beyond: Visions of the Interplanetary Probes, New York, Abrams, 2003, 9–11.
36. MacGowan and Ordway, 233–235. Bob Parkinson visualized the exploration (and exploitation) of the solar system as a joint carbon–silicon enterprise. "The Carbon or Silicon Colonization of the Universe?" JBIS, Vol. 58 (2005), 111–116.
37. Koerner and LeVay, 211; Hans Moravec, Mind Children: The Future of Robot and Human Intelligence, Cambridge, MA, Harvard University Press, 1988, republished 1990.
38. Ray Kurzweil, Age of Spiritual Machines: When Computers Exceed Human Intelligence, New York, Texere, 2001; Michael A. Goldman, "Promises and

Perils of Technology's Future" (review of Kurzweil's Living with the Genie), Science, Vol. 303 (30 January 2004), 629.

39. Rees, Our Final Hour, 19, 167; Barrow and Tipler, 595.

40. Sagan, editor, CETI, 160–161.

41. Clarke, Greetings, Carbon-Based Bipeds, 276; Mark Rose, Alien Encounters: Anatomy of Science Fiction, Harvard University Press, 1981, 152; Shostak and Barnett, Cosmic Company, 70.

42. Frank Herbert, Dune, Radnor, PA, Chilton, 1965; Berkeley Medallion Edition 1977.

43. Steven Dick, "They Aren't Who You Think." Dick had published a more detailed version entitled "Cultural Evolution, the Postbiological Universe and SETI," International Journal of Astrobiology," Vol. 2 (2003), 65–74. Also see Peter J. Richerson and Robert Boyd, Not by Genes Alone: How Culture Transformed Human Evolution, Chicago, University of Chicago Press, 2004.

44. Drake and Sobel, 52.

45. R.E. Spier, review of two books on genetic engineering, Science, Vol. 296 (7 June 2002), 1807–1809; Morrison in Ekers et al., xxx; Richard Lewontin, "The Wars Over Evolution," The New York Review of Books, 20 October 2005, 51–54.

46. Ulmschneider, 212–213.

47. Christian de Duve, "Lessons of Life," in Dick, editor, Many Worlds, 10.

48. Arthur C. Clarke, Interplanetary Flight, New York, Berkley, 1985, 147; originally published by Harper and Row in 1950.

49. Isaac Asimov, A Choice of Catastrophes: The Disasters That Threaten Our World, New York, Fawcett, 1979, passim.

50. L.W. Alvarez, et al., "Extraterrestrial Cause for the Cretaceous-Tertiary Extinction," Science, Vol. 208 (6 June 1980), 1095–1108. Richard A. Kerr, "Whiff of Gas Points to Impact Mass Extinction," Science, Vol. 291 (23 February 2001), 1469–1470; Kenneth Chang, "Meteor Seen As Causing Extinctions on Earth," The New York Times, 21 November 2003; Kenneth Chang, "Scientists Say Crater Is Result of a Killer Meteor," The New York Times, 14 May 2004; L. Becker, et al., "Bedout: A Possible End—Permian Impact Crater Offshore of Northwestern Australia," Science, Vol. 304 (4 June 2004), 1469–1476. For an overall view of the end Permian extinction, see Michael Benton, When Life Nearly Died: The Greatest Mass Extinction of All Time, New York, Thames and Hudson, 2003 and Duncan Steel, Rogue Asteroids and Doomsday Comets, New York, Wiley, 1995, 104–107.

51. David J. Dunlop, "Magnetic Impact Craters," Nature, Vol. 435 (12 May 2005), 156–157, and Laurent Carporzen, et al., "Paleomagnetism of the Vredefort Meteorite Crater and Implications for Craters on Mars," Nature, Vol. 435 (12 May 2005), 198–201.

52. Steel, 20–21; Clarke, Greetings, 505; David Keys, Catastrophe: An Investigation into the Origins of the Modern World, New York, Ballantine, 1999. The story of the long effort to develop and confirm the dinosaur-killing impact theory is told in Walter Alvarez, T. Rex and the Crater of Doom, Princeton, NJ, Princeton University Press, 1997.

53. Steel, 187, 203–204, 247–259; Arthur C. Clarke, Rendezvous with Rama, New York, Bantam, 1990; originally published by Harcourt Brace in 1973.

54. Leonard David, "Homeland Defense for Earth," Aerospace America, August 2004, 41–44; Steel, 7; "Two Telescopes Join Hunt for ET," Vol. 440 (13 April 2006), 853.

55. Alan MacRobert, "Asteroid 2004MN4: A Really Near Miss," Sky and Telescope, May 2005, 16–17. Former astronaut Rusty Schweikart provided a useful overview in "We Must Decide to Do It! The Saga of Asteroid 2004 MN4," The Planetary Report, July–August 2005, 14–18.

56. Bova, 81.

57. Dick and Strick, 126; Ekers, et al., 287; Christopher F. Chyba, "Life Beyond Mars," Nature, Vol. 382 (15 August 1996), 576–577; Steel, 103.

58. Robert S. Dixon, "What Caused the Dinosaurs Extinction?" Cosmic Search, Vol. 2, Number 4 (Year End 1980), 14; Luann Becker, "Repeated Blows," Scientific American, March 2002, 76–83.

59. Joseph A. Burns, "Double Trouble," Nature, Vol. 427 (5 February 2004), 494–495; Roald Tagle and Philippe Claeys, "Comet or Asteroid Shower in the Late Eocene?" Science, Vol. 305 (23 July 2004), 492.

60. Sagan, Pale Blue Dot, 327; Clarke, Greetings, 508; Chyba, "Life Beyond Mars."

61. Fred Hoyle, The Black Cloud, New York, Penguin, 1960 (originally published by Heinemann in 1957); David H. Clark, "Extraterrestrial Climate Threats," Astronomy, August 1980, 66–71; "Our Galaxy's Deadly Spiral Arms," Sky and Telescope, July 1998, 20; John Noble Wilford, "Sun's Cruise Through Space is About to Hit Bumpy Patch," The New York Times, 18 June 1996. The quotation is from "50, 100, and 150 Years Ago," Scientific American, February 2006, 16.

62. Nigel Henbest and Heather Couper, The Guide to the Galaxy, Cambridge, Cambridge University Press, 1994, 93–94.

63. Clark, "Extraterrestrial Climate Threats;" Walter Alvarez, 58, 74; Sagan, Cosmos, 235; John Noble Wilford, "Heavens Send Earth An Enormous Surge of Stellar Radiation," The New York Times, 30 September 1998.

64. See Robert Irion, "A Very Good Year for Explosions," Science, Vol. 311 (6 January 2006), 30–32, and the cluster of reports by E. Berger, et al., N.R. Tanvir, et al., and S.D. Barthelmy, et al., Nature, Vol. 438 (15 December 2005), 988–996.

65. Peter J.T. Leonard and Jerry T. Bonnell, "Gamma-Ray Bursts of Doom," Sky and Telescope, February 1998, 28–34.

66. Neil Gehrels, et al., "The Brightest Explosions in the Universe," Scientific American, December 2002, 85–91; Malcolm W. Browne, "Astronomers Detect Immense Explosion 2d Only to Big Bang," The New York Times, 7 May 1998. S. Yu. Sazonov, et al. reported that, in the brightest gamma-ray bursters, the gamma-rays are so highly collimated that these events can be seen across the observable universe. "An Apparently Normal Gamma Ray Burst with an Unusually Low Luminosity," Nature, Vol. 430 (5 August 2004), 647.

67. Leonard and Bonnell, "Gamma-Ray Bursts;" Clark and Clark, 218.

68. In Allen Tough, editor, When SETI Succeeds: The Impact of High-Information Contact, Foundation for the Future, 2000, 105.

69. Milan M. Cirkovic, "Permanence—An Adaptationist Solution to Fermi's Paradox," JBIS, Vol. 58 (2005), 62–70; Milan M. Cirkovic, "Earths: Rare in

Time, Not Space?" JBIS, Vol. 57 (2004), 53–59. Also see James Annis, "An Astrophysical Explanation for the Great Silence," JBIS, Vol. 52 (1999), 19–22.

70. "Our Galaxy's Stellar Speedsters," Sky and Telescope, May 1996, 15; Robert Naeye, "Extreme Pulsars," Sky and Telescope, December 2005, 20; Sagan, Pale Blue Dot, 393; J. Diemand, et al., "Earth-Mass Dark-Matter Haloes as the First Structures in the Universe," Nature, Vol. 433 (27 January 2005), 389–391; E.P.J. Van den Heuvel, "Pulsar Magnetospheres and Pulsar Death," Science, Volume 312 (28 April 2006), 539–540.

71. Ray Jayawardhana, "Unraveling Brown Dwarf Origins," Science, Vol. 303 (16 January 2004), 322–323; Robert Irion, "Astronomers Shine a Light Upon Dim Nearby Stars," Science, Vol. 304 (11 June 2004), 1587–1589; "Protostellar Ping-Pong," Science, Vol. 303 (2 January 2004), 16.

72. "Silicon Solar System," Science, Vol. 303 (19 March 2004), 1741.

73. Mark Rose, Alien Encounters: Anatomy of Science Fiction, Cambridge, MA, Harvard University Press, 1981, 24.

74. Fred C. Adams and Gregory Laughlin, "The Future of the Universe," Sky and Telescope, August 1998, 32–39; Mark A. Garlick, "The Fate of the Earth," Sky and Telescope, October 2002, 30–35; Jill Tarter, "Implications of Contact with ETI Far Older than Humankind," in Tough, editor, 45.

75. Bova, 286.

76. Quoted by Euan Nisbet in his review of Doug MacDougall's Frozen Earth, Nature, Vol. 432 (9 December 2004), 673.

77. Kevin Krajik, "Tracking Myth to Geological Reality," Science, Vol. 310 (4 November 2005), 762–764.

78. Dick and Strick, 230.

The Drake Equation, Take Two

1. Ward and Brownlee, 274–275.

2. Michael Papagiannis, "Colonies in the Asteroid Belt, or a Missing Term in the Drake Equation," in Hart and Zuckerman, editors, 77–84; John H. Mauldin, Prospects for Interstellar Travel, San Diego, American Astronautical Society, 1992, 260; Simon Goodwin with John Gribbin, XTL: Extraterrestrial Life and How to Find It, London, Seven Dials, 2002, 115.

Should We Continue the Search?

1. Goldsmith and Owen, 432.

2. Casti, 413.

3. For a list, see Ekers, et al., 381–425.

4. Grinspoon, 305; Frank Drake, "On Hands and Knees in Search of Elysium," Technology Review, June 1976, 22–29. Drake repeated that thought in his 1992 autobiographical book. Drake and Sobel, 234.

5. Swift, 107; Achenbach, 280; Michael J. Klein, "Where and What Can We See?"

6. Trefil was quoted in Gregg Easterbrook, "Are We Alone?" The Atlantic Monthly, August 1988, 25–38; Rood and Trefil, 239.

7. Ekers et al., editors, 34.

8. Ekers, et al., editors, 283–301.

9. In Bova and Preiss, editors, 147.

10. Drake, "Reflections on the Modern History of SETI."

11. Achenbach, 361.

12. Paul Horowitz and William R. Alschuler, "The Harvard SETI Search," in Bova and Preiss, editors, 206–214.

13. Paul Horowitz and Ken Clubok, "High Resolution SETI," Acta Astronautica, Vol. 26 (1992), 193–200; Ekers, et al., 7.

14. Ekers, et al., 180; Seth Shostak, "The Future of SETI," Sky and Telescope, April 2001, 50; "In Support of Xeno-Optimism," Nature, Vol. 421 (20 February 2003), 769; Woodruff T. Sullivan III, "Eavesdropping Mode and Radio Leakage from Earth," in Billingham, editor, Life in the Universe, 377–390. Also see W.T. Sullivan III, et al., "Eavesdropping: The Radio Signature of the Earth," Science, Vol. 199 (27 January 1978), 377–388.

15. T.L. Wilson, "The Search for Extraterrestrial Intelligence," Nature, Vol. 409 (22 February 2001), 1110–1114.

16. Jill Tarter, in Dick, editor, Many Worlds, 145; Ekers, et al., 230; Sagan, editor, CETI, 368.

17. Arthur C. Clarke, "Where Are They?" in Bova and Preiss, editors, 308.

18. Ekers et al., 150; Richard A. Kerr, "No Din of Alien Chatter in Our Neighborhood," Science, Vol. 303 (20 February 2004), 1133.

19. Frank Drake, "Promising New Approaches in the Search for Extraterrestrial Intelligence," paper presented at the 2002 International Astronautical Congress (IAC-02-IAA.9.1.01).

20. Beatty and MacRobert, "Why no SETI Signals Yet?"; Drake and Sobel, 69.

21. Drake and Sobel, 233.

22. White, 70–71, 100.

23. Kerr, "No Din of Alien Chatter;" Frank D. Drake, "When the Brightest is the Farthest," Cosmic Search, Vol. 3, Number 2 (Fall 1981), 7.

24. Alan M. MacRobert, "50 Ways to Lose Your Aliens" (review of Stephen Webb's If the Universe is Teeming with Aliens, Where is Everybody?), Sky and Telescope, March 2003, 70–71.

25. McDonough, 209.

26. Frank J. Tipler, "Extraterrestrial Intelligence: A Skeptical View of Radio Searches," letter to Science, Vol. 219 (14 January 1983), 110–112; Crowe, 548; Drake and Sobel, 206; Dennis Overbye, "Under Prodding, Cosmologists Debate, Well, Everything," The New York Times, 29 April 2003; Gordon Kane, "The Mysteries of Mass," Scientific American, July 2005, 41–48; Philip Morrison, "Reflections," in Billingham, editor, Life in the Universe, 421–433.

27. Clark and Clark, 98–99.

28. David Black, et al., Mercury, 1977; Kerr, "No Din of Alien Chatter."

29. Swift, 321: Clarke, Greetings, 480–481.

30. MacGowan and Ordway, 355.

31. Morrison, et al., 1977, 7.

32. Dick, "SETI Before the Space Age," 7; "Back to the Future: SETI Before the Space Age," paper presented at the 1993 International Astronautical Congress (IAA.9.2.93-790); William Sims Bainbridge, The Spaceflight Revolution: A Sociological Study, New York, Wiley, 1976, 3.
33. Rees, 142.
34. Ekers, et al., 3.
35. Achenbach, 355.
36. Sagan, Pale Blue Dot, 365; Achenbach, 62.
37. Roger Angel, "Future Optical and Infrared Telescopes," Nature, Vol. 409 (18 January 2001), 427–430; Michaud, "SETI Hearings in Washington."
38. Quoted by Norman N. Nelson, "Some Diplomatic and Journalistic Implications of SETI," paper presented at the 1985 International Astronautical Congress (IAA-85-469).
39. Peter B. Boyce, "Signal Verification in the Real World: When Is a Signal not a Signal?" Acta Astronautica, Vol. 21 (1990), 81–84; Albert A. Harrison, After Contact: The Human Response to Extraterrestrial Life, New York, Plenum, 1997, 319.
40. Steven J. Dick, "Consequences of Success in SETI," in G. Seth Shostak, editor, Progress in the Search for Extraterrestrial Life (1993 Bioastronomy Symposium), Astronomical Society of the Pacific, 1995, 521–530.
41. Dick, Dibner Lecture, 3.
42. Guthke, 16; Black, et al.
43. Allen Tough, "Positive Consequences of SETI Before Detection," paper presented at the 1995 International Astronautical Congress (IAA-95-IAA.9.2.06).
44. Morrison, et al., editors, 7.

Direct Contact

1. Stapledon, 327–328.
2. In A.G.W. Cameron, editor, Interstellar Communication, Amsterdam, New York, Benjamin, 1963, 144–159.
3. In Cameron, editor, 121–143.
4. Morrison, et al., 1977, 107.
5. Black, et al., 3.
6. Brin, "Xenology," 69; Drake and Sobel, 120.
7. Sullivan, 226; Shklovskii and Sagan, 449.
8. Arthur C. Clarke, The Promise of Space, New York, Harper and Row, 1968, 313.
9. In Cameron, editor, 147.
10. Goldsmith and Owen, 372.
11. Swift, 150.
12. Arthur C. Clarke, Greetings, Carbon-Based Bipeds!, New York, St. Martin's, 1999, 49–50.
13. See, among others, Michael A.G. Michaud, "Spaceflight, Colonization, and Independence: A Synthesis," JBIS, Vol. 30, Number 3 (March, 1977), 83–95 (Part One); Vol. 30, Number 6 (June, 1977), 203–212 (Part Two); Vol. 30, Number 9 (September, 1977), 323–331 (Part Three).

14. A. Bond, et al., Project Daedalus: The Final Report on the BIS Starship Study, special issue of JBIS, 1978.
15. Clarke, Greetings, 52.
16. Clarke, in Bova and Preiss, editors, 307; Ian Crawford, "Where Are They?", Scientific American, July 2000, 38–43.
17. MacGowan and Ordway, 348; D.F. Spencer and L.D. Jaffe, "Feasibility of Interstellar Travel," Astronautica Acta, Vol. 9 (1963), 49–59.
18. Robert L. Forward, A National Program for Interstellar Exploration, Malibu, CA, Hughes Research Laboratories Research Report 492, 1975.
19. Freeman Dyson, "Interstellar Propulsion Systems," in Hart and Zuckerman, editors, 41–45. Forward reviewed propulsion options in his 1986 article "Feasibility of Interstellar Travel," JBIS, Vol. 39 (1986), 379–384, and in "Ad Astra!", JBIS, Vol. 49 (1996), 23–32. Paul Gilster gave us an updated survey in Centauri Dreams: Imagining and Planning Interstellar Exploration, New York, Copernicus, 2004.
20. Eugene F. Mallove and Robert L. Forward, Bibliography of Interstellar Travel and Communication, Malibu, CA, Hughes Research Laboratories Research Report 460, 1972; Robert L. Forward, Bibliography of Interstellar Travel and Communication—April 1977 Update, Malibu, CA, Hughes Research Laboratories Research Report 512, 1977; Gilster, 20; Eugene F. Mallove and Gregory L. Matloff, The Starflight Handbook, New York, Wiley, 1989; Casti, 389.
21. Robert L. Forward, Starwisp, Malibu, CA, Hughes Research Laboratories Research Report 555, 1983; Yoji Kondo, et al., editors, Interstellar Travel and Multi-Generation Space Ships, New York, Apogee Books, 2003, 57.
22. Gilster, 235.
23. MacGowan and Ordway, 253, 268.
24. Goldsmith, editor, 183.
25. Mallove and Matloff, 23; Kondo, 36–37; Gilster, 12; L.D. Jaffe, et al., "An Interstellar Precursor Mission," JBIS, Vol. 33 (1980), 3–26; Richard A. Kerr, "Voyager 1 Crosses a New Frontier and May Save Itself from Termination," Science, Vol. 308 (27 May 2005), 1237–1238; "Voyager Crosses Boundary," Sky and Telescope, September 2005, 22.
26. Morrison, et al., editors, 1977, 108.
27. Project Daedalus, Special Issue of JBIS, 1978.
28. Achenbach, 147–148; Gilster, 11–13; James Glanz, "Engineers Dream of Practical Star Flight," Science, Vol. 281 (7 August 1998), 765–767; Warren E. Leary, "NASA Mission Will Explore Solar System's Frozen Edge," The New York Times, 15 January 2006.
29. Arthur C. Clarke, introduction to Michael Benson, Beyond: Visions of the Interplanetary Probes, New York, Abrams, 2003, 11; Shostak, Sharing the Universe, 107–109.
30. Tipler, quoted in Kondo, et al., editors, 68.
31. Clarke, quoted in Mallove and Matloff, 4; John Kraus, "Gerard K. O'Neill on Space Colonization and SETI (interview)," Cosmic Search, Vol. 1, Number 2 (March 1979), 16–23. Also see Goldsmith, Voyage to the Milky Way, 225–226.
32. Grinspoon, 219.
33. Morrison, et al., editors, 1977, 108.

34. Bracewell, "Communications from Superior Galactic Communities." Also see Ronald N. Bracewell, The Galactic Club: Intelligent Life in Outer Space, San Francisco, Freeman, 1975, 70–83; Morrison, et al., 107; O.G. Villard, Jr., et al., "LDEs, Hoaxes, and the Cosmic Repeater Hypothesis," QST, LV (May 1971), 54–58.

35. Swift, 151; Morrison's remark is in Cameron, editor, 263.

36. Frank J. Tipler, "Alien Life," (review of Davoust's The Cosmic Water Hole), Nature, Vol. 354 (28 November 1991), 334–335.

37. Christopher Rose and Gregory Wright, "Inscribed Matter as an Energy-Efficient Means of Communication with an Extraterrestrial Civilization," Nature, Vol. 431 (3 September 2004), 47–49; Woodruff T. Sullivan, "Message in a Bottle," Nature, Vol. 431 (2 September 2004), 27–28. Bracewell had hinted at such a concept 20 years earlier, writing that a substantial reference library could be compressible into the volume of an interstellar probe. In Cyril Ponnamperuma and A.G.W. Cameron, editors, Interstellar Communication: Scientific Perspectives, Boston, Houghton-Mifflin, 1974, 116.

38. Freeman J. Dyson, Letter to the editor, Scientific American, April 1964. Quoted in MacGowan and Ordway, 347.

39. Leslie R. Shepherd, "Interstellar Flight," JBIS, Vol. 11 (1952), 149–167.

40. Mallove and Matloff, 199; Yoshinari Minami, "Traveling to the Stars: Possibilities Given by a Spacetime Featuring Imaginary Time," JBIS, Vol. 56 (2003), 205–211.

41. Charles Sheffield, "Fly Me to the Stars," in Kondo, et al., editors, 25; Stapledon, 329, 340; Mauldin, 163–164. Anthony R. Martin provided a useful overview of this concept in "World Ships—Concept, Cause, Cost, Construction, and Colonisation," JBIS, Vol. 37 (1984), 243–253.

42. J.D. Bernal, The World, The Flesh, and the Devil: An Inquiry into the Future of the Three Enemies of the Rational Soul, London, Jonathan Cape, 1970, 23–30, originally published in 1929. Also see Mallove and Matloff, 13; Gerard K. O'Neill, "The Colonization of Space," Physics Today, September 1974, 32–40, reprinted in Goldsmith, editor, 283–292; Gerard K. O'Neill, The High Frontier, New York, Morrow, 1977; T.A. Heppenheimer, Colonies in Space, New York, Warner, 1977; G.L. Matloff, "Utilization of O'Neill's Model I Lagrange Point Colony as an Interstellar Ark," JBIS, Vol. 29 (1976), 775–785; Michaud, "Spaceflight, Colonization, and Independence."

43. Sheffield, in Kondo, et al., editors, 20–28; Shostak, 141; Mauldin, 243.

44. Papagiannis, in Hart and Zuckerman, editors, 79; Michaud, "Spaceflight, Colonization, and Independence."

45. Martin, "World Ships," 251.

46. R.W. Moir and W.L. Barr, "Analysis of Interstellar Spacecraft Cycling Between the Sun and Nearby Stars," JBIS, Vol. 58 (2005), 332–341.

47. Sheffield, in Kondo, et al., editors, 26. Mark Ayre, et al. provided a useful overview of hibernation through cooling in "Morpheus—Hypometabolic Stasis in Humans for Long Term Space Flight," JBIS, Vol. 57 (2004), 325–339. Scientists reported in 2005 that hydrogen sulfide induces a state like suspended animation in mice. As core body temperature went down, oxygen consumption and carbon dioxide output declined to about 10% of normal.

Eric Blackstone, et al., "H2S Induces a Suspended Antimation-Like State in Mice," Science, Vol. 308 (22 April 2005), 518.

48. Feinberg and Shapiro, 431–432; Mauldin, 224; Achenbach, 65.
49. Clarke, The Promise of Space, 293; Mauldin, 161, citing a work by Fogg.
50. Mauldin, 225.
51. J. Kelley Beatty, "Distant Planetoid Sedna Baffles Astronomers," Sky and Telescope, June 2004, 14–15. For a general overview of research on trans-Neptunian bodies, see John Davies, Beyond Pluto: Exploring the Outer Limits of the Solar System, Cambridge, Cambridge University Press, 2001.
52. Mark Hopkins, "Future Earth Prosperity Will Depend on Resources in Space," Ad Astra, Vol. 16, Number 2 (April–May–June 2004), 14.
53. Eric M. Jones and Ben R. Finney, "Interstellar Nomads," in James D. Burke and April S. Whitt, editors, Space Manufacturing 1983, San Diego, Univelt (for the American Astronautical Society), 1983, 357–374; Eric Jones and Ben Finney, "Fastships and Nomads: Two Roads to the Stars," in Ben R. Finney and Eric M. Jones, Editors, Interstellar Migration and the Human Experience, Berkeley, CA, University of California Press, 1985, 88–102.
54. Quoted in Sagan, Pale Blue Dot, 382.
55. John MacVey, Interstellar Travel, New York, Stein and Day, 1977, 13; Michaud, "Spaceflight, Colonization and Independence," Part 2, 205. Also see Saul J. Adelman and Benjamin Adelman, Bound for the Stars, Englewood Cliffs, NJ, Prentice-Hall, 1981, 294–298; Yoji Kondo, "Interstellar Travel and Multi-Generation Space Ships: An Overview, in Kondo, et al., editors, 7–18; Robert H. Goddard, "The Ultimate Migration," manuscript dated 14 January 1918. In The Goddard Biblio Log, Friends of the Goddard Library, 11 November 1972; Project Cyclops, 31; Michael Mautner, "Life in the Cosmological Future," JBIS, Vol. 58 (2005), 167–180.
56. Barrow and Tipler, 590–591.
57. Clark and Clark, 208–209.
58. Darling, 102.
59. Rood and Trefil, 204.
60. In Michaud, "SETI Hearings in Washington."
61. Stapledon, 332; Stephen Baxter, "A Human Galaxy: A Prehistory of the Future," JBIS, Vol. 58 (2005), 138–142.
62. Achenbach, 317; Sagan, Pale Blue Dot, 327.
63. Barrow and Tipler, 600.
64. Bracewell, in Hart and Zuckerman, editors, 37–38; R.N. Bracewell, "Man's Role in the Galaxy," Cosmic Search, Vol. 1, Number 2 (March 1979), 48–51.
65. See, for example, Michael Balter, "Ancient DNA Yields Clues to the Puzzle of European Origins," Science, Vol. 310 (11 November 2005), 964–965.
66. Easterbrook, "Are We Alone?"; Asimov, A Choice of Catastrophes, 267; David G. Stephenson, "Extraterrestrial Cultures Within the Solar System?", QJRAS, Vol. 20 (1979), 422–426 (reprinted in Goldsmith, editor, 246–249); Lamb, 180.
67. Rood and Trefil, 206.
68. Jill Tarter, "Planned Observational Strategy for NASA's First Systematic Search for Extraterrestrial Intelligence," in Ben R. Finney and Eric M. Jones,

editors, Interstellar Migration and the Human Experience, Berkeley, CA, University of California Press, 1985, 314–329.

69. Scot Lloyd Stride, "An Instrument-Based Method to Search for Extraterrestrial Interstellar Robotic Probes," JBIS, Vol. 54, (2001), 2–13.

70. Ekers, et al., 2; Jill C. Tarter and Christopher F. Chyba, "Is There Life Elsewhere in the Universe?", Scientific American, December 1999, 118–123.

71. Richard Burke-Ward, "Possible Existence of Extra-Terrestrial Technology in the Solar System," JBIS, Vol. 53 (2000), 2–12; Stride, "An Instrument-Based Method," and Allen Tough, "Small Smart Interstellar Probes," JBIS, Vol. 51 (1998), 167–174.

72. Benford in Bova and Preiss, editors, 172.

73. Hart and Zuckerman, editors, 44–45; Ridpath, 109; Clark and Clark, 242.

74. Burke-Ward, "Possible Existence of Extra-Terrestrial Technology in the Solar System."

75. Barrow and Tipler, 591.

76. Papagiannis in Hart and Zuckerman, editors, 77–84; in Goldsmith, editor, 243–245; Bova and Preiss, editors, 189.

77. Gregory L. Matloff and Anthony R. Martin, "Suggested Targets for an Infrared Search for Artificial Kuiper Belt Objects," JBIS, Vol. 58 (2005), 51–61.

78. Koerner and LeVay, 212.

79. Alan E. Rubin, Disturbing the Solar System: Impacts, Close Encounters, and Coming Attractions, Princeton, NJ, Princeton University Press, 2002, 300.

80. "Astronomers Warm Up to the Infrared Universe," Science, Vol. 304 (18 June 2004), 1740; Rees, Our Final Hour, 172; Freeman J. Dyson, "Looking for Life in Unlikely Places," in Kondo, et al., editors, 105–119. A planned NASA spacecraft that would orbit Mars was described as the harbinger of a network of communications and space-weather outposts throughout the solar system. "Mars 2009," Sky and Telescope, May 2004, 27.

81. Michael C. Malin, "Finding Martian Landers," Sky and Telescope, July 2005, 42–46; Kenneth Chang, "Mars Lander Still Missing," The New York Times, 8 November 2005.

82. Arthur C. Clarke, By Space Possessed: Essays on the Exploration of Space, London, Gollancz, 1993, 126.

83. MacGowan and Ordway, 286.

84. Project Cyclops, 35.

85. MacVey, Interstellar Travel, 112–113.

86. Stephen Baxter, "Mars the Recording Angel: Traces of Extra-Martian Events in the Polar Layered Terrain," JBIS, Vol. 58 (2005), 206–210; Leonard David, "Uncovering the Secrets of Mars," Aerospace America, March 2006, 24–29.

87. Clark and Clark, 239.

88. Clarke, Greetings, 260. Space artist Adolph Schaller, in a book authored by Terence Dickinson, depicted a huge and mysterious alien artifact on Jupiter's moon Callisto. Extraterrestrials: A Field Guide for Earthlings, Buffalo, NY, Camden House, 1994, 58.

89. Tony Reichhardt, "Going Underground," Nature, Vol. 435 (19 May 2005), 266–267.

90. Paul Wason, "Symbolism and the Inference of Intelligence," SETI Institute Explorer, First Quarter 2005, 16–17.

91. Richard Hoagland described speculations about the Face and other mammoth relics in his book The Monuments of Mars, Berkeley, North Atlantic Books, 1987.
92. "Cydonia Defaced," Sky and Telescope, July 1998, 20; E.C. Krupp, "Facing Mars," Sky and Telescope, August 2003, 86–88; "A Face-Off on Mars," Science, Vol. 261 (10 September 1993), 1392. The comment about NASA is from Fitzgerald's Cosmic Test Tube, p. 19. Shostak and Barnett provided a useful summary of the face on Mars in Cosmic Company, 94–95.
93. Malcolm Smith, "Planetary SETI Craves Scientific Credibility," Spaceflight, Vol. 46, No. 1 (January 2004), 35–37.
94. Carl Sagan, "Direct Contact Among Galactic Civilizations by Relativistic Interstellar Spacecraft," Planetary and Space Science, Vol. 11 (1963), 485–490, reprinted in Goldsmith, editor, 205–213; Achenbach, 61.
95. David Michael Jacobs, The UFO Controversy in America, Bloomington, IN, Indiana University Press, 1975, 218.
96. Erich von Daniken, Chariots of the Gods?, New York, G.P. Putnam's, 1970 (originally published in Germany by Econ-Verlag, 1968); John C. Baird, The Inner Limits of Outer Space, Hanover, NH, University Press of New England, 1987, 20.
97. Baird, 17. One interesting spinoff was the International Journal of Paleovisitology, meant to promote scientific research into this question. The journal was active as of 1990.
98. Shklovskii and Sagan, 455–461; White, 45–48.
99. Stern, in Maruyama and Harkin, editors, 48; Chris Boyce, Extraterrestrial Encounter, Secaucus, NJ, Chartwell Books, 1979, 105.
100. Walter Gratzer, review of Robert Ehrlich's Eight Preposterous Propositions, Nature, Vol. 426 (18/25 December 2003), 766.
101. Ulmschneider, 166.

The UFO Controversy

1. Quoted by Randall Fitzgerald in Cosmic Test Tube, 292.
2. Robert L. Hall, "Sociological Perspectives on UFO Reports," in Carl Sagan and Thornton Page, editors, UFO's: A Scientific Debate, New York, Norton, 1972, 213–222; Clark and Clark, 243; Charles Fort, The Book of the Damned, Mineola, NY, Dover, 2002, 20, originally published in 1919 by Boni and Liveright; Hoyle and Wickramasinghe, Cosmic Life Force, 39.
3. Jacobs, 265.
4. Hynek's Foreword to Jacobs, ix–xi. Also see J. Allen Hynek, The UFO Experience: A Scientific Inquiry, London, Corgi, 1974, originally published by Abelard-Schuman in 1972.
5. David W. Swift, "SETI Without Saucers?" Aerospace America, April 1982, 52–53; Michael Kurland, The Complete Idiot's Guide to Extraterrestrial Intelligence, New York, Alpha Books, 1999.
6. Jack Cohen and Ian Stewart, What Does a Martian Look Like?, Hoboken, NJ, Wiley, 2002, 310.
7. Davies, Are We Alone?, 133; The Editors of Time-Life Books, The UFO Phenomenon, New York, Barnes and Noble, 1987, 12; Roulet, 71.

8. A text with commentary can be found in MacVey's book Interstellar Travel, 133–135. A briefer version is in Peter Hough and Jenny Randles, Looking for the Aliens, New York, Barnes and Noble, 1997, 135–136, originally published by Blandford Press in 1991.
9. MacVey, Interstellar Travel, 140.
10. Editors of Time-Life Books, 16.
11. Fitzgerald, Cosmic Test Tube, 277.
12. Jacobs, 5–34; Editors of Time-Life Books, 19–23; Michael Busby, Solving the 1897 Airship Mystery, Gretna, LA, Pelican, 2004.
13. For a general description, see Ann Druffel, "Fatima (Portugal), miracle at," in Ronald D. Story, editor, The Encyclopedia of Extraterrestrial Encounters, New York, New American Library, 2001, 188–191.
14. Fort, 17–23.
15. Loren E. Gross, "Ghost Rockets of 1946," in Story, editor, 217–219; The Editors of Time-Life Books, 26–27.
16. Jacobs, 36–37.
17. William K. Hartmann, "Historical Perspectives," in Sagan and Page, editors, 11–21.
18. Jacobs, 37–41.
19. Curtis Peebles, Watch the Skies! A Chronicle of the Flying Saucer Myth, Washington, DC, Smithsonian Institution Press, 1994, 17. Jacobs documented the shifting opinions among Air Force investigators; although they sometimes were willing to consider the extraterrestrial hypothesis, they dismissed it at other times.
20. Hartmann, in Sagan and Page, editors, 17–18.
21. Clark and Clark, 189.
22. Peebles, 130; Jacobs, 102.
23. Quoted in Ian Favell, "Human Aspirations in the History of Space Flight," JBIS, Vol. 57 (2004), 340–347.
24. Jacobs, 86.
25. Jacobs, 115, 123–125.
26. Edward U. Condon, et al., Scientific Study of Unidentified Flying Objects, New York, New York Times Company, 1969; Hynek, The UFO Experience, 242; Fitzgerald, 299–301.
27. Jacobs, 246.
28. Hynek, foreword to Jacobs, ix–xvi.
29. Jacobs, 248, 251, 259.
30. Peter Sturrock, The UFO Enigma, 156.
31. Carl Sagan, "The Extraterrestrial and Other Hypotheses," in Sagan and Page, 265–274.
32. Jacobs, 296. The poll was reported in Time, January 21, 1974, 74.
33. The Editors of Time-Life, 128; Timothy Good, Above Top Secret: The World-Wide UFO Coverup, New York, William Morrow, 1988, 368–369, originally published by Sidgwick and Jackson in 1987.
34. Peebles, 199–201.
35. Randall Fitzgerald provided a good overview of the Roswell case in Cosmic Test Tube, 75–95. Shostak and Barnett gave a capsule description in Cosmic Company, 77.
36. "The Truth is Out There, but It's Classified" (editorial), The New York Times, 27 September 2003.

37. Jenny Randles, "Crop Circles," in Story, editor, 146–148; Fitzgerald, Cosmic Test Tube, 286–292.

38. Harrison, After Contact, 67–68.

39. John E. Mack, Abduction: Human Encounters with Aliens, New York, Simon and Schuster, 1994.

40. David M. Jacobs, Secret Life, New York, Simon and Schuster, 1993; Lamb, 131.

41. Susan Clancy, Abucted: How People Come to Believe They Were Kidnaped by Aliens, Cambridge, MA, Harvard University Press, 2005; Benedict Carey, "Explaining Those Vivid Memories of Martian Kidnapers," The New York Times, 9 August 2005.

42. Baird, 35.

43. Webb, 4.

44. Linda Billings, "From the Observatory to Capitol Hill," in Bova and Preiss, editors, 223–239; Swift, 83; Sturrock, The UFO Enigma, 160; David W. Swift, "Parallel Universes: A Tale of Two SETIs," Astronomy, October 1981, 24–28.

45. Sagan and Page, xviii.

46. The Editors of Time Life, 9.

47. Hall, in Sagan and Page, editors, 220.

48. Goldsmith and Owen, 409–419.

49. Grinspoon, 379; R.N. Bracewell, "Interstellar Probes," in Ponnamperuma and Cameron, 102–116.

50. Jenny Randles, Alien Contact: The First Fifty Years, London, Collins and Brown, 1997, 7; Story, editor, 459; Grinspoon, 371.

51. Hynek, The UFO Experience, 288.

52. Hynek, Foreword to Jacobs, ix–xvi.

53. Jacobs, 215–216.

54. Fitzgerald, 292.

55. Baird, 40, 123–126.

56. Hall in Sagan and Page, editors, 213–221.

57. C. G. Jung, Flying Saucers: A Modern Myth of Things Seen in the Skies, translated from the German by R.F.C. Hull, New York, Harcourt, Brace, 1959, 14, 21, 147, 150–152. Originally published by Rascher and Cie., Zurich in 1958. Jung had published an earlier article on UFOs in the German language magazine Weltwoche in July 1954.

58. Davies, Are We Alone?, 135; Robert Schaeffer, "An Examination of Claims that Extraterrestrial Visitors to the Earth are Being Observed," in Hart and Zuckerman, editors, 20–27.

59. Swift, SETI Pioneers, 325; White, 49.

60. Davies, Are We Alone?, 133.

61. Dick, Dibner lecture, 40; Steven Dick, "Cosmotheology: Theological Implications of the New Universe," in Dick, editor, Many Worlds, 191–208.

62. Peebles, 241; Grinspoon 372.

63. Albert Harrison, After Contact, 90.

64. Harrison, 89; Glen David Brin, "The Great Silence: The Controversy Concerning Extraterrestrial Intelligent Life," Quarterly Journal of the Royal Astronomical Society, Vol. 24 (1983), 283–309.

65. Quoted by Randall Fitzgerald in Cosmic Test Tube, 127.

66. Lamb, 144.
67. In Bova and Preiss, editors, 310.
68. Swift, 110; Shostak, Sharing The Universe 135.
69. Rood and Trefil, 219; Schaeffer, in Hart and Zuckerman, editors, 20.
70. David W. Schwartzman, "The Absence of Extraterrestrials on Earth and the Prospects for CETI," in Goldsmith, editor, 264–266. Originally published in Icarus, Vol. 32 (1977), 473–475.
71. Cade, Other Worlds Than Ours, 22.
72. Clark and Clark, 188, 190–196, 222, 243, 255.
73. Sturrock, 149; Swift, "Parallel Universes."
74. Quoted in Stephen J. Garber, "Searching for Good Science."
75. Sturrock, 121–122, 125, 153, 169, 255–256; Clark and Clark, 246.
76. Torsten Neubert, "On Sprites and Their Exotic Kin," Science, Vol. 300 (2 May 2003), 747–749.
77. Sturrock, 155.
78. Sagan, editor, CETI, 189, 228.

The Drake Equation: Take Three

1. Dyson, in G. Seth Shostak, editor, Third Decennial US-USSR Conference on SETI, Astronomical Society of the Pacific, 1993, 434.
2. Shostak and Barnett, Cosmic Company, 134.
3. David Viewing, "Directly Interacting Extraterrestrial Technological Communities," JBIS, Vol. 28 (1975) 735–744.
4. Brin, in Bova and Preiss, editors, 130.

Why Don't We See Them?

1. Viewing, "Directly Interacting Extra-Terrestrial Technological Communities."
2. Crowe, 230.
3. Vladimir Lytkin, Ben Finney, and Liudmila Alepko, "Tsiolkovsky, Russian Cosmism, and Extraterrestrial Intelligence," QJRAS, Vol. 36 (1995), 369–376; Albert Harrison, 187–188.
4. Fort, 162–163.
5. Finney and Jones, editors, 298–300; Lytkin, et al., 373; Drake and Sobel, 203.
6. John A. Ball, "The Zoo Hypothesis," Icarus, Vol. 19 (1973), 347–349, reprinted in Goldsmith, editor, 241–242.
7. Rood and Trefil, 248.
8. For an overview of the O'Neill enterprise, see Michael A.G. Michaud, Reaching for the High Frontier: The American Pro-Space Movement, 1972–1984, New York, Praeger, 1986, 57–80.
9. Casti, 398.
10. Brin, "The Great Silence."
11. Mauldin, Prospects for Interstellar Travel, 280.

12. Michael H. Hart, "An Explanation for the Absence of Extraterrestrials on Earth," QJRAS, Vol. 16 (1975), 128–135, reprinted in Goldsmith, editor, 228–231.

13. L. J. Cox, "An Explanation for the Absence of Extraterrestrials on Earth," QJRAS, Vol. 17 (1976), 201–208; Swift, 32; George Abell, "The Search for Life Beyond Earth: A Scientific Update," in James L. Christian, editor, Extraterrestrial Intelligence, Buffalo, NY, Prometheus, 1976, 53–71.

14. Eric M. Jones, "Colonization of the Galaxy," Icarus, Vol. 28 (1976), 421–422.

15. Eric M. Jones, "Estimates of Expansion Time Scales," in Hart and Zuckerman, editors, 66–74.

16. T. B. H. Kuiper and M. Morris, "Searching for Extraterrestrial Civilizations," Science, Vol. 196 (6 May 1977), 616–621, reprinted in Goldsmith, editor, 170–177.

17. David W. Schwartzman, "The Absence of Extraterrestrials on Earth and the Prospects for CETI," Icarus, Vol. 32 (1977), 473–475, reprinted in Goldsmith, editor, 264–266.

18. Webb, 24; Hart and Zuckerman, editors, vii.

19. Michael H. Hart, "Atmospheric Evolution, The Drake Equation, and DNA: Sparse Life in an Infinite Universe," in Hart and Zuckerman, editors, 154–163.

20. Papagiannis, "Colonies in the Asteroid Belt," in Hart and Zuckerman, editors, 77–84.

21. Michael Papagiannis, "Strategies for the Search for Life in the Universe," Cosmic Search, Vol. 2, Number 1(Winter 1980), 24–27; Michael D. Papagiannis, "The Search for Extraterrestrial Civilizations—A New Approach," Mercury, January-February 1982, 12–16, 25.

22. Casti, 401.

23. Frank J. Tipler, "A Brief History of the Extraterrestrial Intelligence Concept," QJRAS, Vol. 22 (1981), 133–145.

24. Frank J. Tipler, "We Are Alone," Discover, March 1983, 56–61; Frank J. Tipler, "Alien Life" (review of Emmanuel Davoust's The Cosmic Water Hole), Nature, Vol. 354 (28 November 1991), 334–335; Frank J. Tipler, "Extraterrestrial Intelligent Beings Do Not Exist," QJRAS, Vol. 21 (1980), 267 (Tipler published an article under the same title in Physics Today, Vol. 34 (1981), 9, 70–71); Barrow and Tipler, 576–601.

25. Frank D. Drake, letter in Physics Today, March 1982, 26–27, 31; Drake and Sobel, 205.

26. Finney, "Exponential Expansion."

27. William I. Newman and Carl Sagan, "Galactic Civilizations: Population Dynamics and Interstellar Diffusion," Icarus, Vol. 46 (1981), 293–327; Carl Sagan and William I. Newman, "The Solipsist Approach to Extraterrestrial Intelligence," in Regis, editor, 151–161. The separate argument by Sagan is from "The Solipsistic Approach to Extraterrestrial Intelligence," QJRAS, Vol. 24 (1983), 113–121, quoted in Lamb, 163.

28. Goldsmith and Owen, 425–431.

29. Clark and Clark, 240; Sagan and Newman, "The Solipsist Approach;" Eric M. Jones and Barham W. Smith, Letter in Physics Today, March 1982, 32–33; Gregory Benford, Letter in Physics Today, March 1982, 33; Tim Weiner, "A New Model Army Soldier Rolls Closer to the Battlefield," The New York Times, 16 February 2005.

30. B. R. Finney, "Exponential Expansion: Galactic Destiny or Technological Hubris?" in Michael Papagiannis, editor, The Search for Extraterrestrial Life: Recent Developments, Dordrecht, Reidel, 1985, 455–463.

31. B. R. Finney, "Exponential Expansion." Also see Rebecca L. Cann, "Talking Trees Tell Tales," Nature, Vol. 405 (29 June 2000), 1008–1009.

32. Former British naval officer Gavin Menzies advocated an expanded version of what the voyages achieved in his book 1421: The Year China Discovered America, New York, William Morrow (Harper Collins), 2003, originally published in Great Britain in 2002 by Transworld Publishers.

33. Claudius Gros, "Expanding Advanced Civilizations in the Universe," JBIS, Vol. 58 (2005), 108–110; Sagan, Cosmos, 310–311.

34. Finney, "Exponential Expansion."

35. For a brief description of the classic agenda, see Michael A.G. Michaud, Reaching for the High Frontier, New York, Praeger, 1986, 7–8.

36. Dole and Asimov, 203–204.

37. See Eugene N. Parker, "Sheilding Space Travelers," Scientific American, March 2006, 40–47, and Joshua Roth, "A Record-Setting Solar Flare," Sky and Telescope, September 2005, 16–17.

38. C.M. Cade, "Communicating with Life in Space," Discovery, May 1963, 36–41; MacGowan and Ordway, 259, 272; A.J. Dunlop, "Galaxy Dynamics and Desirable Real Estate," paper prepared for the 1987 International Astronautic Congress.

39. Bracewell, "Communications from Superior Galactic Communities."

40. Mallove and Matloff, 4.

41. Brin, "The Great Silence." The more popularized version appeared as "Mystery of the Great Silence" in Bova and Preiss, editors, 118–139, and as "Xenology," Analog, May 1983, 64–82.

42. McConnell, 68, 70.

43. John Kraus, "Gerard K. O'Neill on Space Colonization and SETI," Cosmic Search, Vol. 1, Number 2 (March 1979), 16.

44. Stapledon, 292.

45. Grinspoon, 412.

46. Milan M. Cirkovic and Richard B. Cathcart, "Geo-Engineering Gone Awry," JBIS, Vol. 57, (2004), 209–215.

47. James Annis, "An Astrophysical Explanation for the Great Silence," JBIS, Vol. 52 (1999), 19–22.

48. Paul Davies, "A Brief History of the Multiverse," The New York Times, 12 April 2003; Stephen Baxter, "The Planetarium Hypothesis: A Resolution of the Fermi Paradox," JBIS, Vol. 54 (2001), 210–216.

49. Frank Drake, "On Hands and Knees in Search of Elysium," Technology Review, June 1976, 22–29.

50. Drake and Sobel, 161.

51. Frank D. Drake, "A Speculation on the Influence of Biological Immortality on SETI," Cosmic Search, Summer 1980, 9–11.
52. Arthur C. Clarke, The Promise of Space, 293–294.
53. Goldsmith, Voyage to the Milky Way, 240.
54. Claudia Dreifus, "Live Longer With Evolution?" (interview with Michael Rose), The New York Times, 6 December 2005.
55. Webb, 239.

Reformulating the Problem

1. Ivan Almar, "Analogies Between Olbers' Paradox and the Fermi Paradox," Acta Astronautica, Vol. 26 (1992), 253–256; George Musser, "String Instruments," Scientific American, October 1998, 24–28.
2. Shklovskii and Sagan, 418; Asimov, Extraterrestrial Civilizations, 192.
3. Sullivan, 244–245.
4. Heidmann's argument was described in Stuart Clark, Life on Other Worlds and How to Find It, New York, Springer, 2000, 156. For other rejections of the paradox, see Koerner and LeVay, 181, Robert A. Freitas, Jr., "There is No Fermi Paradox," Icarus, Vol. 62 (1985), 518–520, and Robert A. Freitas, Jr., "Fermi's Paradox: A Real Howler," Isaac Asimov's Science Fiction Magazine, 8 September 1984, 30–44.
5. McDonough, 198.
6. Webb, 42; Michael D. Papagiannis, "Bioastronomy: The Search for Extraterrestrial Life," Sky and Telescope, June 1984, 508–511.
7. L. M. Gindilis and G.M. Rudnitskii, "On the Astrosociological Paradox in SETI," in Shostak, editor, 404, 413.

Thinking Outside the Box

1. Stephen Baxter, "The Planetarium Hypothesis: A Resolution of the Fermi Paradox," JBIS, Vol. 54 (2001), 210–216.
2. Padgen, 49.
3. Edward Harrison, review of The Anthropic Cosmological Principle, American Scientist, Vol. 75 (January-February 1987), 72–73. Also see Edward Harrison, Masks of the Universe, New York, MacMillan, 1985, 65.
4. Stephen Eales, "Smoking Supernovae and Dusty Galaxies," Sky and Telescope, August 2004, 36–42.
5. George Musser, "Growing Pains," Scientific American, July 2004, 32–33.
6. J. Michael Shull, "Hot Pursuit of Missing Matter," Nature, Vol. 433 (3 February 2005), 465–466; McConnell, 370.
7. James Glanz, "Is the Dark Matter Mystery Solved?", Science, Vol. 271 (2 February 1996), 595–596; Mark Sincell, "Astronomers Glimpse Galaxy's Heavy Halo," Science, Vol. 291 (23 March 2001), 2293–2294.
8. James Trefil, The Dark Side of the Universe: A Scientist Explores the Mysteries of the Cosmos, Anchor reprint edition, 1989. Also see Robert

Kirshner, The Extravagant Universe: Exploding Stars, Dark Energy, and the Accelerating Cosmos, Princeton, Princeton University Press, 2004, quoted by John Noble Wilford, "From Distant Galaxies, News of a Stop and Go Universe," The New York Times, 3 June 2003, and Daniel Clery, "Dwarf Galaxies May Help Define Dark Matter," Science, Vol. 311 (10 February 2006), 758–759.

9. David B. Cline, "The Search for Dark Matter," Scientific American, March 2003, 50–59. Yale University astronomer Meg Urry described the search for dark energy as arguably the biggest revolution in physics in a century. Andrew Lawler, "Balancing the Right Stuff," Science, Vol. 308 (22 April 2005), 484–487. For an overview of dark matter and dark energy, see the special section in Science, Vol. 300 (30 June 2003), 1893–1918.

10. Sources differ on the exact percentages comprised by dark matter and dark energy. See Alan MacRobert, "Refining the Cosmic Recipe," Sky and Telescope, February 2004, 18–19, Charles Bennett, "Astrophysical Observations: Lensing and Eclipsing Einstein's Theories," Science, Vol. 307 (11 February 2005), 879–884, and Masataka Fukugita, "The Dark Side," Nature, Vol. 422 (3 April 2003), 489–491.

11. Alejandro Gangui, "A Preposterous Universe," Science, Vol. 299 (28 February 2003), 1333–1334. For a dissenting view, see Paul J. Steinhardt and Neil Turok, "A Cyclic Model of the Universe," Science, Vol. 296 (24 May 2002), 1436–1439.

12. Sean Carroll, "Dark Energy & the Preposterous Universe," Sky and Telescope, March 2005, 32–39; Dennis Overbye, "From Space, A New View of Doomsday," The New York Times, 17 February 2004.

13. Davies, "A Brief History of the Multiverse;" Dan Falk, "The Anthropic Principle's Surprising Resurgence," Sky and Telescope, March 2004, 43–47.

14. George Johnson, "A Universe of Universes," (review of Lee Smolin's The Life of the Cosmos), The New York Times, 27 July 1997; Steven Weinberg, "Life in the Universe," Scientific American, October 1994, 44–49.

15. Mario Livio and Martin J. Rees, "Anthropic Reasoning," Science, Vol. 309 (12 August 2005), 1022–1023.

16. Martin J. Rees, "Piecing Together the Biggest Puzzle of All," Science, Vol. 290 (8 December 2000), 1919–1925; Sean Carroll, "Insignificance," Nature, Vol. 429 (6 May 2004), 27.

17. Stapledon, 401.

18. Max Tegmark, "Parallel Universes," Scientific American, May 2003, 41–51; Dick, Dibner Lecture, 27–29.

19. Nima Arkani-Hamed, et al., "The Universe's Unseen Dimensions," Scientific American, August 2000, 62–69; Roland Pease, "Brane New World," Nature, Vol. 411 (28 June 2001), 986–988.

20. Charles Seife, "Physics Enters the Twilight Zone," Science, Vol. 305 (23 July 2004), 464–466.

21. Davies, Are We Alone?, 121; Koerner and LeVay, 235.

22. Roger Newton, "Weird Science" (review of The Fabric of the Cosmos), The New York Times Book Review, 11 April 2004; Davies, "A Brief History of the Multiverse."

23. Barrow and Tipler, 1; Edward Harrison, Masks of the Universe, 250; Rees, Our Final Hour, 135–136; Dan Falk, "The Anthropic Principle's Surprising Resurgence."
24. Barrow and Tipler, 16.
25. James Glanz, "Debating the Big Questions," Science, Vol. 273 (30 August 1996), 1168–1170.
26. Edward Witten, "When Symmetry Breaks Down," Nature, Vol. 429 (3 June 2004), 507–508.
27. Barrow and Tipler, 601.
28. Barrow and Tipler, 21, 23; James N. Gardner, Biocosm: The New Scientific Theory of Evolution—Intelligent Life is the Architect of the Universe, Makawao, Maui, Inner Ocean, 2003, 38–39.
29. John H. Mauldin, Prospects for Interstellar Travel, San Diego, Univelt (for the American Astronautical Society), 1992, 283.
30. Dan Falk, "The Anthropic Principle's Surprising Resurgence," 46.
31. Alan H. Guth and David I. Kaiser, "Inflationary Cosmology: Exploring the Universe from the Smallest to the Largest Scales," Science, Vol. 307 (11 February 2005), 884–890; Livio and Rees, "Anthropic Reasoning."
32. Aveni, Conversing with the Planets, 219.
33. White, 117.
34. The Linde and Smolin theories are summarized in Gardner, 82–83, 155–156. Also see Lee Smolin, The Life of the Cosmos, Oxford, Oxford University Press, 1997.
35. Gardner, Biocosm, passim; James N. Gardner, "Assessing the Computational Potential of the Eschaton: Testing the Selfish Biocosm Hypothesis," JBIS, Vol. 55 (2002), 285–288.
36. Eugene F. Mallove, "Do We Control the Universe's Fate?" The Washington Post, 30 November 1986; Eugene F. Mallove, "The Self-Reproducing Universe," Sky and Telescope, September, 1988, 253–256.
37. J.D. Bernal, The World, The Flesh, and the Devil, London, Jonathan Cape, 1970 (originally published in 1929); 31; Gerald Feinberg, The Prometheus Project, New York, Doubleday, 1968, 142; Michael A.G. Michaud, "Spaceflight, Colonization, and Independence;" Michael A.G. Michaud, "The Final Question: Paradigms for Intelligent Life in the Universe," JBIS, Vol. 35, Number 3 (March, 1982), 131–134; Michael A.G. Michaud, "The Last Question," Cosmic Search, Volume 4, Number 1 (First Half 1982), 6–7.
38. Rees, 147; Gardner, 130, 161–163; "Founder's Message," Science, Vol. 310 (2 December 2005), 1421; "Creator Calling Card?" Sky and Telescope, May 2006, 25.
39. Stapledon, Star Maker, 393.
40. Richard Dawkins, River Out of Eden: A Darwinian View of Life, London, Weidenfeld and Nicholson, 1995. Quoted in a review by Michael Shermer in Science, Vol. 308 (8 April 2005), 205–206.
41. Crowe, 403; Fred Hoyle, The Intelligent Universe, 226.
42. Michael J. Behe, "Design for Living," The New York Times, 7 February 2005.
43. See Kenneth Chang, "In Explaining Life's Complexity, Darwinists and Doubters Clash," The New York Times, 22 August 2005, and Cornelia Dean,

"Scientists Speak Up on Mix of God and Science," The New York Times, 23 August 2005. de Vore was quoted in an unauthored box in Science, Volume 284 (30 April 1999), 9.

44. Claudia Dreifus, "How Culture Pushed Us to the Top of the Food Chain" (interview with Robert Boyd), The New York Times, 10 May 2005; Verlyn Klinkenborg, "Grasping the Depth of Time as a First Step in Understanding Evolution," The New York Times, 23 August 2005.

45. Davies, The Fifth Miracle, 270; In Dick, editor, Many Worlds, 27.

46. Christoph Schonborn, "Finding Design in Nature," The New York Times, 7 July 2005.

47. Guillermo Gonzalez and Jay W. Richards, The Privileged Planet: How Our Place in the Cosmos is Designed for Discovery, Washington, DC, Regnery, 2004.

48. David Tytell, review of The Privileged Planet, Sky and Telescope, March 2005, 108; Brin, "A Contrarian Perspective on Altruism;" Letter to the Editor from Karen B. Rosenberg, Chairwoman of the Department of Anthropology, University of Delaware, The New York Times, 9 February 2005.

49. Douglas Vakoch, review of The Privileged Planet in Nature, Vol. 429 (24 June 2004), 808–809.

50. Grinspoon, 107; Cohen and Stewart, 212.

51. Robert J. Sawyer, "SETI and God," SearchLites, Vol. 8 (Autumn 2002), 2.

52. Darling, 115.

53. George Musser, "String Instruments," Scientific American, October 1998, 24–28; Gardner, 8.

54. Quoted in Kevin Krajick, "Winning the War Against Island Invaders," Science, Vol. 310 (2 December 2005), 1410–1413.

55. Gardner, Bicosm, 89.

56. Quoted in Richard Lewontin, "The Wars over Evolution," The New York Review of Books, 20 October 2005.

57. Vaclav Havel, "The New Measure of Man." Excerpts from this speech were published in The New York Times, 8 July 1994.

58. Philip Blond and Adrian Pabst, "Why the West Gets Religion Wrong," International Herald Tribune, 8 July 2005.

SETI and Religion

1. Stapledon, 292.

2. Dick, editor, Many Worlds, 205. George Basalla addressed this linkage at length in his book Civilized Life in the Universe, published by Oxford University Press in 2006. Unfortunately, I did not see this book until after I sent my manuscript to my publisher.

3. Guthke, ix; Sagan, editor, CETI, 344–345.

4. Gerritt L. Verschuur, "If We Are Alone, What on Earth Are We Doing?" Sky and Telescope, November 1989, 452.

5. Swift, 57; S.A. Kaplan, editor, Extraterrestrial Civilizations, 215.

6. Benjamin, 184, 187.

7. Mark Rose, Alien Encounters: Anatomy of Science Fiction, Cambridge, MA, Harvard University Press, 1981, 50.

8. Tarter, in Regis, editor, 170, and Dick, editor, 143; Tarter, in White, 204.

9. Clarke, Profiles of the Future, 94; Clarke, Greetings, 230.

10. Jung, Flying Saucers, 48; Pickover, 173.

11. Gardner, Biocosm, 150; MacGowan and Ordway, 270.

12. Dick, editor, Many Worlds, 202–204. The website for this organization was at www.bibleufo.com/isso.

13. Robert J. Sawyer, "SETI and God," in SearchLites, Vol. 8, Number 4 (Autumn 2002), 2.

14. Stapledon, 291; Gardner, 93.

15. Douglas Vakoch, "Roman Catholic Views of Extraterrestrial Intelligence," in Tough, editor, 165–174; Sagan and Page, editors, 272.

16. Baird, 116.

17. Davies, Are We Alone?, 136–137.

18. Brian Aldiss, "Desperately Seeking Aliens," Nature, Vol. 409 (22 February 2001), 1080–1082.

19. Cohen and Stewart, 183; Puccetti, Persons, 144.

20. Shostak, Sharing the Universe, 99–100; Paul Wasson, "Symbolism and the Inference of Intelligence," SETI Institute Explorer, First Quarter 2005, 16–17.

21. Natalie Angier, "The Origin of Religions, From a Distinctly Darwinian View," The New York Times, 24 December 2002.

22. Tarter, in Tough, editor, 45.

23. Dick, editor, Many Worlds, 145–146; Tough, editor, 45.

24. Davies, Are We Alone?, 50–51.

25. Musso, op. cit.

26. Vakoch, "Roman Catholic Views;" Douglas A. Vakoch, "Framing Spiritual Principles for Interstellar Communications," Science and Spirit, Vol. 10 (November-December 1999), 21.

27. Fort, 162.

28. Guthke, 119.

29. Shostak, 120; Don Lago, letter to the editor, Cosmic Search, Vol. 3, Number 1 (Winter 1981), 14.

30. Sagan, editor, CETI, 345; Timothy Ferris, "Seeking and End to Cosmic Loneliness," The New York Times Magazine, 23 October 1977, 31–32, 92.

31. For a powerful example, see Richard Fletcher, The Barbarian Conversion: From Paganism to Christianity, Berkeley, University of California Press, 1997.

The Consequences of Contact

1. Ben Finney, "The Impact of Contact," in Tarter and Michaud, editors, Acta Astronautica, Vol. 21 (1990), 117–121; quoted in White, 137.

2. Swift, 406; Sagan, The Cosmic Connection, 218; Sagan, "The Quest for Intelligent Life."

3. Sagan, editor, CETI, 353; Carl Sagan and Frank Drake, "The Search for Extraterrestrial Intelligence," Scientific American, May 1975, 80–89; Sagan, "The Quest for Intelligent Life;" Drake and Sobel, 159.

4. Clarke, Greetings, 230; Billingham, et al., Social Implications, 50; Richard Berendzen, editor, Life Beyond Earth and the Mind of Man, Washington, DC, NASA (SP-328), 1973, 17; Harrison, After Contact, 315; Michael A.G. Michaud, The Consequences of Contact," AIAA Student Journal, Vol. 15, Number 4 (Winter, 1977–78), 18–23.
5. Frank Drake, "In Which Klingons Become Chimeras," Cosmic Search, Vol. 3, Number 1 (Winter 1981), 9.
6. Dyson, in Project Cyclops, 177; Albert A. Harrison, "Slow Track, Fast Track, and the Galactic Club," paper prepared for the Foundation for the Future, 1998.
7. Arnold J. Toynbee, A Study of History, Vol. 9, Oxford, Oxford University Press, 1954; McDonough, 216.
8. Steven J. Dick, "Consequences of Success in SETI: Lessons from the History of Science," in G. Seth Shostak, editor, Progress in the Search for Extraterrestrial Life (1993 Bioastronomy Symposium), San Francisco, Astronomical Society of the Pacific, 1995, 521–532; Morrison, et al., editors, SETI, 8; Harrison, After Contact, 280.
9. Harrison, "Slow Track, Fast Track, and the Galactic Club."
10. Brin, "A Contrarian Perspective on Altruism;" Chris Boyce, Extraterrestrial Encounter, Seacaucus, NJ, Chartwell, 1979, 100.
11. Billingham, et al., Social Implications, 62.
12. Brin, "A Contrarian Perspective."
13. White, 156.
14. Bracewell in Ponnamperuma and Cameron, editors, 116.
15. Personal communication, 18 September 2003.
16. Jean Heidmann, Intelligences Extra-Terrestres, Paris, Editions Odile Jacob, 1992, 11.
17. Trudy E. Bell, "The Grand Analogy: History of the Idea of Extraterrestrial Life," Cosmic Search, Vol. 2, Number 1 (Winter 1980), 2–10. This article appeared previously in the August 1978 issue of the Griffith Observer, a publication of the Griffith Observatory.
18. Billingham, et al., editors, Social Implications, 34.
19. Billingham, et al., editors, 32.
20. MacGowan and Ordway, 183, 235. Jack Cohen and Ian Stewart also concluded that we will meet mostly or entirely mechanical–electronic creatures. What Does a Martian Look Like? The Science of Extraterrestrial Life, Hoboken, NJ, Wiley, 2002, 308.
21. A.A. Harrison and J.T. Johnson, "ETI: Our First Impressions."
22. Francois Jacob, "Evolution and Tinkering," Science, Vol. 196 (10 June 1977), 1161–1166; Darling, 118.
23. Casti, 392–393; Swift, 125, 273.
24. Swift, 83; Clarke, Greetings, 480. For galleries of alien types, see Ronald D. Story, The Encyclopedia of Extraterrestrial Encounters, New York, New American Library, 2001, 22–25, 27–28, 39–45.
25. Quoted by Clarke in Bova and Preiss, editors, 311.
26. Cohen and Stewart, 100–106, 297.
27. Swift, 322.
28. Billingham, et al., editors, Social Implications, 71–72, 116; Harrison, After Contact, 214; John Clute and Peter Nicholls, editors, The Encyclopedia of

Science Fiction, New York, St Martin's Griffin, 1993, 15–19. Wayne Barlowe and others depicted imaginary aliens from popular science fiction works in Barlowe's Guide to Extraterrestrials, New York, Workman, 1979, reprinted 1987.

29. Beck, in Regis, editor, 13; Boyce, 60.
30. MacGowan and Ordway, 235; Arthur C. Clarke, foreword to MacGowan and Ordway, v.
31. Koerner and LeVay, 213; Davies, Are We Alone?, 51–53; Grinspoon, 399.
32. Shostak, Sharing The Universe, 109.
33. Albert Harrison, After Contact, 222.
34. Billingham, et al., Social Implications, 70, 116.

Hopes

1. Berendzen, editor, 49–50; McConnell, 383.
2. "Interview: Robert E. Edelson," Mercury, July-August 1977, 8–12; Swift, 270.
3. Sagan, The Cosmic Connection, 219–220; Swift, 375.
4. Puccetti, Persons, 118.
5. Crowe, 394.
6. Swift, 406; "What If?" Sky and Telescope, November 1992, 515; Arthur C. Clarke, The Exploration of Space, 1951, 195.
7. Ian Ridpath, Messages from the Stars, London, Fontana Books, 1978, 25.
8. Project Cyclops, 29.
9. Grinspoon, 224; Stapledon, 302, 368.
10. Bracewell, in Goldsmith, editor, 105–106; Ronald N. Bracewell, The Galactic Club, San Francisco, Freeman, 1974. Bernal is quoted in David W. Schwartzman, "The Absence of Extraterrestrials on Earth and the Prospects for CETI," in Goldsmith, editor, 264–266.
11. Harrison, "Slow Track, Fast Track;" Harrison, After Contact, 174–180.
12. White, 103–104.
13. Sagan, "The Quest for Intelligent Life;" Sagan, Cosmos, 339.
14. Sagan, The Cosmic Connection, 242.
15. Harrison, "Slow Track, Fast Track."
16. McDonough, 230; Sagan and Shklovskii, 393.
17. Davies, Are We Alone?, 129; Sullivan, 309.
18. Launius and McCurdy, 38; Jung, Flying Saucers, 98.
19. White, 138, 199; Sagan, "The Quest for Intelligent Life;" Drake and Sobel, 159.
20. Billingham, et al., editors, 86.
21. MacVey, Interstellar Travel, 228; Roberto Pinotti, "ETI, SETI, and Today's Public Opinion," paper presented at the 1988 International Astronautical Congress (IAF/IAA.88.532); Roberto Pinotti, "Contact: Releasing the News," Acta Astronautica, Vol. 21 (1990), 109–115.
22. White, 151.
23. Tarter, in Dick, editor, Many Worlds, 148; Swift, 271.
24. Swift, 190.

25. Drake and Sobel, 159–160 Sagan, "The Quest for Extraterrestrial life." Sagan, Cosmos, 314.
26. MacGowan and Ordway, 249–250, 269.
27. Project Cyclops, 36.
28. Sagan, The Cosmic Connection, 236–239.
29. Sagan, Cosmos, 291–315; Carl Sagan, Other Worlds, New York, Bantam, 1975, 13; McConnell, 384.
30. Project Cyclops 30–31, 60, 169.
31. Frank Drake, "On Hands and Knees in Search of Elysium," Technology Review, June 1976, 22–29; Drake and Sobel, 159–160; Sullivan, 308; Shostak, Sharing The Universe, 195; Billingham, et al., editors, 59–60; Chloe Zerwick and Harrison Brown, The Cassiopeia Affair, New York, Doubleday, 1968; Robert J. Sawyer, "SETI and God," SearchLites, Vol. 8, Number 4 (August, 2002), 2.
32. Drake, "On Hands and Knees;" Black, et al.
33. Sagan, Cosmos, 314.
34. Sagan, editor, CETI, 337–338.
35. Beck, in Regis, editor, 15.
36. Baird, 25, 213.
37. Samuel P. Huntington, The Clash of Civilizations and the Remaking of World Order, New York, Simon and Schuster, 1996, 54–55. Sullivan, 304; John W. MacVey, Whispers from Space, New York, Macmillan, 1973, xvii; MacVey, Interstellar Travel, 4.
38. Harrison and Dick, in Tough, editor, 9, 16; S. A. Kaplan, editor, Extraterrestrial Civilizations: Problems of Interstellar Communication, Glavnaya Redaktsiya (Russian), Moscow, 1969, translated into English in 1971 by the Israel Program for Scientific Translations.
39. White, 158–159; Harrison, After Contact, 281.
40. Albert A. Harrison, et al., "The Role of the Social Sciences in SETI, in Tough, editor, 71–84.
41. MacGowan and Ordway, ix; Jill Tarter, "What if Everybody is Listening and Nobody is Transmitting?" SETI Institute Explorer, Third Quarter 2004, 18–19.
42. James Gunn, Alternate Worlds: The Illustrated History of Science Fiction, Englewood Cliffs, NJ, A&W Visual Library (Prentice-Hall), 1975, 169.
43. Sullivan, 308–309; James Gunn, The Listeners, New York, Scribners, 1972, 212.
44. Stapledon, 321.
45. Crowe, 88, 558; Stapledon, Star Maker, 354; Sagan, The Cosmic Connection, 241.
46. Swift, 85; Drake, "On Hands and Knees;" McDonough, 224.
47. Sagan, "The Quest for Intelligent Life;" Diane Richards, "Interview with Dr. Frank Drake," SETI Institute News, Vol. 12 (First Quarter 2003), 5–6; Goldsmith, editor, 269.
48. McDonough, 206–207; Rubin, 311.
49. Drake, "On Hands and Knees;" McDonough, 230; Koerner and LeVay, 245.
50. Project Cyclops, 181; "Project Cyclops," Newsweek, 29 April 1974.

51. Goldsmith and Owen, 352; Sagan, The Cosmic Connection, 220; Bracewell, The Galactic Club, 66.

52. Peter Schenkel, "The Nature of ETI, Its Longevity and Likely Interest in Mankind: The Human Analogy Re-examined," JBIS, Vol. 52 (1999), 13–18.

53. Black, et al.; Harrison, After Contact, 302.

54. Billingham, et al., editors, 52; Brian W. Aldiss, "Desperately Seeking Aliens," Nature, Vol. 409 (22 February 2001), 1080–1082.

55. Project Cyclops, 169; David Brin, Otherness, New York, Bantam, 1994, 132.

56. Crowe, 558.

57. Story, editor, The Encyclopedia of Extraterrestrial Encounters, 485–486.

58. C.M. Cade, "Communicating with Life in Space," Discovery, May 1963, 36–41 (he repeated this thought in his 1966 book Other Worlds than Ours, 228).

59. Stephen J. Pyne, "Voyage of Discovery," in Valerie Neal, editor, Where Next, Columbus? The Future of Space Exploration, New York, Oxford University Press, 1994, 9–39.

60. Gregory Benford, Deep Time: How Humanity Communicates Across Milennia, New York, Harper Collins Perennial, 1999, and Avon, 1999.

61. Norris, in Tough, editor, 105; Swift, 113.

62. Peter G. Stanley, "Prospects for Extraterrestrial Life," JBIS, Vol. 50 (1997), 243–248.

63. Stewart Brand, The Clock of the Long Now, London, Weidenfeld & Nicholson, 1999, and John Casti's review in Nature, Vol. 400 (1 July 1999), 33.

64. Don Lago, "In the Time Machine," Cosmic Search, Vol. 2, Number 2 (Spring 1980), 30–31.

Fears

1. Arthur C. Clarke, Profiles of the Future, New York, Bantam, 1963, 94.

2. Billingham, et al., editors, Social Implications, 57.

3. Crowe 395–396.

4. Clarke, Greetings, 480; Shapley, 107; William J. Broad, "Life on Mars: What Would Neighbors Mean for Earthlings," International Herald Tribune, 11 Jan 2004; David Brin, "Xenology: The New Science of Asking Who's Out There," Analog Science Fiction/Science Fact, May 1983, 64–82.

5. Sagan and Shklovskii, 22.

6. Davies, Are We Alone?, 50.

7. Jung was quoted by Sullivan, 294, and by Stern in Maruyama and Harkins, editors, 50; White, 149.

8. Davies, Are We Alone?, 54; "Interview: Robert E. Edelson," Mercury, July-August 1977, 8–12.

9. Quoted in William J. Broad, "Be Careful What You Look For on Mars," The New York Times, 11 January 2004; Drake and Sobel, 210.

10. Billingham, et al., editors, 56.

11. Anthony DePalma, Here: A Biography of the New American Continent, New York, Public Affairs Press, 2001, 35.
12. Drake, "On Hands and Knees."
13. Donald N. Michael, "Proposed Studies on the Implications of Peaceful Space Activities for Human Affairs: A Report Prepared for the Committee on Long-Range Studies of the National Aeronautics and Space Administration by the Brookings Institution," 1960; Sullivan, 293.
14. Morrison in Sagan, editor, CETI, 337, 342.
15. Sagan, The Cosmic Connection, 219; McDonough, 221.
16. Kuiper and Morris in Goldsmith, editor, 175; Paolo Musso, "Philosophical and Religious Implications of Extraterrestrial Intelligent Life," paper presented at the 2004 International Astronautical Congress (IAC-04-IAA.1.1.2.11).
17. Sagan, editor, CETI, 345; Swift, 193; "von Hoerner on SETI" (interview), Cosmic Search, Vol. 1, Number 1 (January 1979), 40–45; Shostak, 196.
18. S.A. Kaplan, editor, Extraterrestrial Civilizations, 234; Ulmschneider, 226; Harrison, "Slow Track, Fast Track."
19. Roberto Pinotti, "Contact: Releasing the News," Acta Astronautica, 1990, 107–114; Donald E. Tarter, "Security Considerations in Signal Detection," paper presented at the 1997 International Astronautical Congress (IAA-97-IAA.9.2.05).
20. Ben Finney, "The Impact of Contact,"Acta Astronautica, Vol. 21 (1990), 117–121; Billingham, et al., 52.
21. Robert Jastrow, "What Are the Chances for Life?" (review of Dick's The Biological Universe), Sky and Telescope, June 1997, 62–63.
22. Eric J. Chaisson, "Null or Negative Effects of ETI Contact in the Next Millennium," in Tough, editor, 59; Foundation for the Future News, Winter 1999, 5.
23. Sagan, editor, CETI, 345; Richard G. Klein, "Whither the Neanderthals?" Science, Vol. 299 (7 March 2003), 1525–1526.
24. Allen Tough, "A Critical Examination of the Factors that Might Encourage Secrecy," Acta Astronautica, Vol. 21 (1990), 97–101; Huntington, 91.
25. Alan Moorhead, The Fatal Impact: The Invasion of the South Pacific, New York, Harper and Row, 1966, 86–87.
26. White, 153; Shostak, Sharing The Universe, 196.
27. Sullivan, 294; Stern, in Maruyama and Harkin, editors, 56; Ulmschneider, 226.
28. Rood and Trefil, 240.
29. Harrison, After Contact, 170.
30. Huntington, 302; Clarke, Profiles of the Future, 91.
31. Padgen, xv–xvi.
32. Berendzen, editor, 17, 60.
33. Rees, 170; Harrison, After Contact, 284.
34. Dick, "SETI Before the Space Age."
35. Huntington, 67, 271; Alexander Stille, "Historians Trace an Unholy Alliance," The New York Times, 31 May 2003.
36. Arne Ohman, "Conditioned Fear of a Face: A Prelude to Ethnic Enmity?", Science, Vol. 309 (29 July 2005), 711–713; Andreas Olsson, et al., "The Role of Social Groups in the Persistence of Learned Fear," Science, Vol. 309 (29

July 2005), 785–786; David Bereby, Us and Them: Understanding Your Tribal Mind, Boston, Little, Brown, 2005, quoted by Henry Gee in Scientific American, December 2005, 124–126.

37. Shostak, Sharing The Universe, 121.
38. In Berendzen, editor, 87.
39. See, for example, the quotations in Erik Hildinger, Warriors of the Steppe, Cambridge, MA, Da Capo, 1997, 133, 136.
40. British classicist J.V. Luce, as quoted in John Noble Wilford, "Will We Ever Find Atlantis?", The New York Times, 11 November 2003.
41. Robert L. Hall, "Sociological Perspectives on UFO Reports," in Sagan and Page, editors, 215–216.
42. In Goldsmith, editor, 245.

Dangers

1. Quoted by Dick, The Biological Universe, 515, and by Billingham, et al., 53.
2. Zdenek Kopal, Man and His Universe, London, New York, William Morrow, 1972, 306. Also quoted in MacVey, Interstellar Travel, 226, and in Sullivan, 297.
3. As quoted by Nigel Calder in Spaceships of the Mind, New York, Viking, 1978, 10.
4. Purcell, in Goldsmith, editor, 188–196; Morrison in Sagan, editor, CETI, 337; Drake, "On Hands and Knees."
5. Sagan, The Cosmic Connection, 179–180, 215.
6. Joseph F. Goodavage, "An Interview with Carl Sagan," Analog Science Fiction/Science Fact, August 1976, 92–101; Sagan, Cosmos, 311; Sagan, Pale Blue Dot, 394, 398.
7. Sagan, editor, CETI, 347–348.
8. Harrison, "Slow Track, Fast Track."
9. MacGowan and Ordway, 330.
10. Asimov, Catastrophes, 265.
11. MacGowan and Ordway, 243, 268, 271, 366; Cade, Other Worlds than Ours, 224–225.
12. Project Cyclops, 31.
13. Donald Tarter, "Security Considerations in Signal Detection."
14. Republished as "Advice for Astronomers," International Herald Tribune, 30 December 1982; Clark and Clark, 106.
15. Swift, 166.
16. "Possibility of Intelligent Life Elsewhere in the Universe" (report prepared by the Congressional Research Service for the Committee on Science and Technology, U.S. House of Representatives), Washington, DC, U.S. Government Printing Office, 1977, 68.
17. William R. Polk, Neighbors and Strangers: The Fundamentals of Foreign Affairs, Chicago, University of Chicago Press, 1997.
18. Shostak, Sharing the Universe, 123; Clark and Clark, 107.
19. Ronald Bracewell, "Man's Role in The Galaxy," Cosmic Search, Vol. 1, Number 2 (March, 1979), 48–51; Sullivan, 295.

20. Rood and Trefil, 241.
21. Brin, "A Contrarian Perspective."
22. C.M. Cade, "Communicating with Life in Space," Discovery, April 1963, 26–41; MacVey, Interstellar Travel, 228; Grinspoon, 290.
23. Jared Diamond, "To Whom It May Concern," The New York Times Magazine, 5 December 1999, 68–69; Jared Diamond, The Third Chimpanzee, New York, Harper Collins, 1992, 214.
24. Sullivan, 295–296; Fitzgerald, 350.
25. Fred Hoyle, A for Andromeda, New York, Avon, 1975, originally published in 1962; Carl Sagan, Contact, New York, Pocket, 1997.

Mixed Emotions

1. Dick, editor, Many Worlds, 195–196; Dick, The Biological Universe, 58; Dick, "Consequences of Success in SETI." 529.
2. Shapley, 104–114; Swift, 47.
3. Swift, 270.
4. Edward Harrison, Masks of the Universe, 109–110.
5. Davies, Are We Alone?, 42–43.
6. Drake and Sobel, 233; Swift, 85.
7. Baird, 11.
8. White, 199.
9. Mary M. Connors, "The Role of the Social Scientist in the Search for Extraterrestrial Intelligence," NASA Ames Research Center, 1976, and Mary M. Connors, "The Consequences of Detecting Extraterrestrial Intelligence for Telecommunication Policy," NASA Ames Research Center, 1977; Rees, 64.
10. Donald E. Tarter and Walter G. Peacock, "The Consequences of Contact: Views of the Scientific Community and the Science Media," paper presented at the International Bioastronomy Symposium, 1990; Donald E. Tarter, "SETI and the Media," Acta Astronautica, Vol. 26 (1992), 281–289; Harrison, After Contact, 211.
11. McDonough, 221; Swift, 304.
12. Billingham, et al., editors, Social Implications, 65–79.
13. A. A. Harrison and J. T. Johnson, "ETI: Our First Impressions."
14. Quoted in Norman N. Nelson, "Some Diplomatic and Journalistic Implications of SETI," paper presented at the 1985 International Astronautical Congress (IAA-85-469).
15. Douglas A. Vakoch and Yuh-shiow Lee, "Reactions to Receipt of a Message from Extraterrestrial Intelligence: A Cross-Cultural Empirical Study," Acta Astronautica, Vol. 46 (2000), 737–744.
16. William S. Bainbridge, "Attitudes Toward Interstellar Communication," JBIS, Vol. 36 (1983), 298–304; Douglas A. Vakoch, "Roman Catholic Views of Extraterrestrial Intelligence," in Tough, editor, 165–174.
17. Harrison, After Contact, 198, 223.
18. Berendzen, editor, Life Beyond Earth and the Mind of Man, 25.
19. Albert A. Harrison, "Slow Track, Fast Track, and the Galactic Club," paper prepared for the Foundation for the Future, 1998.

20. Mary M. Connors, "The Consequences of Detecting Extraterrestrial Intelligence," unpublished document, NASA Ames Research Center, 1978; Timothy Ferris, "An End to Cosmic Loneliness," The New York Times Magazine, 23 October 1977, 30–32, 92, 97.
21. James V. Scotti, "The Tale of an Asteroid," Sky and Telescope, July 1998, 30–34.
22. Lee Clarke, "Social Science and Near Earth Objects," paper prepared for a conference in 2004.
23. Donald E. Tarter and Walter G. Peacock, "The Consequences of Contact."
24. Billingham, et al., editors, "Social Implications," 68–69.
25. Clarke, The Exploration of Space, 191.
26. Dick, The Biological Universe, 526; McDonough, 228.
27. George V. Coyne, "The Evolution of Intelligent Life on the Earth and Possibly Elsewhere," in Dick, editor, Many Worlds, 177–188; Davies in Tough, editor, 51.
28. Dick, editor, Many Worlds, 146–148, 202–208.
29. Paolo Musso, "Philosophical and Religious Implications of Extraterrestrial Intelligent Life." Davies reported that the Vatican had embarked on an evaluation of the significance for Christianity of the discovery of alien intelligent life. Davies, Are We Alone?, 44.
30. Sullivan, 302; Lamb, 191.
31. Ashkenazi, op. cit.
32. Vakoch, "Roman Catholic Views," in Tough, editor, 165–174; Roland Puccetti, Persons: A Study of Possible Moral Agents in the Universe, London, Macmillan, 1968, 125–126, 140–141.
33. Tenzin Gyatso, "Our Faith in Science," The New York Times, 12 November 2005.
34. Jack Jennings, "Impact of Contact on Religion," Second Look, Vol. 2, Number 2 (January-February 1980), 11–14, 41.
35. Shapley, 148–149.
36. Michael Ruse, Darwin and Design: Does Evolution Have a Purpose? Harvard University Press, 2003; Dick, editor, Many Worlds, 191–208.
37. Robert Jastrow, Until The Sun Dies, Norton, 1977.
38. Dick, editor, Many Worlds, 146–148.
39. Dick, editor, Many Worlds, 200.
40. Vladimir Lytkin, Ben Finney, and Liudmilla Alepko, "Tsiolkovsky, Russian Cosmism, and Extraterrestrial Intelligence," QJRAS, Vol. 36 (1995), 369–376; Grinspoon, 223–224.
41. Huntington, 59; White, 36; Hall, in Sagan and Page, editors, 215.
42. Bainbridge, "Attitudes Toward Interstellar Communication;" McDonough, 228.
43. Freeman, The Closing of the Western Mind, 154–177; Roberts, The Triumph of the West, 71.

Some Assumptions Examined

1. White, 138.

Before Contact

1. Achenbach, Captured by Aliens, 286, 294.
2. Fort, 162–163.
3. Jill Tarter, "What if Everybody is Listening and Nobody is Transmitting?" SETI Institute Explorer, Third Quarter 2004, 18–19; Richard Burke-Ward, "Possible Existence of Extra-Terrestrial Technology in the Solar System," JBIS, Vol. 53 (2000), 2–12.
4. Arthur C. Clarke, Rendezvous with Rama, New York, Bantam, 1990; originally published by Harcourt Brace in 1973.
5. Martin Rees, Our Final Hour, 187; Shostak and Barnett, 156.
6. N. Chandra Wickramasinghe, "Why Alien Intelligence May Not be So Alien," Searchlites, Vol. 8, Number 4 (Autumn 2002), 3–7.
7. Benjamin, 160.
8. Mayr, in Regis, editor, 29; Klein, in Bova and Preiss, editors, 148.
9. Clark and Clark, 236.
10. Ulmschneider, 207; Interview with Robert Edelson, Mercury.
11. Harrison, "Slow Track, Fast Track, and the Galactic Club," 1998.
12. Clark and Clark, 235.
13. Sagan, Pale Blue Dot, 373.
14. John Kraus, "Gerard K. O'Neill on Space Colonization and SETI," Cosmic Search, Vol. 1, Number 2 (March 1979), 16–23.
15. Sagan, editor, CETI, 254.
16. Gardner, 44.
17. Ekers, et al., 2, 70.
18. Drake and Sobel, 206; Ferris, op. cit.
19. Drake, in Ponnamperuma and Cameron, editors, 119; Frank D. Drake, "A Speculation on the Influence of Biological Immortality on SETI," Cosmic Search, Vol. 2, Number 3 (Summer 1980), 9–11; Drake, "Promising New Approaches;" Joseph E. Pingree, "One Creature's Signal Is Another One's Noise," Letter to Sky and Telescope, May 1999, 14; Ekers, et al., 123; Koerner and LeVay, 164.
20. Ekers, et al., 133.
21. Harrison, After Contact, 44; Bova and Preiss, editors, 200.
22. Sagan, The Cosmic Connection, 224; Drake, "A Speculation on the Influence of Biological Immortality on SETI."
23. Interview with Robert Edelson, Mercury, July-August 1977, 8–12; Sagan, editor, CETI, 213.
24. R.N. Bracewell, "Interstellar Probes," in Cyril Ponnamperuma and A.G.W. Cameron, editors, Interstellar Communication: Scientific Perspectives, Boston, Houghton-Mifflin, 1974, 107.
25. McConnell, 67; Tough, "Small, Smart Interstellar Probes."
26. Harrison, After Contact, 193.
27. Gregory Benford, Beyond Infinity, New York, Aspect (Warner), 2004; Project Cyclops, 180.
28. Quoted by Dennis Overbye, "Mirror, Mirror," The New York Times, 30 August 2005.
29. Sagan, The Cosmic Connection, 232; Sagan, "The Quest for Intelligent Life."

30. Burke-Ward, "Possible Existence of Extraterrestrial Technology in the Solar System;" Swift, 188–189.
31. Sagan, The Cosmic Connection, 229–231.
32. Grinspoon, 277; White, 55.
33. Rees, 168.
34. In Shostak, editor, Third Decennial US-USSR Conference, 411; McDonough, 218.
35. Hart and Zuckerman, editors, 33.
36. Grinspoon, 323; Clarke, Greetings, 399. Sagan repeated that thought: "Civilizations hundreds or thousands or millions of years beyond us should have sciences and technologies so far beyond our present capabilities as to be indistinguishable from magic." The Cosmic Connection, 222.
37. Rees, 167.
38. Sagan, editor, CETI, 189–190; Achenbach, 73.
39. Shostak, 200; Sagan, editor, CETI, 8.
40. Kondo, et al., editors, 118–119.
41. Bracewell, in Goldsmith, editor, 105–106.
42. In Minutes of the Workshop on Cultural Evolution, Center for Advanced Study in the Behavioral Sciences, Stanford, CA, November 24–25, 1975; Baird, 115–116.
43. J. M. Roberts, The Triumph of the West, 118.
44. Quoted by Steven J. Dick, "Exploration in Historical Context," paper presented at a Conference on International Cooperation in Space, June 21, 2004, partly reprinted in NASA History Notes, Vol. 21, Number 3 (August 2004), 1–2; Ben R. Finney, "From Africa to the Stars: The Evolution of the Exploring Animal," in James D. Burke and April S. Whitt, Space Manufacturing 1983, San Diego, Univelt, 1983, 85–104.
45. Joseph Kahn, "China Has an Ancient Mariner to Tell You About," The New York Times, 20 July 2005.
46. Philip Morrison, "Interstellar Communication," in Cameron, editor, 249–271; Sagan, editor, CETI, 214–215.
47. McDonough, 208–209; Sagan, The Cosmic Connection, 222.
48. Sagan, The Cosmic Connection, 216.
49. Sagan, Other Worlds, 22; Peter Boyce, "Planetary Radars Have Announced Our Existence," paper presented at the 1991 International Astronautical Congress (IAA-91-611); McDonough, 230.
50. Woodruff Sullivan III, et al., "Eavesdropping: The Radio Signature of the Earth," Science, Vol. 199 (27 January 1978), 377–387; John Billingham and Jill Tarter, "Detection of the Earth with the SETI Microwave Observing System Assumed to be Operating Out in the Galaxy," Acta Astronautica, Vol. 26 (1992), 185–188.
51. Jill Tarter, in Regis, editor, 187; Michael D. Papagiannis, "The Hunt is On: SETI in Action," in Bova and Preiss, editors, 181–195.
52. Morrison, et al., editors, SETI, 51–52; MacGowan and Ordway, 259; Harrison, After Contact, 140.
53. George W. Swenson, Jr., "Intragalactically Speaking," Scientific American, July 2000, 44–47. Shostak was quoted by C. Claiborne Ray in The New York Times, 31 July 2001.

54. Swift, 409.
55. Project Cyclops, 28.
56. Fred Hoyle, The Black Cloud, 185.
57. Kerr, "No Din of Alien Chatter."
58. Stern, in Maruyama and Harkins, editors, 30; Ben Finney, "SETI and the Two Terrestrial Cultures," Acta Astronautica, Vol. 26 (1992), 263–265.
59. Aveni, 212; Beck, in Regis, editor, 10; Finney, "SETI and the Two Terrestrial Cultures."
60. Rood and Trefil, 97; Shostak, Sharing The Universe, 112.
61. Goldsmith and Owen, 351–352.
62. Goldsmith and Owen, 346; Zuckerman, in Hart and Zuckerman, editors, 9; Bernard M. Oliver, "Radio Search for Distant Races," International Science and Technology, Vol. 10 (October 1962), 55–60.
63. Jill Tarter, "What if Everybody is Listening and Nobody is Transmitting?" SETI Institute Explorer, Third Quarter 2004, 18–19.
64. Rood and Trefil, 98–99.
65. Regis, in Regis, editor, 240. Also see Stern, in Maruyama and Harkins, editors, 31.
66. Zsolt Hetesi and Zsolt Regaly, "A New Interpretation of Drake Equation," JBIS, Vol. 59 (2006), 11–13.
67. Carl Sagan and William I. Newman, "The Solipsist Approach to Extraterrestrial Intelligence," in Regis, editor, 151–161; Sagan, editor, CETI, 210, 254; Baird, 203; Sagan, The Cosmic Connection, 222, 225; Carl Sagan, "On the Detectivity of Advanced Galactic Civilizations," Icarus, Vol. 19 (1973), 350–352 (also in Goldsmith, editor, 140–141); Regis, in Regis, editor, 240; Achenbach, 279.
68. "What If?" Sky and Telescope, November 1992, 515.
69. Robin H. D. Corbet, "SETI at X-Ray Energies—Parasitic Searches from Astrophysical Observations," JBIS, Vol. 50 (1997), 253–257.
70. Shostak, Sharing The Universe, 192–193; Seth Shostak, "The Future of SETI," Sky and Telescope, April 2001, 42–53.
71. Bova and Preiss, editors, 208; McDonough, 219–220.
72. Project Cyclops, 60, 170, 171.
73. Ekers, et al., 172.
74. Jill Tarter, "What if Everybody is Listening."
75. Drake, "On Hands and Knees;" Sagan, editor, CETI, 301; Swift, 115.
76. Davies, Are We Alone?, 71; Bracewell, The Galactic Club, 54–55, 128.
77. Seth Shostak, "Transmitting to a Million Worlds," SETI Institute home page, 12 January 2006; Rubin, 301–305.
78. Horowitz, in Bova and Preiss, editors, 209; Brin, "A Contrarian Perspective on Altruism."
79. Ekers, et al., 230–231.
80. McDonough, 224–226.
81. Louis Berman, "What it Was Like Before Ozma," Cosmic Search, Vol. 1, Number 4 (Fall 1979), 17–20.
82. Andrew MacFadyen, "Long Gamma-Ray Bursts," Science, Vol. 303 (2 January 2004), 45–46.
83. McDonough, 226; Brin, "A Contrarian Perspective."

84. Jill Tarter, "What if Everybody is Listening;" Morrison in Berendzen, editor, 93–94; Bracewell, in Ponnamperuma and Cameron, editors, 116; Ronald N. Bracewell, "Communications from Superior Galactic Communities," in Donald Goldsmith, editor, The Quest for Extraterrestrial Life, Mill Valley, CA, University Science Books, 1980, 105–106.

85. Albert A. Harrison, "Slow Track, Fast Track, and the Galactic Club," paper published by the Foundation for the Future, 1998.

Assumptions: After Contact

1. Sullivan, 160.
2. Frank D. Drake, "Methods of Communication," in Ponnamperuma and Cameron, editors, 139.
3. Sagan, "The Quest for Intelligent Life;" Sagan, Other Worlds, 21.
4. Sagan, The Cosmic Connection, 217; Tarter, in Dick, editor, Many Worlds, 147.
5. McDonough, 228; Baird, 194.
6. Sagan, editor, CETI, 334–335; Bracewell, The Galactic Club, 79.
7. Douglas A. Vakoch, "Constructing Messages to Extraterrestrials: An Exosemiotic Approach," paper delivered at the 1995 International Astronautical Congress (IAA-05-IAA.9.2.05), and Douglas A. Vakoch, "The Dialogic Model: Representing Human Diversity in Messages to Extraterrestrials," Acta Astronautica, Vol. 42 (1998), 705–710.
8. Sagan, editor, CETI, 320, 343.
9. Andrew Lawler, "The Indus Script—Write or Wrong?" Science, Vol. 306 (17 December 2004), 2026–2029.
10. Baird, 205; Guillermo Lemarchand, "Who Will Speak in the Name of Earth?" Bioastronomy News, Vol. 8 (Third Quarter 1996), 1–3.
11. Baird, 93, 117, 190, 193.
12. John Elliott and Eric Atwell, "Is Anybody Out There? The Detection of Intelligent and Generic Language-Like Features," JBIS, Vol. 53 (2000), 13–22.
13. Kuniyoshi L. Sakai, "Language Acquisition and Brain Development," Science, Vol. 310 (4 November 2005), 815–819.
14. Crowe, 27; Michael Ruse, "Is Rape Wrong on Andromeda?", in Regis, editor, 43–72.
15. Regis, editor, 240; Goldsmith, editor, 208; Drake, "On Hands and Knees;" Stanislav Lem, Solaris, New York, Berkley, 1976.
16. Baird, 178; Fred Hoyle, The Black Cloud, 206; Michael Arbib, "Minds and Millennia," Cosmic Search, Vol. 1, Number 3 (Summer 1979), 21–25, 47–48.
17. McConnell, 275.
18. Regis, in Regis, editor, 238; Arbib, "Minds and Millennia."
19. Sagan, Other Worlds, 21.
20. Carl L. De Vito, "Languages Based on Science," Acta Astronautica, Vol. 26 (1992), 267–271.
21. Rescher in Regis, editor, 89–90, 105, 113–114, 242; Baird, 178, "Hawaii Seminar Focuses on the Cultural Impact of Extraterrestrial Contact," Foundation for the Future News, Winter 1999, 3–5.

22. Ruse, in Regis, editor, 53; Douglas A. Vakoch, "Pictorial Messages to Extra-terrestrials, Part 1," SETI Quest, Vol. 4 (First Quarter 1998), 8–10.

23. White, 84; Nicholas Rescher, "Extraterrestrial Science," in Regis, editor, 105.

24. David N. Livingstone, Putting Science in its Place: Geographies of Scientific Knowledge, Chicago, University of Chicago Press, 2003; reviewed by Cristina Gonzalez in Science, Vol. 302 (5 December 2003) 1683–1684.

25. Sagan, editor, CETI, 344.

26. Rescher, in Regis, editor, 85–87, 114; Regis, in Regis, editor, 242.

27. Regis, editor, 81; Rood and Trefil, 155.

28. Musso, "Philosophical and Religious Implications of Extraterrestrial Intelligence."

29. Regis, editor, 81, 117–128.

30. Sagan, editor, CETI, 346; Burke-Ward, "Possible Existence of Extra-Teres-trial Technology;" Vakoch, "Constructing Messages to Extraterrestrials."

31. George Johnson, "Useful Invention or Absolute Truth: What is Math?" The New York Times, 10 February 1998.

32. Baird, 44, 133.

33. Puccetti, 115.

34. Philip Morrison, "Conclusion: Entropy, Life, and Communication," in Ponnamperuma and Cameron, editors, 168–186.

35. Raup in Regis, editor, 38.

36. White, 158.

37. Ulmschneider, 220–221.

38. Regis, editor, 9.

39. Regis, in Regis, editor, 230.

40. Bova, Faint Echoes, Distant Stars, 251.

41. Sagan, The Cosmic Connection, 216.

42. Morrison in Cameron, editor, 317, and in Sagan, editor, CETI, 335.

43. Project Cyclops, 60, 62, 64.

44. Harrison, After Contact, 284.

45. Jean Heidmann, "A Reply from Earth: Just Send Them the Encyclopedia," Acta Astronautica, Vol. 29 (1993), 233–235; McDonough, 229–230.

46. Brin, "A Contrarian Perspective."

47. Achenbach, 283.

48. John C. Lilly, Man and Dolphin, New York, Pyramid, 1969, 8, originally published by Doubleday in 1961.

49. McDonough, 220.

50. This story has been told many times. For an authoritative version including the original paper in Nature, see Marcia Bartusiak, Archives of the Universe: A Treasury of Astronomy's Historic Works of Discovery, New York, Pantheon, 2004, 513–518. Also see Walter Sullivan, 211–213.

51. Dick and Strick, 189.

52. Govert Schilling, "Catching Gamma-Ray Bursts on the Wing," Sky and Telescope, March 2004, 33–40.

53. Fitzgerald, 333.

54. Allen Tough, "A Critical Examination of Factors that Might Encourage Secrecy," Acta Astronautica, Vol. 21 (1990), 97–101; Harrison, After Contact, 270–272.

55. Donald E. Tarter, "Security Considerations in Signal Detection," paper presented at the 1997 International Astronautical Congress (IAA-97-IAA.9.2.05); Bracewell, the Galactic Club, 75; Baird, 11; White, 63.
56. Swift, 84–85.
57. White, 64.
58. Kyle Jansen and Fiona Murray, "Intellectual Property Landscape of the Human Genome," Science, Vol. 310 (14 October 2005), 239–240.
59. "Team is Ready to Publish Full Set of Dead Sea Scrolls," The New York Times, 15 November 2001.
60. Harrison, "Slow Track, Fast Track."
61. Robert Wright, Non-Zero: The Logic of Human Destiny, New York, Vintage, 2000, 104; Jane R. McIntosh, A Peaceful Realm: The Rise and Fall of the Indus Civilization, New York, Boulder, CO, Westview, 2002, 209.
62. Baird, 25.
63. Baird, 216.
64. Dick, "Consequences of Success in SETI," 525.
65. Huntington, 101, 184; Ian Buruma and Avishai Margalit, Occidentalism: The West in the Eyes of its Enemies, New York, Penguin, 2004.
66. Quoted in "Confounding Machines: How the Future Looked," The New York Times, 28 August 2005.
67. Billingham, et al., editors, Social Implications, 86–87.
68. Regis in Regis, editor, 234–237.
69. MacGowan and Ordway, 270.
70. Rubin, 311.
71. Alan S. Blinder, "Offshoring: The Next Industrial Revolution," Foreign Affairs, Vol. 85, Number 2 (March/April 2006), 113–128; Harrison, After Contact, 278, 287, 293, 299.
72. Frank Tipler, "Alien Life" (review of Davoust's The Cosmic Water Hole," Nature, Vol. 354 (28 November 1991), 334–335.
73. Wasim Maziak, "Science in the Arab World," Science, Vol. 308 (3 June 2005), 1416–1418.
74. Bernard Lewis, The Muslim Discovery of Europe, London, Weidenfeld and Nicolson, 1982.
75. White, 32; Dick, editor, Many Worlds, 148.
76. Bracewell, The Galactic Club, 75.
77. Harrison and Dick in Tough, editor, 10.
78. Drake, "On Hands and Knees;" Cameron, editor, Introduction, 1.
79. Edward Rothstein, "Paradise Lost: Can Mankind Live Without Utopias?" The New York Times, 5 February 2000.
80. James C. Scott, Seeing Like a State: How Certain Schemes to Improve the Human Condition Have Failed, New Haven, CT, Yale University Press, 1998.
81. Robert Conquest, Reflections on a Ravaged Century, London, John Murray, 1999, 1, 13, 15. It is disturbing to see that some scientists still flirt with totalitarian ideas. While recognizing that policing in human societies has been used by repressive regimes to sustain inequalities, Francis L. W. Ratnieks and Tom Wenseleers wrote that "a human society in which policing is used to promote greater equality and justice may not be an unattractive

prospect." "Policing Insect Societies," Science, Vol. 307 (7 January 2005), 54–56.

82. Crowe, 51–52; Harrison, After Contact, 187–188; Launius, "Perfect Worlds, Perfect Societies;" Favell, "Human Aspirations in the History of Space Flight."

83. MacGowan and Ordway, 265–268.

84. Crowe, 197.

85. Fred Hoyle, The Black Cloud, 186.

86. Clarke, The Exploration of Space, 193; Clarke, The Promise of Space, 306.

87. Grinspoon, 412.

88. McDonough, 202; Tough, editor, 45.

89. Quoted by James W. Deardorff, "Possible Extraterrestrial Strategy for Earth," QJRAS, Vol. 27 (1986), 94–101.

90. Lamb, 193; Shostak, 102.

91. Richard Burke-Ward, "Possible Existence of Extra-Terrestrial Technology in the Solar System," JBIS, Vol. 53 (2000), 2–12; Pickover, 172.

92. Bruce E. Fleury, "The Aliens in Our Oceans."

93. Musso, "Philosophical and Religious Implications."

94. Puccetti, 117–118.

95. Michael S. Gazzaniga, The Ethical Brain, Washington, DC, Dana Press, 2005. Also see the review by Paul Bloom in Nature, Vol. 436 (14 July 2005), 178–179.

96. Lilly, 123; Ruse, in Regis, editor, 71.

97. In Regis, editor, 238.

98. McKay in Dick, editor, Many Worlds, 56, Dick in Dick, editor, Many Worlds, 205.

99. Charles Darwin, On The Origin of Species, as quoted by Robert Ardrey in The Territorial Imperative, New York, Laurel, 1971, 260, originally published in 1966.

100. Drake, in Ponnamperuma and Cameron, editors, 122.

101. Claudia Dreifus, "The Congressman Who Loved Science," (interview with George Brown), The New York Times, 9 March 1999.

102. Project Cyclops, 31.

103. Rood and Trefil, 240; Brin, "A Contrarian Perspective;" Shostak, Sharing The Universe, 100.

104. Joan B. Silk, et al., "Chimpanzees are indifferent to the welfare of unrelated group members," Nature, Vol. 437 (27 October 2005), 1357–1359.

105. Brin, "A Contrarian Perspective."

106. Robert Trivers, "Mutual Benefits at All Levels of Life" (review of Peter Hammerstein's Genetic and Cultural Evolution of Cooperation), Science, Vol. 304 (14 May 2004), 964–965; Paul Seabright, The Company of Strangers: A Natural History of Economic Life, Princeton, NJ, Princeton University Press, 2004, reviewed by Herbert Gintis in Nature, Vol. 431 (16 September 2004), 245–246.

107. Andrew Haley, Space Law and Government, New York, Appleton-Century-Crofts, 1965; Clive D. L. Wynne, "Willy Didn't Yearn to Be Free," The New York Times, 27 December 2003.

108. MacVey, Interstellar Travel, 225, 231.

109. Dick, editor, Many Worlds, 148.

110. Arthur C. Clarke, Childhood's End, London, Pan Books, 1956, 59, originally published by Sidgwick and Jackson in 1954; Cohen and Stewart, 59.

111. Barrow and Tipler, 597–598; Grinspoon, 394.

112. Brin, "A Contrarian Perspective."

113. See Michael A.G. Michaud, "Organizing Ourselves for Contact," Analog Science Fiction/Science Fact, Vol. CXVIII, No. 1 (January 1998), 51–63.

114. Quoted in Douglas Vakoch, "The Dialogic Model: Representing Human Diversity in Messages to Extraterrestrials," Acta Astronautica, Vol. 42 (1998), 705–710.

115. Douglas Vakoch, "Messages to the Cosmos?" SETI Institute News, Vol. 12 (Second Quarter 2003), 14–15.

116. McDonough, 230; Goldsmith was quoted in Douglas A. Vakoch, "The Dialogic Model."

117. "The War of the Worlds," in The Complete Science Fiction Treasury of H.G. Wells, New York, Avenel, 1978, 266.

118. Steven A. LeBlanc (with Katherine E. Register), Constant Battles: The Myth of the Peaceful, Noble Savage, New York, St. Martin's, 2003, 166.

119. Huntington, 129; Felipe Fernandez-Armesto, Humankind: A Brief History, Oxford, Oxford University Press, 2004, 91.

120. Anthony Padgen, Peoples and Empires, New York, Modern Library, 2003, 12; Fernandez-Armesto, 68.

121. Sagan, Cosmos, 311; Harrison and Dick, in Tough, editor, 20; Harrison, After Contact, 166.

122. Harrison, After Contact, 167; David Christian, Maps of Time: An Introduction to Big History, Berkeley, CA, University of California, Press, 2004, 482.

123. Quoted in William R. Polk, Neighbors and Strangers: The Fundamentals of Foreign Affairs, Chicago, University of Chicago Press, 1997, 273.

124. Edward Grendon, "Crisis," in Groff Conklin, editor, Invaders of Earth, New York, Pocket Books, 1955, 167–177.

125. Sagan and Newman, in Regis, editor, 159–160.

126. Horowitz was quoted in Easterbrook, "Are We Alone?"; Clark and Clark, 107.

127. Gregory L. Matloff, Peter Schenkel, and Jaime Marchan, "Direct Contact with Extraterrestrials: Possibilities and Implications," paper presented at the 1999 International Astronautical Congress (IAA-99-IAA.8.2.02); MacGowan and Ordway, 272.

128. Stern, in Maruyama and Harkins, editors, 49.

129. Easterbrook, "Are We Alone?"; Casti, 393; Guillermo Lemarchand, "Who Will Speak in the Name of Earth?"; MacVey, Interstellar Travel, 230; Baldwin, "Keeping the ET's Away;" Yaoming Hu, et al., "Large Mesozoic Mammals Fed on Young Dinosaurs," Nature, Vol. 433 (13 January 2005), 149–152; Anne Weil, "Living Large in the Cretaceous," Nature, Vol. 433 (13 January 2005), 116–117; Erik Stokstad, "New Fossils Show Dinosaurs Weren't the Only Raptors," Science, Vol. 307 (14 January 2005), 192.

130. Nicholas Wade, "A Course in Evolution, Taught by Chimps," The New York Times, 25 November 2003; Frans B.M. de Waal, "A Century of Getting to Know the Chimpanzee," Nature, Vol. 437 (1 September 2005), 56–58; Steven A. LeBlanc, "Prehistory of Warfare," Archaeology, May–June 2003, 18–25.

131. Robert L. Carroll, "Between Water and Land," Nature, Vol. 437 (1 September 2005), 38–39; LeBlanc, Constant Battles, 54; Hart and Zuckerman, editors, 37.

132. Drake, in Ponnamperuma and Cameron, editors, 120; Harrison, After Contact, 15.

133. Sagan, editor, CETI, 105.

134. Peter D. Walsh, et al., "Catastrophic Ape Decline in Western Equatorial Africa," Nature, Vol. 422 (10 April 2003), 611–613.

135. Rood and Trefil, 241.

136. Robert Freitas, "Metalaw and Interstellar Relations," Mercury, Number 6 (March-April 1977), 15–17. Michaud developed this theme in several articles, beginning with "Interstellar Negotiation" in 1972.

137. Brin, A Contrarian Perspective; Achenbach, 290.

138. The Complete Science Fiction Works of H.G. Wells, 266. Robert Markley developed this theme at some length in Dying Planet: Mars in Science and the Imagination, Raleigh, NC, Duke University Press, 2005.

139. Harrison, After Contact, 122, 194, 312; Harrison, "Slow Track, Fast Track;" Albert Harrison, "The Relative Stability of Belligerent and Peaceful Societies," Acta Astronautica, Volume 46 (2000), 707–712.

140. William Pfaff, "Europe Pays the Price for Cultural Naivete," International Herald Tribune, 25 November 2004.

141. Puccetti, 105.

142. Sullivan, 296.

143. Murray Leinster, "First Contact," in Damon Knight, editor, First Contact, Pinnacle, 1971, 9–44.

144. Barry Gewen, review of Victor Davis Hanson's Ripples of Battle, The New York Times, 28 September 2003.

145. Lawrence Keeley, War Before Civilization: The Myth of the Peaceful Savage, Oxford, Oxford University Press, 1997 (reprint edition).

146. Natalie Angier, "Is War Our Biological Destiny?" The New York Times, 11 November 2003; LeBlanc, "Prehistory of Warfare;" John Noble Wilford, "Archaeologists Unearth a War Zone 5,500 Years Old," The New York Times, 16 December 2005. Researchers have found a genetic variant in primates, including chimpanzees and humans, that predisposes males to aggressive, impulsive, and even violent behavior. "Tracking the Evolutionary History of a Warrior Gene," Science, Vol. 304 (7 May 2004), 818.

147. LeBlanc, Constant Battles, 3, 8, 12, 167, 185, 192, 220.

148. LeBlanc, Constant Battles, 84, 89, 225; Asimov, A Choice of Catastrophes, 268.

149. LeBlanc, "Prehistory of Warfare."

150. David Christian, 458.

151. Harrison, After Contact, 184; David Christian, 458.

152. Hart and Zuckerman, editors, 32.

153. Sullivan, 295–296; Brin in Bova and Preiss, editors, 128; McConnell, 370.

154. Brin, "A Contrarian Perspective."

155. Clarke, The Exploration of Space, 192; Clarke, Greetings, 39.

156. In Regis, editor, 263.

157. Drake and Sobel, xiii.

158. William I. Newman and Carl Sagan, "Galactic Civilizations: Population Dynamics and Interstellar Diffusion," Icarus, Vol. 46 (1981), 293–327.
159. Swift, 112.
160. Shklovskii and Sagan, 464; Congressional Research Service, "Possibility of Intelligent Life Elsewhere in the Universe."
161. Clark and Clark, 212–213.
162. Asimov, A Choice of Catastrophes, 265; Ulmschneider, 214. This Asimov estimate seems inconsistent with the much larger distance he proposed in his book Extraterrestrial Civilizations, page 192.
163. Barrow and Tipler, 585; Newman and Sagan, "Galactic Civilizations."
164. Barrow and Tipler, 586, 594.
165. Robert Ardrey, The Territorial Imperative: A Personal Inquiry into the Animal Origins of Property and Nations, New York, Dell, 1966 (Laurel Edition, 1971), 295.
166. Project Cyclops, 31.
167. MacVey, Interstellar Travel, 232.
168. Sebastian von Hoerner, Population Explosion and Interstellar Expansion, Gottingen, Germany, Vandenhoek and Ruprecht, as paraphrased by Alfred Roulet, The Search for Intelligent Life in Outer Space, 121–124.
169. Harrison, "Slow Track, Fast Track."
170. Sagan, Cosmos, 308.
171. Barrow and Tipler, 596.
172. John Kraus, "Gerard K. O'Neill on Space Colonization and SETI," Cosmic Search, Vol. 1, Number 2 (March 1979), 16–23.
173. Richard Burke-Ward, "Possible Existence of Extra-Terrestrial Technology in the Solar System," JBIS, Vol. 53 (2000), 2–12; Rood and Trefil, 241.
174. Charles Pellegrino and George Zebrowski, The Killing Star, New York, William Morrow, 1995.
175. Ulmschneider, 226.
176. Mark Pagel and Ruth Mace, "The Cultural Wealth of Nations," Nature, Vol. 428 (18 March 2004), 275–278.
177. Harrison, After Contact, 182.
178. White, 103–104.
179. Harrison, "Slow Track, Fast Track."
180. Bracewell, The Galactic Club, 75.
181. Clarke, Profiles of the Future, 118–119; Greetings, 207.
182. Arthur C. Clarke, Foreword to 2001: A Space Odyssey, Hutchison Radius, 1968; reissued by Roc (New American Library) in 2000, xx.
183. Stapledon, 356.
184. James Gunn, The Illustrated History of Science Fiction, 169.
185. Groff Conklin, Invaders of Earth, New York, Pocket Books, 1955, ix.
186. Gunn, Alternate Worlds, 225–226.
187. Harrison, "Slow Track, Fast Track."
188. Harrison, After Contact, 175.
189. Huntington, 51.
190. Padgen, xxi; Henry A. Kissinger, Diplomacy, New York, Simon and Schuster, 1994, 21.
191. Dmitri Simes, "America's Imperial Dilemma," Foreign Affairs, Vol. 83, Number 2 (November-December 2003), 91–102; Kissinger, 21.

192. Eliot A. Cohen, "History and the Hyperpower," Foreign Affairs, July–August 2004, 49–63.
193. Padgen, 26, 29, 64.
194. Deepak Lal, In Praise of Empires, New York, Palgrave Macmillan, 2004.
195. Niall Ferguson, Empire: The Rise and Demise of the British World Order and the Lessons for Global Power, New York, Basic, 2003; John Lewis Gaddis, "The Last Empire, for Now," review of Niall Ferguson's Colossus, The New York Times Book Review, 25 July 2004; Arthur Herman, To Rule the Waves: How the British Navy Shaped the Modern World, New York, Harper Collins, 2004.
196. Padgen, xxiii, xxiv.
197. Henry Kamen, Empire: How Spain Became a World Power, New York, Harper Collins, 2003. First published in Great Britain by Penguin, 2002.
198. See Edward N. Luttwak, The Grand Strategy of the Roman Empire, Baltimore, Johns Hopkins University Press, 1976, particularly the diagrams on pages 22 and 23.
199. G. John Ikenberry, "Illusions of Empire" (book essay), Foreign Affairs, Vol. 83, Number 2 (March-April 2004), 144–151.
200. Paul Kennedy, Review of Hugh Thomas' Rivers of Gold, The New York Times Book Review, 25 July 2004; Cohen, "History and the Hyperpower;" David Abernethy, The Dynamics of Global Dominance: European Overseas Empires, 1415–1980, New Haven, CT, Yale University Press, 2001.
201. Padgen, 20, 132.
202. Shostak, Sharing The Universe, 175.
203. Gilster, 19.
204. Lilly, 67.
205. Baird, 35, 133.; Clark and Clark, 242.
206. Stern, in Maruyama and Harkins, editors, 54; Harrison and Dick in Tough, editor, 9; Harrison, After Contact, 198; David Sivier, "SETI and the Historian," JBIS, Vol. 53 (2000), 23–25; E.J Coffey, quoted in Harrison, After Contact, 145.
207. Shostak, 113, 201.
208. Jill Tarter, "Searching for Them: Interstellar Communications," Astronomy, October 1982, 6–20.
209. Steven J. Dick, "Cultural Evolution, the Postbiological Universe and SETI," International Journal of Astrobiology, Vol. 2 (2003), 65–74.

What Is Missing

1. Project Cyclops, 32.
2. Morrison, et al., editors, 8; "Should Mankind Hide?" The New York Times, 22 November 1976, republished in Goldsmith, editor, 269.
3. Brin, "A Contrarian Perspective;" Burke-Ward, "Possible Existence of Extra-Terrestrial Technology in the Solar System;" MacVey, Interstellar Travel, 226.
4. Ivan Almar and Jill Tarter, "The Discovery of ETI as a High-Consequence, Low-Probability Event," paper presented at the 2000 International Astronautical Congress (IAA-00-IAA.9.2.01); Seth Shostak and Ivan Almar,

"The Rio Scale Applied to Fictional SETI Detections," paper presented at the 2002 World Space Congress (IAC-02-IAA.9.1.06).

5. von Hoerner, in Hart and Zuckerman, editors, 29.
6. Leonardo Maugeri cited this theorem in his article "Oil: Never Cry Wolf—Why the Petroleum Age Is Far From Over," Science, Vol. 304 (21 May 2004), 1114.
7. White described a sample scenario from a contact conference in The SETI Factor, 145–148. Harrison gave a shorter description in After Contact, 29.
8. Dole and Asimov, 197.
9. Robert Freitas, "Metalaw and Interstellar Relations," Mercury, March-April 1977, 15–17.
10. D.T. Michael, Proposed Studies on the Implications of Peaceful Space Activities for Human Affairs, Washington, DC, Brookings Institution, 1960.
11. MacGowan and Ordway, 261, 377.
12. Black, et al., 1977; David C. Black and Mark A. Stull, "The Science of SETI," in Morrison, et al., editors, 95–100.
13. Ben Finney, "SETI and the Two Terrestrial Cultures," Acta Astronautica, Vol. 26 (1992), 263–265; Finney, quoted in Douglas A. Vakoch, "The Dialogic Model," Acta Astronautica, Vol. 42 (1998), 705–710.
14. Grinspoon, 257; Sivier, "SETI and the Historian."
15. John Billingham, Scientific and Cultural Aspects of SETI, paper prepared as background material for a meeting of the International Academy of Astronautics SETI Committee in 1994 (the meeting was canceled).
16. Billingham, et al., editors, Social Implications, 80–81.
17. Ekers, et al., 236; Dick and Strick, 222.
18. This is not identical with behaviorism, a particular school of psychological research that was succeeded by cognitive psychology.
19. Harrison, et al., "The Role of the Social Sciences in SETI," in Tough, editor, 71–85.
20. Melvin Konner, "Why We Did It" (review of Michael Cook's A Brief History of the Human Race), Nature, Vol. 428 (11 March 2004), 123–124; Cohen, "History and the Hyperpower".
21. Paolo Musso, "Philosophical and Religious Implications;" Joseph N. Tatarewicz, review of Dick's The Biological Universe," Science, Vol. 275 (21 March 1997), 1748–1749.
22. Freeman, The Closing of the Western Mind, xvi.
23. Crowe, 559.
24. Sagan and Newman in Regis, editor, 161.

Some Conclusions Drawn

1. Govert Schilling, "The Chance of Finding Aliens," Sky and Telescope, December 1998, 36–42.
2. Lamb, 3.
3. Quoted in Dick, The Biological Universe, 537.

4. Edward Purcell, "Radio Astronomy and Communication Through Space," in Cameron, editor, 121–143.
5. Jack Cohen and Ian Stewart, What Does a Martian Look Like?, Hoboken, NJ, Wiley, 2002, 209.
6. Crowe, 559.
7. Quoted by George Ellis in a review of Leonard Susskind's The Cosmic Landscape, Nature, Vol. 438 (8 December 2005), 739–740.
8. Project Cyclops, 171; Achenbach, 50.
9. Martin Rees, "Cosmological Challenges: Are We Alone, and Where?", in John Brockman, editor, The Next Fifty Years: Science in the First Half of the Twenty-First Century, New York, Vintage, 2002, 18–28.
10. Fitzgerald, Cosmic Test Tube, 292.
11. James L. Christian, Preface, in James L. Christian, editor, Extra-Terrestrial Intelligence, Buffalo, NY, Prometheus, 1976, 2.
12. Joel Achenbach, "Life Beyond Earth," National Geographic, January 2000, 24–51.
13. Beck, in Regis, editor, 14.
14. Francois Jacob, "Evolution and Tinkering," Science, Vol. 196 (10 June 1977), 1161–1166.
15. Ridpath, 140.
16. Sullivan, 295.
17. Clark and Clark, 107; Brin, "A Contrarian Perspective;" Achenbach, 288–292.
18. This perspective was suggested by Robert DeBiase in a personal communication, 2005.

Paradigm Shifts

1. Mark Rose, Alien Encounters, Cambridge, MA, Harvard University Press, 1981, 53.
2. Swift, 229.
3. Sagan, editor, CETI, ix–x; Sagan, The Cosmic Connection, 59; Davies, Are We Alone?, 136.
4. White, 3, 18.
5. Benjamin, 177.
6. Ned Martel, "Doomsayers of All Stripes, Now Together in One Show," The New York Times, 24 December 2004.
7. Gardner, 125.
8. Baird, 214.
9. Harrison "Slow Track, Fast Track."
10. Black, et al., in Mercury.
11. Dick and Strick, 222; White, 113.
12. White, 182.
13. Easterbrook, "Are We Alone?"; William J. Broad, "Maybe We Are Alone in the Universe, After All," The New York Times, 8 February 2000.
14. Achenbach, Captured by Aliens, 314.
15. Edward Regis Jr., "SETI Debunked," in Regis, editor, 234–235.
16. Morris, Life's Solution, 328.

The Human Role

1. Stapledon, Star Maker, 290.
2. Steven Weinberg, The First Three Minutes: A Modern View of the Origin of the Universe, Glasgow, William Collins, 1977, 148.
3. Davies, Are We Alone?, 128.
4. Fred Hoyle and Chandra Wickramasinghe, Cosmic Life Force, New York, Paragon, 1990, 143, originally published in the United Kingdom by Dent in 1988.
5. Sagan, Pale Blue Dot, 403. The Fromm quote is from Richard Lewontin, "The Wars Over Evolution," The New York Review of Books, 20 October 2005.
6. Hoyle and Wickramasinghe, Cosmic Life Force, 143.
7. Olaf Stapledon, "Interplanetary Man?" JBIS, Vol. 7 (1948), 213–224.
8. J.D. Bernal, The World, The Flesh, and the Devil, London, Jonathan Cape, 1970, 74, originally published in 1929.
9. Bernal, 11.
10. J.M. Roberts, 36, 286.
11. Seielstad, 278.
12. Seielstad, 240, 272.
13. M. Mautner and G.L. Matloff, "Directed Panspermia: A Technical and Ethical Evaluation of Seeding Nearby Solar Systems," JBIS, Vol. 32 (1979), 419–423. Also see Michael N. Mautner, "Life in the Cosmological Future," JBIS, Vol. 58 (2005), 167–180. As of 1996, a New York company was proposing to launch more than a million samples of human biological material beyond the solar system. "Dust to Stardust," Science, Vol. 272 (14 June 1996), 1593.
14. Fred Hoyle, The Intelligent Universe, 224; Seielstad, 190; Aldiss, "Desperately Seeking Aliens;" Barrow and Tipler, 614.
15. Seielstad, 191.
16. Steven Wolfe, "Space Settlement: The Journey Inward," Ad Astra, Vol. 16, Number 1 (January-February-March 2004), 30–33.
17. "A Conversation with Carl Sagan," U.S. News and World Report, 1 December 1980, 62–63.
18. Kenneth Heuer, Men of Other Planets, New York, Pellegrini and Cudahy, 1951, 23.
19. William J. Broad, The Universe Below, New York, Simon and Schuster, 1997, 22.
20. These concepts are developed in Michael A.G. Michaud, "The Extraterrestrial Paradigm," Interdisciplinary Science Reviews, Vol. 4 (September, 1979), 177–192.
21. Space optimists believe that expansion will remove the problem of limits. Consider Gregg Easterbrook's declaration: "Above us in the Milky Way are essentially infinite resources and living space. If the phase of fossil-driven technology leads to discoveries that allow *Homo sapiens* to move into the galaxy, then resources, population pressure ... will be forgotten. Most of the Earth may even be returned to primordial stillness, and the whole thing would have happened in the blink of

an eye by nature's standards." New York Times Book Review, 30 January 2005.

22. See Michael A.G. Michaud, "Towards a Grand Strategy for the Species," Earth-Oriented Applications of Space Technology, Vol. 2 (1982), 213–219. For a more popularized version, see Michael A.G. Michaud, "Sharing the Grand Strategy," Space World, Vol. U-8-248 (August 1984), 4–9.

23. For another version of the reconnaissance argument, see Finney's views in White, 3.

24. Carl Sagan, Pale Blue Dot, 385.

25. Freeman Dyson, Disturbing the Universe, New York, Harper, 1979, 234. Also quoted in John Noble Wilford, Mars Beckons, New York, Knopf, 1990, 215.

26. Sagan, Pale Blue Dot, 327, 371, 375, 405.

27. Nigel Calder, Spaceships of The Mind, New York, Viking, 1978, 141.

28. Rodney Brooks, "The Merger of Flesh and Machines," in John Brockman, editor, The Next Fifty Years: Science in the First Half of the Twenty-First Century, New York, Vintage, 2002, 183–193.

29. Michael A.G. Michaud, "An Extraterrestrial Ethos," America, Vol. 139 (11 November 1978), 330–331; Ivan Almar, "Protection of the Planetary Environment," paper presented at the 1999 Astronautical Congress (IAA-99-IAA.7.1.02); Martyn J. Fogg, Terraforming: Engineering Planetary Environments, Warrendale, PA, Society of Automotive Engineers, 1995, 496. Also see James E. Oberg, New Earths, Harrisburg, PA, Stackpole, 1981.

30. Dave Brody, "Terraforming: Human Destiny or Hubris?" Ad Astra, Vol. 17, Number 2 (Spring 2005), 29–33; Oliver Morton, Mapping Mars, New York, Picador USA, 2002, 313; Martyn J. Fogg, Terraforming: Engineering Planetary Environments, Society of Automotive Engineers, 1995, 496. Also see James E. Oberg, New Earths, Harrisburg, PA, Stackpole, 1981.

31. Mautner and Matloff, "Directed Panspermia."

32. William H. McNeill, The Rise of the West, New York, New American Library, 1963, 877.

33. Berendzen, editor, 26.

34. Gerald Feinberg, The Prometheus Project, Garden City, New York, Doubleday, 1968, 142.

35. Stapledon, Star Maker, 302, 368.

Annex: Preparing

1. Steven J. Dick, "Cosmotheology," in Dick, editor, Many Worlds, 205.

2. Quoted in MacGowan and Ordway, 359 .

3. John Billingham, "Scientific and Cultural Aspects of SETI," paper prepared as background material for a meeting of the International Academy of Astronautics SETI Committee in 1994 (the meeting was canceled).

4. Michael A.G. Michaud, "On Communicating with Aliens" (book essay on five works about SETI), Foreign Service Journal, Vol. 51, Number 6 (June 1974), 33–34, 40.

5. Stephen E. Doyle, "Cultural and Institutional Aspects" (for roundtable on SETI and Society), paper presented at the 1997 International Astronautical Congress (IAA-97-IAA.7.1.04).

6. Drake and Sobel, xii–xiii.
7. White, 67, 175.
8. Beck, in Regis, editor, 15.
9. Andrew G. Haley, "Space Law and Metalaw: A Synoptic View," in Proceedings of the 1956 Astronautical Congress, Rome, Associazione Italian Razzi, 1956; Andrew G. Haley, Space Law and Government, New York, Appleton-Century-Crofts, 1965; Ernst Fasan, Relations with Alien Intelligences, Berlin, Berlin Verlag Arno Spitz, 1970.
10. Michael A.G. Michaud, "Interstellar Negotiation," Foreign Service Journal, Vol. 49 (December 1972), 10–14, 29–30; "Negotiating With Other Worlds," The Futurist, Vol. VII (April 1973), 71–77; "There's Somebody Out There We're Going to Have to Deal With," Los Angeles Times, May 27, 1973.
11. Allan E. Goodman, "The Diplomatic Implications of Discovering Extraterrestrial Intelligence," Mercury, Vol. XVI (March-April 1987), 14–16; "Diplomacy and the Search for Extraterrestrial Intelligence," in Tarter and Michaud, editors, 137–141; Colin Campbell, "Protocol of Calls From Distant Space," The New York Times, 10 June 1985; Paul Ney, "Extraterrestrial Intelligence Contact Treaty?" JBIS, Vol. 38 (1985), 521–522.
12. For a brief history of this effort, see Michael A.G. Michaud, "Organizing Ourselves for Contact," Analog Science Fiction/Science Fact, Vol. CXVIII (January 1998), 51–63.
13. Koerner and LeVay, 169.
14. Koerner and LeVay, 169–170; Regis, editor, 175–176.
15. Scot Lloyd Stride, "An Instrument-Based Method to Search for Extraterrestrial Interstellar Robotic Probes," JBIS, Vol. 54 (2001), 2–13; Allen Tough, "Small Smart Interstellar Probes," JBIS, Vol. 51 (1998), 167–174.
16. Michael A.G. Michaud, "Broadening and Simplifying the First SETI Protocol," JBIS, Vol. 58, Number 12 (January–February 2005), 40–42.
17. Gregory Matloff, Peter Schenkel, and Jaime Marchan, "Direct Contact with Extraterrestrials: Possibilities and Implications," paper presented at the 1999 International Astronautical Congress (IAA-99-IAA.8.2.02); Peter Schenkel, "Legal Frameworks for Two Contact Scenarios," JBIS, Vol. 50 (1997), 258–262.
18. Margaret S. Race, "Communicating About the Discovery of Extraterrestrial Life: Different Searches, Different Issues," paper presented to the 2004 International Astronautical Congress (IAC-04-IAA.1 .1.2.02).
19. David H. Levy, "The Rules of the Game," Sky and Telescope, February 1998, 69–70; Gretchen Vogel, "Asteroid Scare Provokes Soul-Searching," Science, Vol. 279 (20 March 1998), 1843–1844; James V. Scotti, "The Tale of an Asteroid," Sky and Telescope, July 1998, 30–34; "After Asteroid Episode, Scientists Agree to Agree," The New York Times, 20 March 1998. The revised Torino scale can be found at http://neo.jpl.nasa.gov/torino_scale.html.
20. Billingham, et al., "Social Implications," 78; John Billingham, "SETI and Society: Decision Trees," Acta Astronautica, Vol. 51 (2002), 667–672.
21. Ivan Almar and Jill Tarter, "The Discovery of ETI as a High-Consequence, Low-Probability Event," paper presented at the 2000 International Astronautical Congress; Ivan Almar, "How the Rio Scale Should be Improved," paper presented at the 2001 International Astronautical Congress (IAA-01-IAA.9.2.03). Also see Seth Shostak, "The Rio Scale Applied to Fictional

SETI Detections," paper presented at the 2002 International Astronautical Congress (IAA-02-IAA.9.1.06).

22. David Morrison, "Are Astronomers Crying Wolf," Mercury, Vol. 32, Number 6 (November-December 2003), 15.

23. Billingham, et al., editors, "Social Implications,"117–119.

24. Project Cyclops, 32.

25. Michael A.G. Michaud, "Detection of ETI—An International Agreement," Space Policy, Vol. 5, Number 2 (May, 1989), 103–106; Donald Goldsmith, "Who Will Speak for Earth?" Acta Astronautica, Vol. 21 (1990), 149–151; Michael A.G. Michaud, "An International Agreement Concerning the Detection of Extraterrestrial Intelligence," Acta Astronautica, Vol. 26 (March/April, 1992), 291–294; Michael A.G. Michaud, John Billingham, and Jill Tarter, "A Reply from Earth?" Acta Astronautica, Vol. 26, No. 3/4 (March/April, 1992), 295–297.

26. See the International Academy of Astronautics website at www.iaanet.org, link to Position Papers.

27. Doyle, "Cultural and Institutional Aspects."

28. Billingham, et al., editors, Social Implications," 94.

29. A. Harrison, "Confirmation of ETI: Initial Organizational Response," paper presented at the 1999 International Astronautical Congress (IAA-99-IAA.9.2.02) Later published in Acta Astronautica, Vol. 53 (2002), 229–236.

30. Fred Hoyle, The Black Cloud, 199.

31. Douglas Vakoch, "Messages to the Cosmos?," SETI News, Second Vol. 12, Number 2 (Quarter 2003), 14–15.

32. Drake and Sobel, 228.

33. D.E. Tarter, "Reply Policy and Signal Type," paper presented to the 1995 International Astronautical Congress (IAA-95-IAA.9.2.12).

34. Wright, Non-Zero, 170.

35. As of 2005, information on the San Marino Scale could be found on the SETI League website, www.setileague.org/iaaseti/smscale.html.

36. Douglas Vakoch, "The Dialogic Model," Acta Astronautica, Vol. 42 (1998), 705–710.

37. Michael A.G. Michaud, "Policy Issues in Communicating with ETI," Space Policy, Vol. 14 (1998), 173–178; Michael A.G. Michaud, "An International Agreement Concerning the Detection of Extraterrestrial Intelligence," Acta Astronautica, Vol. 26, No. 3/4 (March/April, 1992), 291–294.

38. Douglas Vakoch, "The Dialogic Model."

39. Drake and Sobel, 228.

40. Douglas Vakoch, "Messages to the Cosmos?"; Michaud, "Policy Issues in Communicating with ETI."

41. F. Galton, "Intelligible Signals Between Neighboring Stars," Fortnightly Review, Vol. 60 (1896), 657–664; H.W. Nieman and C. Wells Nieman, "What Shall We Say to Mars?," Scientific American, 122:12 (March 20, 1920), 298–312; Lancelot Hogben, "Astroglossa or First Steps in Celestial Syntax," JBIS, Vol. 11 (1952), 258–274.

42. Hans Freudenthal, "Lincos—Design of a Language for Cosmic Intercourse," Amsterdam, North Holland, 1960 (Freudenthal provided an update on his thinking 20 years later in "Towards a Cosmic Language," Cosmic Search,

Vol. 2, Number 1 (Winter 1980), 35–39; S.W. Golomb, "Extraterrestrial Linguistics," Astronautics, Vol. 6 (January–June 1961, 46–47, 95.

43. Michael A. Arbib, "Minds and Millennia," Cosmic Search, Vol. 1, Number 3 (Summer 1979), 21–25, 47.

44. Carl L. De Vito, "Languages Based on Science," Acta Astronautica, Vol. 26 (1992), 267–271. Brian McConnell gave us an overview of the technical issues involved in his 2001 book Beyond Contact: A Guide to SETI and Communicating with Other Civilizations, Sebastopol, CA, O'Reilly, 2001.

45. Douglas A. Vakòch, "Constructing Messages to Extraterrestrials: An Exosemiotic Perspective," Acta Astronautica, Vol. 42 (1998), 697–704.

46. Albert A. Harrison, "Slow Track, Fast Track, and the Galactic Club," paper published by the Foundation for the Future, 1998.

47. George W. Swenson, Jr., "Intragalactically Speaking," Scientific American, July 2000, 44–47.

48. Ernst Fasan, "Legal Consequences of a SETI Detection," Acta Astronautica, Vol. 42 (1998), 677–679.

49. Haley, "Space Law and Metalaw—A Synoptic View;" Haley, Space Law and Government; Robert A. Freitas, Jr., "The Legal Rights of Extraterrestrials," Analog Science Fiction/Science Fact, April 1977, 54–67.

50. Ernst Fasan, Relations with Alien Intelligences, Berlin, Arno Spitz, 1970.

51. Tough, editor, 18; Freitas, "The Legal Rights of Extraterrestrials."

52. Asimov, A Choice of Catastrophes, 268.

53. Henry Fountain, "Armageddon Can Wait: Stopping Killer Asteroids," The New York Times, 19 November 2002; Russell L. Schweikart, et al., "The Asteroid Tugboat," Scientific American, November 2003, 54–61; Rusty Schweickart, "We Must Decide to Do It!," The Planetary Report, July-August 2005, 14–18; Henry Fountain, "Asteroid Poses Tiny Danger, But It May Be Lured Away," The New York Times, 22 November 2005; "Tilting at Asteroids," Scientifc American, December 2005, 36; Claudio Maccone, "Planetary Defense in Space," SETI Institute Explorer, second Quarter 2005, 10–13.

54. "NASA Told to Think Small in Hunt for Asteroids," Nature, Vol. 435 (26 May 2005), 398.

55. Edward T. Lu and Stanley G. Love, "Gravitational Tractor for Towing Asteroids," Nature, Vol. 438 (10 November 2005), 177–178.

56. Paul Gilster, Centauri Dreams, New York, Copernicus, 2004, 236.

57. Billingham, et al., editors, Social Implications, 117–119.

58. Peter Schenkel, ETI: A Challenge for Change, New York, Vantage, 1988, reviewed by Michael Papagiannis in Sky and Telescope, March 1990, 281–283.

Index

Printed in the United States of America.